Developmental Biology

A COMPREHENSIVE SYNTHESIS

Volume 6

Genomic Adaptability in Somatic Cell Specialization

Developmental Biology
A COMPREHENSIVE SYNTHESIS

Editor
LEON W. BROWDER
University of Calgary
Calgary, Alberta, Canada

Developmental Biology

A COMPREHENSIVE SYNTHESIS

Volume 6

Genomic Adaptability in Somatic Cell Specialization

Edited by

MARIE A. DiBERARDINO
The Medical College of Pennsylvania
Philadelphia, Pennsylvania

and

LAURENCE D. ETKIN
University of Texas M.D. Anderson Cancer Center
Houston, Texas

PLENUM PRESS • NEW YORK AND LONDON

Library of Congress Cataloging in Publication Data

(Revised for vol. 6)

Developmental biology.

Vol. 3 edited by Malcolm S. Steinberg; v. 4 edited by Ralph B. L. Gwatkin; v. 6
edited by Marie A. DiBerardino and Laurence D. Etkin.
Includes bibliographies and indexes.
Contents: v. 1. Oogenesis — v. 2. The cellular basis of morphogenesis — (etc.]—v.
6. Genomic adaptability in somatic cell specialization.
1. Developmental biology. I. Browder, Leon W.
QH491.D426   1985   574.3   85-3406
ISBN-13: 978-1-4615-6822-3   e-ISBN-13: 978-1-4615-6820-9
DOI: 10.1007/978-1-4615-6820-9

Cover illustration: Artist's rendition of the genetic control of
development juxtaposed with modulations of the cell phenotype.

Contributors

S. C. Barton Department of Molecular Embryology, Institute of Animal Physiology and Genetics Research, Babraham, Cambridge CB2 4AT, England

Leon W. Browder University of Calgary, Calgary, Alberta, Canada T2N 1N4

Marie A. DiBerardino Department of Physiology and Biochemistry, The Medical College of Pennsylvania, Philadelphia, Pennsylvania 19129

Christine Dreyer Max Planck Institute for Developmental Biology, Department for Cell Biology, D-7400 Tübingen, Federal Republic of Germany

Laurence D. Etkin Department of Molecular Genetics, University of Texas M.D. Anderson Cancer Center, Houston, Texas 77030

Morgan Harris Department of Zoology, University of California–Berkeley, Berkeley, California 94720

John J. Heikkila University of Waterloo, Waterloo, Ontario, Canada N2L 3G1

S. K. Howlett Department of Molecular Embryology, Institute of Animal Physiology and Genetics Research, Babraham, Cambridge CB2 4AT, England

David S. McDevitt Department of Animal Biology, School of Veterinary Medicine, University of Pennsylvania, Philadelphia, Pennsylvania 19104-6045

Robert Gilmore McKinnell Department of Genetics and Cell Biology, University of Minnesota, Saint Paul, Minnesota 55108-1095

Robert W. Nickells University of Calgary, Calgary, Alberta, Canada T2N 1N4

M. L. Norris Department of Molecular Embryology, Institute of Animal Physiology and Genetics Research, Babraham, Cambridge CB2 4AT, England

Michael Pollock University of Calgary, Calgary, Alberta, Canada T2N 1N4

David M. Prescott Department of Molecular, Cellular and Developmental Biology, University of Colorado, Boulder, Colorado 80309-0347

W. Reik   Department of Molecular Embryology, Institute of Animal Physiology and Genetics Research, Babraham, Cambridge CB2 4AT, England

M. A. H. Surani   Department of Molecular Embryology, Institute of Animal Physiology and Genetics Research, Babraham, Cambridge CB2 4AT, England

Robert S. Winning   University of Waterloo, Waterloo, Ontario, Canada N2L 3G1

Foreword

I am pleased to include this book as Volume 6 of *Developmental Biology: A Comprehensive Synthesis*. It has been edited by two of the foremost investigators in the study of genomic adaptability. I owe a special debt of gratitude to Dr. Marie A. DiBerardino, who developed the concept of the volume. Dr. DiBerardino is also a very active member of the editorial board for this series. Much of the success of this series is due to her valuable advice.

This series was established to create comprehensive treatises on specific topics in developmental biology. Such volumes serve a useful role in developmental biology, since it is a very diverse field that receives contributions from a wide variety of disciplines. This series is a meeting ground for the various practitioners of this science, facilitating an integration of heterogeneous information on specific topics.

Each volume is intended to provide the conceptual basis for a comprehensive understanding of its topic as well as analysis of the key experiments upon which that understanding is based. The specialist in any aspect of developmental biology should understand the experimental background of the field and be able to place that body of information in context to ascertain where additional research would be fruitful. At that point, the creative process generates new experiments. This series is intended to be a vital link in that process of learning and discovery.

Leon W. Browder

Preface

The intent of this volume is to focus on the genetic mechanisms of somatic cell specialization in animals and the ability to modulate their differentiated cellular phenotypes. We considered it timely to undertake this project because the fundamental question of genetic totipotency of somatic cells has never been answered; in fact, with the application of molecular techniques, more cases of irreversible genetic changes are being revealed. Thus, the extent of irreversible genetic changes concomitant with cell differentiation in animals remains unknown.

The contributors to this volume show that various genetic mechanisms have evolved that control somatic cell specialization (Chapters 1–4). While some cases involve alterations in the genome, most cell specializations appear not to involve genomic changes. This latter hypothesis is examined in various experimental systems (Chapters 5–9) that show that many silent genes can be activated in normal and cancer cells and that, in special cases, even nearly an entire genome can be activated. The collective studies indicate that genomic multipotentiality is widespread among the animal phyla. Future exploitation of the systems described, as well as others, should permit analysis of the regulatory factors involved in activating not only single genes but also those responsible for activating an entire genome. We thank the authors for contributing to our current knowledge of genomic adaptability in somatic cell specialization.

Marie A. DiBerardino
Laurence D. Etkin

Contents

Chapter 4 • Genomic Imprinting in the Mouse

S. K. Howlett, W. Reik, S. C. Barton, M. L. Norris, and
M. A. H. Surani

Chapter 5 • Phenotypic Changes in Cell Culture

Morgan Harris

Chapter 6 • Developmental Regulation of the Heat-Shock Response

Leon W. Browder, Michael Pollock, Robert W. Nickells,
John J. Heikkila, and Robert S. Winning

Chapter 1

Introduction
Early Development and Cell Commitment

LAURENCE D. ETKIN

1. Introduction

The genome of most eukaryotic organisms is constant and perpetuates itself through numerous cell generations by replication. Yet, during embryogenesis, there is increasing cellular diversity, yielding a multitude of cell-specific phenotypes. Cells become different from one another, and their developmental potential is progressively restricted. The determined state and the overtly differentiated state are exceedingly stable and heritable during normal development. However, various perturbations, both natural and artificially induced, can result in a modulation of cellular phenotypes. It is the purpose of this volume to analyze the nature of the stability of the differentiated state and the ability to reverse the differentiated phenotype. It is not intended to be all-inclusive and, in fact, we have chosen certain biological systems that require further exploration but that have the potential for helping us to understand the cellular and molecular mechanisms underlying these processes.

2. The Genome during Development

Several lines of evidence suggest that the genome of most cells is not dramatically altered or rearranged during development in most eukaryotic organisms. These include biochemical analyses showing that the genome size remains quantitatively equivalent during development and differentiation, and functional studies, including nuclear transplantation and somatic cell hybrids. Nuclear transplantation studies show that the organism maintains a functionally near-complete genome in somatic cells throughout development, although there may be other restrictions imposed on the ability to activate all these genes in the proper sequence (for a review of this topic, see Chapter 8, *this*

LAURENCE D. ETKIN • Department of Molecular Genetics, University of Texas M.D. Anderson Cancer Center, Houston, Texas 77030.

volume). Somatic cell hybrid studies have also demonstrated that it is possible to activate many genes that are not active in a differentiated cell type, providing further evidence that these genes are maintained and neither eliminated nor permanently altered.

There are, however, exceptions to the maintenance of an unaltered genome during development. Prescott (Chapter 2, *this volume*), describes many of these examples, including amplification of specific DNA sequences, such as chorion genes in insects and ribosomal RNA (rRNA) genes in many organisms. There are also numerous examples of rearrangements in the genome. An extreme example discussed in Chapter 2 is the developmental alteration of the macronuclear genome in ciliates. Other examples of rearrangements include the highly specific recombinational events in immunoglobulin (Ig) genes, which in themselves have varied mechanisms of achieving diversity (reviewed by Maizels, 1987; Okazaki *et al.*, 1987; Hesse *et al.*, 1987). Also, gene-conversion events take place during the switching of yeast mating types (Klar, 1987*a*), as well as trypanosome surface antigenic variation (Van der Ploeg, 1987). Thus, at the gross level in most eukaryotic organisms, the genome appears unchanged, but there are instances of rearrangements and gene-conversion events such as those cited above and in Chapter 2.

The process of gene conversion in the switching of yeast mating types is reversible, since a cell can switch back to the original mating type; thus, these types of changes can occur in the germ line without heritable consequences. Recombinational events and amplification that are part of the normal developmental program are not reversible and are probably limited to somatic cells. One exception is the amplification of rRNA genes in amphibian oocytes. However, the amplified copies of the genes are maintained as extrachromosomal nucleoli and are degraded during early embryogenesis. This would not affect the integrity of the germ-line genome.

In addition to the above-mentioned changes in the genome, methylation of CpG pairs occurs to varying degrees throughout most of the animal kingdom. Hypomethylation of sequences is associated with actively transcribing domains, whereas hypermethylation is associated with inactive domains. It is likely that the gross pattern of methylation is not influencing the expression of individual genes but that methylation of specific sites in and around a gene is important in its regulation (Busslinger *et al.*, 1983; Chandler and Jones, 1988).

3. Regulation of Early Development and Cell Specialization

A particularly attractive strategy related to the generation of cellular diversity is the prelocalization of either messenger RNA (mRNA) or proteins in the oocyte or egg. Classic experimental approaches to study this question involved deletion, ablation, or injury to specific regions or blastomeres of developing embryos and subsequent analysis of developmental defects (reviewed by Davidson, 1986). Recently, however, it has been possible to use modern technology in an attempt to assess the molecular nature of such localizations.

Amphibian oocytes and embryos have been used extensively for both embryological and molecular studies of developmental phenomena and more recently for molecular studies dealing with localization. Melton's group (Weeks et al., 1985) have isolated complementary DNA (cDNA) clones for mRNAs localized in either the vegetal or animal pole regions of *Xenopus* oocytes. One of these mRNAs has been identified through sequence comparison encoding as a mitochondrial ATPase (Weeks and Melton, 1987*b*) and is localized in the animal pole region; another mRNA (vg 1) is localized in the vegetal pole region and codes for a transformation growth factor ('TGF$_\beta$)-like factor (Weeks and Melton, 1987*a*). The activity of this growth factor-like molecule may synergistically amplify the mesoderm inducing activity of another maternal molecule, an fibroblast growth factor (FGF)-like molecule (Kimelman and Kirschner, 1987). *In situ* hybridization analysis of vg1 mRNA during development shows a localization of the mRNA in cells of the vegetal tier that are involved in mesoderm induction in the early frog embryo. Thus, although there is no formal evidence per se, it is possible that the localized TGF$_\beta$-like factor, along with other growth factors acting in synergism, may be involved in regional mesoderm induction.

Along with localized mRNA, there are localized distributions of proteins in eggs and oocytes of amphibians. Most proteins appear to be localized in an animal-to-vegetal hemisphere gradient (Moen and Namenworth, 1977; Dreyer et al., 1982; King and Barklis, 1985). Dreyer et al. (1982) produced antibodies against proteins that originated in the germinal vesicle (GV; i.e., nucleus) of *Xenopus* oocytes and examined their fate and distribution during development and differentiation. Dreyer et al. detected unique patterns of tissue specific nuclear localization of some of these proteins much later in development (Chapter 3, *this volume*). It is quite possible that some of these proteins may be involved in the differentiation of these specific cell types.

One of the best examples of a localized cytoplasmic determinant is the bicoid gene product in *Drosophila*. Mutations of the bicoid gene result in perturbations in the position and establishment of anterior structures. The product of the bicoid gene is distributed in an exponential concentration gradient with the maximum at the anterior tip of the embryo. Genetic and experimental manipulations which affect the formation or the shape of the gradient result in corresponding changes in the position of anterior structures (Driever and Nusslein-Volhard, 1988).

Conceptually, localization of macromolecules is the most logical solution to the problem of how the embryo generates cell diversity necessary to establish embryonic cell lineages. An important question is the mechanism by which macromolecules become localized in oocytes and eggs. Recently, Kemphues et al. (1988) identified four genes (par 1, par 2, par 3, and par 4) in the nematode C. *elegans* that are required for normal cytoplasmic localization. Mutations in the genes result in phenotypes such as defects in localization of germ line-specific P granules, abnormal cleavage patterns, and abnormal timing of cleavage. In addition, there are indications that the cell cytoskeleton may be a fundamental component in the establishment of cytoplasmic localizations. The localization phenomenon has begun to yield to molecular and genetic approaches but, for the most part, these analyses are still descriptive in nature.

4. Specification and Stability of Cell Lineages in Early Embryogenesis

An important component in a developing system is the perpetuation and transmission of an established cell phenotype or a determined state to daughter cells. Thus, once a cell is committed to a specific developmental pathway, a molecular mechanism must exist for heritable transmission of the phenotype to daughter cells and, in some cases, to permit one daughter cell to initiate a new cell lineage. An interesting model system for studying cell lineages that may be applicable to higher eukaryotic organisms is that of the mating type switch in yeast (Klar, 1987a). Single-celled yeast that divide by budding differentiate by producing developmentally nonequivalent daughter cells that have different mating types. There is a specific hierarchy or pattern to the switching in which only mother cells (older cells) produce switched progeny, whereas daughter cells (new buds) always remain the same. The switching or mating-type inter-conversion has a genetic basis involving a transposition substitution event from a silent a or α-locus (*HML* α or *HMRa*) to a site in the genome where it will be expressed (*MAT* locus). This gene conversion event involves recombination of DNA at the *MAT* locus by a double-stranded DNA break resulting from the action of a site-specific endonuclease encoded by the *HO* locus (Strathern *et al.*, 1982; Kostriken *et al.*, 1983). In essence, the daughter cells transmit information in a stable heritable manner, while the mother cell is capable of switching to a different cell lineage. This difference between mother and daughter cells is attributable to several factors, one of which is that the competence for *HO* expression is inherited by the mother cell, but not the daughter (bud). The property of competence is due to the differential operation of several genes that inhibit *HO* expression. In yeast that reproduces by fission instead of budding, it appears that the property of competence to switch is actually imprinted in the genome at the *MAT* locus during previous generations (Klar, 1987b). This suggests that the sister cells do not contain different cytoplasmic or nuclear factors, but that asymmetry is established through inheritance of nonequivalent parental chromosomes. Such an imprinting mechanism probably exists in mammalian cells (see Chapter 4, *this volume*).

Cell transfer experiments in early embryos show that, once established, the fate of a cell is determined and the cell will differentiate in a position-independent manner maintaining its original identity and fate. It is possible that the original determined state was initiated through the action of a localized determinant. In *Drosophila* there is an example in which the perpetuation of a determined state may be accomplished by an autoregulatory mechanism. This involves the autoregulation of a homeotic selector gene called deformed (Dfd) which functions in specifying the identity of the mandibular and maxillary segments (Kuziora and McGinnis, 1988).

The differentiation of the blood cell lineage during development provides an example demonstrating the effects of various humoral agents, such as growth inducers, on growth and differentiation (reviewed by Sachs, 1987). Numerous proteins have been identified as playing a role in the differentiation of the myeloid cell lineage. These include various colony-stimulating factors (CSF), which are involved either directly or indirectly in the differentiation of macrophages or granulocytes, and interleukin-3 (IL-3), which is involved in the differentiation of macrophages, granulocytes, and other cell types. Several of these genes have been cloned, and it appears that their products represent a cascade or hierarchy of growth-inducing substances that act at specific stages of blood cell differentiation, biasing cells to follow specific cell lineages (Sachs, 1987).

There are also examples of the effects of cell–cell interactions on the establishment of cell lineages in the nematode *Caenorhabditis elegans*. During the second cleavage of *C. elegans* embryos, the AB blastomere divides into two sister blastomeres ABa and ABp that have different fates (Sulston *et al.*, 1983). ABa produces pharyngeal cells, such as muscle, glands, and neurons, while ABp produces γ-aminobutyric acid (GABA)-containing motor neurons in the ventral nerve cord (Sulston *et al.*, 1983). When the ABa cells are cultured alone, they do not produce pharyngeal muscle cells, suggesting that cell–cell interactions are necessary for proper differentiation. On the other hand, the ABp cell is able to produce pharyngeal cells after the positions of ABp and ABa are interchanged (Priess and Thompson, 1987). This demonstrates that cell–cell inductive interactions are important in the specification of cell lineages and that at least at this early developmental stage the fate of the ABp blastomere may be altered by exposing it to the influence of the neighboring cells normally needed to induce the ABa blastomere. It is quite apparent that the establishment of cell lineage is under genetic control involving specific molecular mechanisms, cell–cell interactions, and exposure to various growth inducers. In embryonic cells, it is still possible to modulate the developmental pathways by various experimental manipulations.

5. Changes or Modulations of Cell Phenotype in Determined and Differentiated Cells

The examples discussed so far all involve the establishment of determined or differentiated cell lineages during development. There are also numerous examples in which differentiated and determined cell phenotypes have been modulated, e.g., via transdifferentiation (see Chapter 7, *this volume*). These include such phenomena as lens regeneration involving the repopulation of crystallin producing lens cells from the dorsal iris epithelium, transdetermination of imaginal discs in *Drosophila*, and several other documented cases. These systems have been studied for many years and may only now be approachable at either the molecular or genetic level, or both. This is especially true of transdetermination of imaginal discs in light of the availability of molecular and genetic tools in *Drosophila*.

Cell lineages may be altered or changed during embryogenesis in the sea urchin. In the sea urchin, the mesoderm forms the primary mesenchyme cells (PMC), which later form the larval skeleton, and the secondary mesenchyme cells (SMC), which form numerous other cell types. Primary mesenchyme cells are derived from the micromeres of the 16-cell embryo, whereas secondary mesenchyme cells are derived from a tier of vegetal cells (veg2) formed at the 64-cell stage (Horstadius, 1939). Ettensohn and McClay (1988) showed that a normal skeleton forms after surgical removal of the PMC and that the skeleton is formed by conversion of SMC into the skeletogenic phenotype. The converted cells not only form the skeleton but also exhibit cell-surface properties typical of PMC. Also, the number of SMC cells converted is directly proportional to the number of PMC that are surgically removed. This suggests that, under normal developmental conditions, the differentiating cell phenotype is extremely stable and heritable but that under the appropriate experimental conditions the SMC can convert to the function of the PMC.

In tissue culture cells, it is well documented that differentiated phenotypes may be lost for many cell generations but are reexpressed under the proper culture conditions (see Chapter 5, *this volume*). The molecular basis for this phenomenon is not known, but it probably involves heritable states of either DNA modification or chromatin conformation, or both.

The ability to modulate the differentiated phenotype is evident from studies using somatic cell hybrids and heterokaryons. When a hybrid is produced between a differentiated and undifferentiated partner (e.g., a hepatocyte cell producing liver-specific products and a fibroblast cell), the differentiated functions may be repressed. When this occurs, it has been termed extinction. In fact, extinction of the liver-specific product tyrosine aminotransferase has been correlated with the retention of a specific chromosome in the undifferentiated partner (Killary and Fournier, 1984).

Studies in heterokaryons, in which the nuclei remain independent in the same cytoplasm of fused cells, indicate that it is possible to activate muscle-specific functions of the nonmuscle partner in some heterokaryon fusions. This suggests that in addition to the negatively acting factors responsible for the extinction phenomenon, there are positively acting components to the system (Blau et al., 1985). This also demonstrates the ability to modulate the phenotype of various cell types under the proper experimental conditions.

Modulation of the differentiated or determined cell phenotype is evident in the synthesis of heat-shock proteins under conditions of stress (see Chapter 6, *this volume*, for a detailed discussion of this topic). Browder et al. discuss how, under heat-shock stress, the heat-shock genes, which are normally repressed at the ambient temperature, are activated at either the transcriptional or translational level, or both. Under these conditions, a small number of heat-shock genes contribute to the primary expression pattern of the cell. Thus, the cellular phenotype is transiently modulated by an external factor.

Cancer cells usually are less differentiated than their normal counterparts and, under appropriate experimental conditions, may be induced to differentiate along different pathways. McKinnell (Chapter 9, *this volume*) discusses a number of these examples such as plant tumors, amphibian neoplasms, and various germ layer tumors of mammals and the different inducers that elicit these differentiations.

Perhaps the most rigorous test of the ability to modulate a cellular phenotype is that of nuclear transplantation. The studies described in Chapter 8 (*this volume*) show that the genome of a terminally differentiated cell type such as an amphibian erythrocyte may be reactivated under the appropriate experimental conditions to support development through embryogenesis and early larval development.

6. Molecular Mechanisms of Gene Regulation and Heritability of the Determined or Differentiated State

Further elucidation of developmental principles will involve an understanding of the molecular basis by which genes are regulated both temporally and spatially in embryogenesis. The past 10 years have yielded a plethora of information regarding the molecular machinery involved in transcription. This includes identification of numerous *cis*-acting DNA elements that modulate the rate and specify the start and stop sites of transcription, and numerous *trans*-acting factors that interact with many of the *cis*-acting elements. The greatest challenge, however, is to begin to identify the factors and elements involved in regulating genes during development and during cellular differentiation and to understand the molecular basis by which the determined or differentiated state is inherited between cell generations (for review of transcription, see Maniatis *et al.*, 1987).

In *Drosophila*, the gene coding for alcohol dehydrogenase (ADH) has been cloned, and the role of both *cis*- and *trans*-acting elements in the temporal and spatial regulation have been partially characterized (Benyajati *et al.*, 1983; Posakony *et al.*, 1986; Heberlein and Tjian, 1988). The ADH gene is regulated by a two-promoter system. The proximal promoter is used during early embryogenesis with transcription increasing during the first and second larval instars and then decreasing; the distal promoter is expressed at low levels in mid-staged embryos and third-instar larvae. In adults, most of the transcription is directed by the distal promoter and is restricted to just a few tissues such as malphigian tubules and the fat body. The distal and proximal promoters are separated by a 700-base-pair (bp) region containing a large intron that is spliced from the mature transcript (Benyajati *et al.*, 1983). The *cis*-acting sequences have been identified by mutating the DNA and introducing the modified sequences into the germ line using p-element-mediated transformation (Posakony *et al.*, 1986). This analysis demonstrated that the control elements of the distal promoter are contained within a region 660 bp upstream of the distal start site, whereas proximal promoter function is regulated by sequences 400 bp upstream of the proximal start site. However, it appears that wild-type transcriptional levels by the proximal promoter are affected by sequences as far away as 2 kilobase (kb).

Using *in vitro* transcription extracts produced from different-staged *Drosophila* embryos, Heberlein and Tjian (1988) showed that the function of the distal promoter depends on the presence of a specific transcription factor (Adf-1). Interestingly, this factor also binds to the proximal ADH promoter as well as promoters from other *Drosophila* genes, but in their assay system does not affect transcription of these other genes.

In the sea urchin, there are two sets of genes coding for histones: an early set, which is active in preblastula embryos, and a late set, which is maximally expressed from the gastrula stage onward. Maxson *et al.* (1988) used the *X. laevis* oocyte as an assay system for *trans*-acting factors involved in this differential regulation. These investigators found that a protein fraction consisting of a 0.45–1 M salt wash of sea urchin gastrula chromatin preferentially stimulated the late sea urchin H2b histone gene when a mixture of the early and late genes along with the salt wash was injected into *Xenopus* oocytes. It appeared that the early and late genes were competing for a common transcription factor. This factor also appeared to decrease in abundance on a per-cell basis during sea urchin embryogenesis. Thus, it is quite possible that part of the mechanism of the developmental switch may be stimulated by the competition for limiting transcription factor.

The phenomenon of competition for a common transcription factor and the change in factor concentration during development as a means of differentially regulating genes was first postulated for the 5S rRNA genes in *Xenopus* (Brown, 1984). The 5S genes are transcribed by RNA polymerase III and do not produce protein products. They are organized as two distinct families: the oocyte type, which are transcribed during oogenesis and are organized as a multigene family consisting of 20,000 members, and the somatic genes, which are expressed during oogenesis and in somatic cells and consist of 400 members. These two gene families use a common transcription factor (TFIIIA), which has a differential affinity for the somatic and oocyte genes and decreases in concentration during development. This establishes a competitive situation in which the somatic genes may be able to outcompete the oocyte genes for the TFIIIA.

Recently, several examples have been discovered in which common or ubiquitous transcription factors may be involved in the differential or tissue specific regulation of gene expression. For example, Becker *et al.* (1987) demonstrated the presence of similar sequence-specific DNA-binding proteins in cell lines that were either expressing or not expressing the tyrosine aminotransferase (TAT) gene. The difference is that specific CpG dinucleotides are methylated in nonexpressing cells, which may inhibit binding of the common transcription factor. This suggests that the presence of a factor in a cell type may not be the only prerequisite for transcription of a particular gene. A similar situation exists for the immunoglobin (Ig) heavy gene enhancer (Gerster *et al.*, 1987). The activation of prolactin and growth hormone genes in different cell types of the anterior pituitary gland depends on the presence of a common *trans*-acting factor (Pit-1), which binds to the cell-specific elements of both genes (Nelson *et al.*, 1988). Thus, a single factor activates both genes in two different cell types.

There are also instances in which proteins may interact with one another to facilitate binding or influence activity of other proteins (Maniatis *et al.*,

1987). In fact, it has been shown that in yeast a second protein (a1) can alter the DNA-binding specificity of the α_2 protein, so that it now recognizes a completely different operator sequence (Goutte and Johnson, 1988).

In the sea urchin, in addition to the early and late histone genes referred to above, there are several tissue-specific variants. One of these, the sperm-specific histone H2B-1, contains two octamer (ATTTGCAT) (Parslow et al., 1984) and two CCAAT motifs 5′ of the gene. Barberis et al. (1987) identified a common binding factor in nuclear extracts from blastula, gastrula, and testes that binds to the CCAAT sequence. There is also a testis-specific octamer-binding factor. Unexpectedly, these workers found that the CCAAT-binding factor from only the testis extract was able to interact with the sperm H2b promoter. Extracts from either the blastula or gastrula stage embryo (where the testis-specific H2b histone is not expressed) contained a CCAAT displacement factor, which bound with high affinity to sequences overlapping the CCAAT sequence and interfered with the binding of the CCAAT-binding factor to the testis H2b gene in these extracts. This protein may be responsible for repressing transcription of the sperm-specific H2b histone gene during embryogenesis.

It is apparent that the phenomenon of transcriptional regulation is exceedingly complex and not only involves the direct interaction of single proteins with specific DNA motifs but also takes on the added complexity of protein–protein interactions, protein modification, multiple proteins binding to the same sequence, and single proteins binding to multiple sequences. It is quite possible that transcription in some cases may be regulated by influencing the ionic conditions of the intracellular milieu, permitting changes in binding affinities, thereby activating a transcriptional unit with the same proteins present in a cell not transcribing that gene.

Once a transcriptionally active complex is established during embryogenesis, it is necessary to transmit the information specifying this pattern to daughter cells through numerous rounds of DNA replication. Attempts to understand and reproduce this phenomenon have been only partially successful. There is evidence that once an active transcription complex is established it is stable through many rounds of transcription (Bogenhagen et al., 1982; Mattaj et al., 1985). In fact, once established, it is difficult to compete for transcription factors from an established complex (Mancebo and Etkin, 1988). However, these complexes, at least in the case of Xenopus 5S genes, tend to dissociate when replication takes place (Wolffe and Brown, 1986; Wolffe et al., 1986).

The chromatin conformation surrounding active genes is marked by the presence of DNase-hypersensitive sites, and these may be heritable. It was found that DNase I-hypersensitive sites, once induced in the globin genes in Roux sarcoma-infected chicken embryo fibroblasts, were maintained for more than 20 cell generations, even in the absence of conditions inducing the expression of these genes (Groudine and Weintraub, 1982). This suggests that at least the chromatin conformation once established around an active gene may be maintained and transmitted through many replication cycles.

An example by which information specifying an active gene or chromatin domain may be inherited is by modification of the DNA involving methylation of CpG sequences. The case of genomic imprinting observed in mice may be due to the methylation states of individual genes or regions of chromosomes. Thus, the imprinted pattern (perhaps due to methylation) is inherited from generation to generation of somatic cells in an embryo but is subject to a mechanism of erasure on passage through the germ line (see Chapter 4, *this volume*). This has far-reaching implications with regard to mechanisms of determination and differentiation, for it provides a remarkably simple method of inheritance and modulation of patterns of gene expression during development.

The collected studies reviewed by the contributors in this volume show that the genomes of eukaryotic animal cells have evolved various genetic mechanisms that control cell specialization. Although some cell specializations involve gains, losses, or rearrangements of DNA sequences and are unique to specific organisms or cell types, most cells do not appear to exhibit irreversible changes in the DNA. In these latter cases, inactive genes in differentiated cells can be activated under appropriate experimental conditions and, in the case of nuclear transplantation in amphibians, nearly an entire genetic program can be activated. The collective studies demonstrate that genomic multipotentiality is widespread among the animal phyla. Future exploitation of the systems described should permit analysis of regulatory factors involved in activating not only single genes but also those responsible for activating an entire genome in both normal and cancer cells.

References

Barberis, A., Superti-Furga, G., and Busslinger, M., 1987, Mutually exclusive interaction of the CCAAT-binding factor and of a displacement protein with overlapping sequences of a histone gene promoter, *Cell* **50**:347–359.

Becker, R., Ruppert, S., and Schutz, G., 1987, Genomic footprinting reveals cell type-specific DNA binding of ubiquitous factors, *Cell* **51**:435–443.

Benyajati, C., Spoerel, N., Haymerle, H., and Ashburner, M., 1983, The messenger RNA for alcohol dehydrogenase in *Drosophila melanogaster* differs in its 5′ end in different developmental stages, *Cell* **33**:125–133.

Blau, H., Pavlath, G., Hardeman, E., Chiu, C. P., Siferseteen, L., Welster, S., Miller, S., and Welster, C., 1985, Plasticity of the differentiated state, *Science* **230**:758–766.

Bogenhagen, D. F., Wormington, M., and Brown, D. D., 1982, Stable transcription complexes of *Xenopus* 5S RNA genes: A means to maintain the differentiated state, *Cell* **28**:413–421.

Brown, D. D., 1984, The role of stable complexes that repress and activate eukaryotic genes, *Cell* **37**:359–364.

Busslinger, M., Hurst, J., and Flavell, R. A., 1983, DNA methylation and the regulation of globin gene expression, *Cell* **34**:197–206.

Chandler, L. A., and P. A. Jones, 1988, Hypomethylation of DNA in the regulation of gene expression, in: *Developmental Biology: A Comprehensive Synthesis*, Vol. 5: *The Molecular Biology of Cell Determination and Cell Differentiation* (L. W. Browder, ed.), pp. 335–349, Plenum, New York.

Davidson, E., 1986, *Gene Activity in Early Development*, Academic, New York.

Dreyer, C., Scholz, E., and Hausen, P., 1982, Oocyte nuclear proteins in early development of *Xenopus* laevis, *Wilhelm Roux Arch.* **191**:228–233.

Driever, W. and Nusslein-Volhard, D., 1988, A gradient of bicoid protein in *Drosophila* embryos, *Cell* **54:**83–93.

Ettensohn, C. A., and McClay, D., 1988, Cell lineage conversion in the sea urchin embryo, *Dev. Biol.* **125:**396–409.

Gerster, T., Matthias, P., Thali, M., Jiricny, J., and Schaffner, W., 1987, Cell type specificity elements of the immunoglobin heavy chain gene enhancer, *EMBO J.* **6:**1323–1330.

Goutte, C., and Johnson, A., 1988, A protein alters the DNA binding specificity of α2 repressor, *Cell* **52:**875–882.

Groudine, M., and Weintraub, H., 1982, Propagation of globin DNAase I-hypersensitive sites in absence of factors required for induction: A possible mechanism for determination, *Cell* **30:** 131–139.

Heberlein, U., and Tjian, R., 1988, Temporal pattern of alcohol dehydrogenous gene transcription reproduced by *Drosophila* stage-specific embryonic extracts, *Nature (Lond.)* **331:**410–415.

Hesse, J. E., Lieber, M. R., Gellert, M., and Mizuuchi, K., 1987, Extrachromosomal DNA substrates in Pre-b cells undergo inversion or deletion at immunoglobulin V-(D)-J joining signals, *Cell* **49:** 775–783.

Horstadius, S., 1939, The mechanisms of sea urchin development, studied by operative methods, *Biol. Rev.* **14:**132–179.

Kemphues, K., Priess, J., Morton, D., and Niansheng, C., 1988, Identification of genes required for cytoplasmic localization in early *C. elegans* embryos, *Cell* **52:**311–320.

Killary, A., and Fournier, K., 1984, A genetic analysis of extinction: Transdominant loci regulate expression of liver-specific traits in hepatoma hybrid cells, *Cell* **38:**523–534.

Kimelman, D., and Kirschner, M., 1987, Synergistic induction of mesoderm by FGF and TGF β and the identification of an mRNA coding for FGF in the early *Xenopus* embryo, *Cell* **51:**869–878.

King, M. L., and Barklis, E., 1985, Regional distribution of maternal messenger RNA in the amphibian oocyte, *Dev. Biol.* **112:**203–212.

Klar, A. J. S., 1987a, Determination of yeast cell lineage, *Cell* **49:**433–435.

Klar, A. J. S., 1987b, Differential parental DNA strands confer developmental asymmetry on daughter cells in fission yeast, *Nature (Lond.)* **326:**466–470.

Kostriken, R., Strathern, J., Klar, A. J. S., Hicks, J., and Heffron, F., 1983, A site-specific endonuclease essential for mating-type switching in *S. cerevisiae*, *Cell* **35:**167–174.

Kuziora, M. A. and McGinnis, W., 1988, Autoregulation of a *Drosophila* homeotic selector gene, *Cell* **55:**477–485.

Maizels, N., 1987, Diversity achieved by diverse mechanisms: Gene conversion in developing B cells of the chicken, *Cell* **48:**359–360.

Mancebo, M., and Etkin, L., 1988, Competition between injected DNA templates in injected *Xenopus* oocytes, *Exp. Cell Res.* **178:**469–478.

Maniatis, T., Goodbourn, S., and Fischer, J., 1987, Regulation of inducible and tissue-specific gene expression, *Science* **236:**1237–1245.

Mattaj, I., Lienhard, S., Jirieny, J., and DeRobertis, E., 1985, An enhancer-like sequence within the *Xenopus* U2 gene promoter facilitates the formation of stable transcription complexes, *Nature (Lond.)* **316:**163–167.

Maxson, R., Ito, M., Balcells, S., Thayer, M., French, M., Lee, F., and Etkin, L., 1988, Differential stimulation of sea urchin early and late H2b histone gene expression by a gastrula nuclear extract after injection into *Xenopus laevis* oocytes, *Mol. Cell. Biol.* **8:**1236–1246.

Moen, T., and Namenworth, M., 1977, The distribution of soluble proteins along the animal-vegetal axis of frog eggs, *Dev. Biol.* **58:**1–10.

Nelson, C., Albert, V., Elsholtz, H., Lu, L. I.-W., and Rosenfeld, M. G., 1988, Activation of cell-specific expression of a rat growth hormone and prolactin genes by a common transcription factor, *Science* **239:**1400–1405.

Okazaki, K., Davis, D. D., and Sakano, A., 1987, T cell receptor β gene sequences in the circular DNA of thymocyte nuclei: Direct evidence for intramolecular DNA deletion in V–D–J joining, *Cell* **49:**477–485.

Parslow, T. G., Blair, D. L., Murphy, W. J., and Granner, D. K., 1984, Structure of the 5′ end of immunoglobulin genes: A novel conserved sequence, *Proc. Natl. Acad. Sci. USA* **81**:2650–2654.

Posakony, J. N., Fischer, J. A., and Maniatis, T., 1986, Identification of DNA sequences required for the regulation of Drosophila alcohol dehydrogenase gene expression, *Cold Spring Harbor Symp. Quant. Biol.* **50**:515–520.

Priess, J. R., and Thompson, S. N., 1987, Cellular interaction in early *C. elegans* embryos, *Cell* **48**:241–250.

Sachs, L., 1987, The molecular control of blood cell development, *Science* **238**:1374–1379.

Strathern, J. N., Klar, A. J. S., Hicks, J. B., Abraham, J. A., Ivy, J. M., Nasmyth, K. A., and McGill, C., 1982, Homothallic switching of yeast mating type cassettes is initiated by a double-stranded cut in the MAT locus, *Cell* **31**:183–192.

Sulston, J., Schierenberg, E., White, J., and Thomson, N., 1983, The embryonic cell lineage of the nematode *Caenorhabditis elegans*, *Dev. Biol.* **100**:67–119.

Van der Ploeg, L. H. T., 1987, Control of variant surface antigen switching in trypanosomes, *Cell* **51**:159–161.

Weeks, D. L., and Melton, D. A., 1987a, A maternal mRNA localized to the vegetal hemisphere in *Xenopus* eggs codes for a growth factor related to TGF β, *Cell* **51**:861–867.

Weeks, D. L., and Melton, D. A., 1987b, An mRNA localized to the animal pole of *Xenopus* encodes a subunit of a mitochondrial ATPase, *Proc. Natl. Acad. Sci. USA* **84**:2798–2802.

Weeks, D. L., Rebagliati, M. R., Harvey, R. P., and Melton, D. A., 1985, Localized maternal mRNAs in *Xenopus laevis* eggs, *Molecular Biology of Development*, Vol 50, Cold Spring Harbor Symposia on Quantitative Biology, Cold Spring Harbor, MA, pp. 21–30.

Wolffe, A. P., and Brown, D. D., 1986, DNA replication *in vitro* erases a *Xenopus* 5S RNA gene transcription complex, *Cell* **47**:217–227.

Wolffe, A., Jordan, E., and Brown, D. D., 1986, A bacteriophage RNA polymerase transcribes through a *Xenopus* 5S gene transcription complex without disrupting it, *Cell* **44**:381–389.

Chapter 2

DNA Gains, Losses, and Rearrangements in Eukaryotes

DAVID M. PRESCOTT

1. Introduction

Cells undergo various stable changes in structural and functional properties in a variety of situations. Several chapters in this volume discuss such changes and the circumstances and mechanisms associated with their occurrence. Stable changes in the structural and functional properties of cells are generally based on modifications at the level of DNA. Five kinds of modifications are known: (1) elimination of DNA sequences; (2) amplification of selected sequences; (3) rearrangement of DNA segments, with no loss or gain of sequence; (4) methylation of DNA; and (5) changes in transcriptional activity of DNA mediated by transcription factors.

Most stable changes in cells come about by the turning on and off of the expression of particular genes by *trans*-acting transcriptional factors. Some modulations are reversible and some are normally permanent, as in the terminal differentiation of cells in multicellular organisms. These kinds of changes, as well as changes in transcription achieved through methylation of DNA, are discussed in other chapters. This chapter is concerned with change in cells brought about by the gain, loss, or rearrangement of DNA sequences in the eukaryotes of animal cells.

2. Gain of DNA Sequences

For genes that encode proteins, gene expression is normally amplified in two successive stages, which form a cascade: transcription and translation. Many primary transcripts are produced from a single gene; each messenger RNA (mRNA) molecule formed from a primary transcript is, in turn, amplified into many copies of an encoded protein. By these two amplifications, sufficient copies of a polypeptide can be produced from a single pair of alleles in most

DAVID M. PRESCOTT • Department of Molecular, Cellular and Developmental Biology, University of Colorado, Boulder, Colorado 80309-0347.

kinds of diploid cells, even in cases in which an enormous amount of a single protein type is required in a short time, e.g., hemoglobin in differentiating red blood cells and silk protein in silk gland cells.

A few exceptions to this generality are known. In *Drosophila*, ovarian follicle cells synthesize large amounts of proteins that form the chorion of the egg in just a few hours during egg formation. The genes encoding these proteins are expressed at high levels in follicle cells at the appropriate time (Spradling and Mahowald, 1980). Before synthesis of the chorion proteins, genes encoding them increase in copy number by differential replication. The genes occur in two clusters, which, between them, contain single copies of genes for the six major chorion proteins. One cluster, located in the X chromosome, undergoes a 15-fold amplification, and the second cluster, in chromosome III, is amplified 60-fold (Kalfayan *et al.*, 1985). The amplified copies in both clusters are produced by extra rounds of bidirectional replication beginning in one or more origins of replication and extending for variable distances (in successive rounds) up to 40–50 kilobase pairs (kbp). The amplified DNA remains in the chromosome in an onion skin arrangement (Osheim and Miller, 1983). This is in contrast to amplification of genes encoding ribosomal RNA (rRNA) in amphibian and certain insect oocytes, in which the extra gene copies are extrachromosomal and form multiple nucleoli (reviewed by Gall, 1976; see also Miller, 1981).

The *trans*-acting molecular signals that trigger and regulate the number of extra replications and the signals themselves have not been identified. Such signals must recognize specific *cis* elements in the two sets of genes, presumably located in the origins of replication. Production of the *trans* signals must be developmentally regulated within the ovarian follicle cells. Two genes have been identified by mutation that are involved in generating the *trans* signal to both chorion gene clusters (Orr *et al.*, 1984). This work is an important step toward deciphering the developmental mechanism that regulates the specific amplification of the chorion genes.

Specific amplification of the chorion genes does not occur in the silk moth (Kafatos *et al.*, 1985). Instead, in the moth, formation of the chorion takes place more slowly and the moth germ-line genome also contains a multiplicity of genes for chorion proteins. These two factors presumably obviate the need for differential replication of genes to meet the protein requirement of choriogenesis.

Amplification of specific DNA sequences was described many years ago in polytene chromosomes of the midge *Rynchosciara* by Pavan and Da Cunha (1969). Polytene chromosomes in insects, plants, and protozoa provide individual cells with multiple copies of all genes. In insects and presumably in plants (but probably not in protozoa), polytene chromosomes represent a strategy for quickly accommodating demands for large amounts of gene products. In polytene chromosomes of *Rynchosciara*, however, further amplification of a few specific genes is superimposed on the generalized amplification of polytenization. The selective amplifications are recognized as puff formation in certain chromosome bands in which DNA replication precedes RNA synthesis in the puff. A protein produced in a large amount after these amplifications has been identified as a candidate for the coding function of a DNA puff.

The first example of specific gene amplification to be discovered is the differential replication of DNA encoding rRNA in amphibian oocytes, a phenomenon subsequently found to be widespread, but not universal, among animals, including insects (Gall, 1976; see also the discussion and references on pages 334–337 in Bostock and Sumner, 1978). Replication occurs just before meiosis or in meiotic prophase. At least one copy of the DNA segment encoding 28S, 18S, and 5.8S rRNA plus the DNA spacers between coding regions is produced from the chromosome. The extra DNA separates from the chromosome and is somehow circularized. It then replicates many times by the rolling circle mechanism, producing many additional DNA circles of various contour lengths. Each circle contains integral multiples of the basic unit encoding one each of the three rRNAs and DNA spacers. The extrachromosomal circles form separate nucleoli, in which rRNA is synthesized and assembled into ribosomes. In this way, a large store of ribosomes (about 10^{12} ribosomes in a frog oocyte) is built up, to be used in embryonic development. In some animal species lacking amplification of rRNA genes, large amounts of rRNA are synthesized in nurse cells surrounding an oocyte, and the RNA is transferred in the oocyte. This apparently accomplishes the same end as amplification of rRNA genes.

In frog oocytes, amplified rDNA accounts for 75% of the total DNA in the oocyte. After fertilization of the egg, the chromosomal DNA replicates with each cell cycle but the extrachromosomal rDNA does not. Therefore, the extra copies of rDNA become rapidly diluted to an undetectable level early in embryonic development.

Differential amplification of rDNA and other genes occurs in the macronucleus in ciliated protozoa, as discussed in Section 5.1.5. These several examples (chorion genes in *Drosophila*, rDNA in oocytes, and rDNA and a few other genes in ciliates) demonstrate that gene amplification has evolved several times as a possible way to meet a demand for large amounts of a gene product. However, differential amplification of genes is clearly not a general strategy used by somatic nuclei to achieve *transient* increases in gene expression. Permanent amplifications of a few genes in the germ lines of most eukaryotes, i.e., genes encoding rRNA, transfer RNA (tRNA), and histones, are well known.

At least two cases of experimentally forced amplification of specific genes are known. When mammalian cells are grown in culture in the presence of methotrexate, an occasional cell (perhaps 1 in 10^3) undergoes amplification of the single copy of the gene encoding dihydrofolate reductase (Schimke et al., 1979). Dihydrofolate reductase catalyzes synthesis of methylene tetrahydrofolate, the methyl-contributing coenzyme required for synthesis of thymidine monophosphate (TMP) from deoxyuridine monophosphate. Methotrexate inhibits dihydrofolate reductase, blocking DNA replication and cell reproduction. In the occasional cell that undergoes amplification of the gene encoding dihydrofolate reductase, the inhibition of methotrexate is overcome by expression of the amplified copies of the gene. Up to 1000 extra copies of the gene may be built up. The amplified copies of the gene can occur as segments of

DNA integrated into a chromosomal DNA molecule at the site of the original gene. In such case, they form a homogeneously staining region in the chromosome. The amplified genes may also be present in an unstably maintained form as small chromosome fragments called double minutes. Since double minutes lack centromeres, they cannot be evenly distributed during mitosis and are rapidly lost from a cell clone when the selective pressure (methotrexate) is removed (reviewed by Cowell, 1982).

Selective amplification of other genes in cultured cells has also been induced by treatment with other toxic substances. For example, amplification of genes encoding the protein metallothionein occurs in cells chronically exposed to a toxic concentration of Cd^{2+} ions (Koropatnick *et al.*, 1985). However, amplification of genes in response to such drugs as methotrexate (mTx) or poisons such as Cd^{2+} ions has not been reported for the germ line of animals.

3. Rearrangement of DNA Sequences

Change in arrangements of DNA sequences occurs widely among eukaryotes but, with few exceptions, these are not achieved by physical shifting of DNA segments. Most rearrangements are mediated by transposable elements, most or all of which are transcribed into RNA, reverse transcribed into DNA, and inserted in chromosomal DNA at new locations. Genes at the point of insertion may be mutated or otherwise inactivated by the insertion. Removal of a transposable element by deletion (excision) may result in reactivation of gene expression at the site of the excision. Gene conversion can also provide a change in DNA sequence arrangement without actual physical shifting of DNA segments. Examples are switching of mating type in yeast (Klar *et al.*, 1984) and switching of genes encoding surface antigens in trypanosomes (Borst, 1986; De Lange, 1986). Since neither transposable elements nor gene conversion involves rearrangement of DNA segments directly, they are not considered further here.

Two phenomena that involve cutting and splicing of DNA have crucial genetic consequences. During the development of an animal, the diversity of genes encoding heavy and light chains of antibodies is generated by combining segments of DNA. A given V segment is spliced to a given J segment following deletion of all intervening V and J segments. In this situation, the process is regular in that deletions occur as specific blocks of multiple V and J segments, rather than as deletion of completely random-sized stretches of DNA.

The formation of antibody genes by deletion is a normal developmental process. The second instance of cutting and splicing of DNA is the more or less random breakage and rejoining of chromosome segments known as translocations. Translocation is probably always the effect of agents that break DNA directly or that cause breakage by interference with DNA replication. Therefore, translocations are generally viewed as abnormal events. In no case do they represent a normal or regulated rearrangement of DNA sequences.

Translocations can produce various effects including mutation of genes at break points, fusion of genes, and changes in gene expression by placing genes under the influence of a different promoter. The latter effect contributes to the transformation of normal cells into transformed cells through enhancement of transcriptional activity or proto-oncogenes. Such enhancement of activity is one of several ways in which proto-oncogenes are converted to oncogenes.

Most kinds of transformed cells have elevated rates of spontaneous trans-location, producing a variety of aneuploid phenotypes. In addition to transloca-tions, many types of transformed cells lose and gain chromosomes through a propensity for mitotic nondisjunction, which further contributes to karyotype variation. The reasons for the elevated rates of translocation and nondisjunc-tion are not known.

4. Loss of DNA Sequences

Loss of entire chromosomes, parts of chromosomes, or DNA sequences dispersed within chromosomes occurs in a few animal species during their development, but it is a relatively rare phenomenon. The losses occur in very early development as cell lineages for future somatic nuclei and future germ-line nuclei are established; losses are restricted to the somatic cell lineage. Ammermann (1985) distinguishes three kinds of DNA in these situations: G DNA is present in nuclei destined to serve as the germline; S DNA is the part of G DNA present in somatic nuclei; and E DNA is the part of G DNA that is eliminated in the formation of somatic nuclei.

Elimination of DNA is a normal part of development in a few species in four taxonomic groups: arthropods, nematodes, insects, and ciliated protozoa. Chromatin diminution was described in the nematode *Ascaris* by Boveri (1887) a century ago. The eliminated DNA is in large heterochromatic blocks; elimina-tion occurs during the first cleavage divisions. Measurements of how much DNA is eliminated are not in agreement. The estimates range from 22% of G DNA (Moritz and Roth, 1976), to 27% (Tobler *et al.*, 1972), to 34% (Pasternak and Barrell, 1976), to 56% (Davies and Carter, 1980). The estimates of remain-ing DNA (S DNA) range from 0.15 to 0.46 pg. In *Parascaris equorum* (a horse parasite), 85% of the DNA is eliminated (Moritz and Roth, 1976).

In *Ascaris*, the eliminated DNA (E DNA) is made up at least in part of repetitive sequences that form a satellite in a CsCl isopyknic gradient (Roth and Moritz, 1981; Mueller *et al.*, 1982). As much as 50% of the E DNA could be composed of unique sequences according to Tobler *et al.* (1972) and to Gold-stein and Straus (1978).

According to Roth and Moritz (1981), about 90% of the S DNA is made up of unique sequences and 10% of repetitive sequences, but only a tiny fraction of the germ-line repetitive sequences correspond to the E sequences that undergo elimination. Mueller *et al.* (1982) conclude from hybridization experiments that E DNA (before elimination) is not transcribed. Whether E DNA in *Ascaris* has any function is not known.

Chromatin elimination also occurs in several species of copepod crustaceans. In the fourth through seventh cleavages in the early blastula of three species of *Cyclops*, about 50% of the G DNA is eliminated (Beermann, 1977). In one species, the E DNA is present in large heterochromatic clumps, which in another species are at the ends of chromosomes and in another are present both at ends and at kinetochores; in a third species, the heterochromatic E DNA blocks are scattered throughout the chromosomes. Thus, the normal elimination of E DNA requires regulated breakage and rejoining of chromosomes. During the stages of elimination, circular strings of nucleosomes of <1 μm to many microns in length appear in the cells (Beermann, 1984). This suggests that elimination involves an intrachromosomal recombination event in which E DNA forms circles during its excision from chromosomes.

Among some insects of *Diptera*, entire chromosomes are lost from somatic cells during early cleavage but are retained in germ-line cells (White, 1954). In gall midges, these chromosomes, composed entirely of E DNA, are necessary for development of spermatocytes and oocytes (Bantock, 1961; Geyer-Duszynska, 1966). E chromosomes are apparently active in RNA synthesis during oogenesis, in keeping with the idea that E chromosomes in insects contain genes essential for gametogenesis.

The gains and losses of DNA among multicellular organisms (animal) discovered to date are listed in Table I. There are no proven cases of direct rearrangement of DNA sequences; in general, DNA-sequence rearrangement involves transcription of DNA into RNA and reverse transcription into DNA or gene conversion. By far the most extensive changes in DNA occur in ciliated protozoa.

5. The Ciliate Genome

Ciliates contain two kinds of nuclei, a micronucleus and macronucleus, which are quite different from each other. Some species contain two or more of each kind. For example, *Urostyla grandis* (a hypotrich) contains more than 100 macronuclei and many micronuclei. The multiple micronuclei in a cell are, in principle, all identical to each other in structure and function, and the multiple macronuclei are all structurally and functionally the same.

The genome of a micronucleus is made up of a set of chromosomes (usually diploid) and divides by mitosis. The genes in a micronucleus are contained in high-molecular-weight chromosomal DNA. However, micronuclear genes are not expressed (e.g., no RNA synthesis) in vegetative cells. The macronucleus (or macronuclei) provides all the RNA (except RNA made in mitochondria) needed for cell metabolism and reproduction. In sharp contrast to micronuclear DNA, all the DNA in a macronucleus is in subchromosomal molecules; i.e., DNA molecules have lower molecular weights than do micronuclear (chromosomal) DNA. Also, the macronucleus divides amitotically. Amitosis is imprecise; one daughter macronucleus usually receives a little more DNA than the other. This random mode of nuclear division is successful because every DNA molecule is present in many identical copies (sometimes thousands), and

Table I. Gains and Losses of DNA

Organism	DNA changes
	Normal changes
Some nematodes	Elimination of DNA from somatic cells during early cleavage
Some *Diptera*	Elimination of DNA from somatic cells during early cleavage
Copepods	Elimination of DNA from somatic cells during early cleavage
Drosophila	Amplification of chorion genes in follicle cells
Many animals	Amplification of rDNA sequences in oocytes
Lymphocytes	Deletion of DNA sequences in formation of immuno-globulin genes
Ciliated protozoa	Elimination of DNA sequences, amplification of DNA sequences, and rearrangement of DNA sequences during macronuclear development
	Abnormal changes
All eukaryotes	Chromosomal translocations and deletions
Vertebrates	Amplification of various genes (e.g., gene-encoding dihydrofolate reductase, metallothionein, protoon-cogenes), usually under some cytotoxic or genotoxic environmental condition

each daughter macronucleus receives some copies of every molecule, hence receives some copies of every DNA sequence to be found in macronuclei of the species.

The micronucleus is a germ-line nucleus. It undergoes meiosis during cell mating (conjugation). In cell mating, haploid micronuclei are exchanged through a cytoplasmic bridge between the two cells of a mating pair. After it is exchanged, a haploid micronucleus fuses with a stationary haploid micronucleus in the recipient cell, forming a diploid micronucleus. About this time, the mating cells separate. The new diploid micronucleus in each exconjugant cell divides mitotically without accompanying cell division, and one of the daughter micronuclei develops into a new macronucleus. While this is occurring, the old macronucleus (or macronuclei) and all unused haploid micronuclei are destroyed. Variations of this pattern occur in different species, but the overall principles are the same.

5.1. DNA Changes in Ciliates

Extensive changes in genomic DNA occur in the development of the macronucleus from a micronucleus in ciliates. Most of the studies have been done on a few species among the holotrichs, chiefly in the genus *Tetrahymena* and on a few species of hypotrichs in the genera *Euplotes, Stylonychia,* and *Oxytricha.* A consistent picture has emerged, with some major differences between the holotrichs and the hypotrichs. The main features of macronuclear development are outlined here to illustrate how these organisms manipulate their genomes. Details of the processes are described in reviews (Brunk, 1986; Karrer, 1986) and the original research reports.

Changes in the ciliate genome during formation of a macronucleus include the following:

1. Elimination of chromosomes
2. Conversion of chromosomal DNA to subchromosomal molecules
3. Elimination of repetitive and unique sequences between genes
4. Splicing of DNA segments
5. Elimination of intron-like sequences
6. Rearrangement of sequences into different orders
7. Addition of repeat sequences (telomeres) at the ends of the subchromosomal molecules

These changes are discussed briefly in the next several sections.

5.1.1. Elimination of Chromosomes

The micronucleus of the hypotrich *Stylonychia lemnae* contains a variable number of chromosomes (Ammermann, 1987) estimated to range between 100 and 180, with an average of 140, (haploid numbers). Only 35 or 36 of these chromosomes participate in formation of a macronucleus; most chromosomes (containing exclusively E DNA) are eliminated from the micronucleus just as it embarks on its development into a macronucleus (Ammermann et al., 1974). The eliminated chromosomes are condensed (heterochromatic) at the time of elimination. This elimination is quite similar to the elimination seen in *Ascaris* and copepods, in which germ-line DNA (G DNA) is composed partly of DNA that is eliminated (E DNA) and partly of DNA that is retained (S DNA) in formation of somatic cells. The situation in these ciliates is more complex, however, since most of the DNA sequences in the retained 35 or 36 chromosomes (putative S DNA) are eliminated later in macronuclear development when individual genes are excised from the chromosomes.

Chromosome elimination has not been documented in other hypotrichs, although it might occur but not be prominent enough to be readily seen cytologically. Some sequences are known not to participate in the subsequent DNA replication that characterizes macronuclear development (Spear and Lauth, 1976), and these may have been eliminated. They may in fact be eliminated in a manner similar to elimination in *Stylonychia*.

In the holotrich *Tetrahymena*, DNA representing about 15% of the DNA sequence complexity in the micronucleus is eliminated (reviewed by Gorovsky, 1980). This diminution does not occur as a visible cytological event in the form of chromatin elimination as in *Stylonychia*, but it does occur very early during the rounds of DNA replication that are part of macronuclear development (Brunk and Conover, 1985; Yokoyama and Yao, 1982). The sequences in question are not simply diluted out by underreplication but are actively and specifically destroyed.

Thus, both hypotrichs and holotrichs undergo DNA diminution at the start of macronuclear development, an event that is extensive and visible as chromosome elimination in *Stylonychia* but much less extensive and not cytologically discerned in *Oxytricha* or *Tetrahymena*. The molecular mechanism by which the E DNA is recognized and then destroyed is completely unknown.

5.1.2. The Second Elimination of DNA Sequences in Macronuclear Development in Hypotrichs

After the first elimination of DNA in hypotrichs, the remaining DNA undergoes repeated replication to form polytene chromosomes. By contrast, the early elimination of sequences in the holotrich *Tetrahymena* apparently leaves the DNA in subchromosomal fragments. These replicate repeatedly (after telomere addition; see Section 5.1.3) but do not form polytene structures. These events in holotrichs are summarized in the diagram in Fig. 1. A more detailed account of sequence elimination in holotrichs (and hypotrichs) is given in Section 5.1.4.

The formation of polytene chromosomes in hypotrichs is complete about 2 days after cell mating. The chromosomes then become fragmented. Fragmentation in its earliest stage is detectable by electron microscopy (Kloetzel, 1970). A septum, believed to be composed of protein, forms through every interband of every polytene chromosome. The septa subsequently extend to form separate envelopes around each individual chromosomal band. In effect, vesicles are

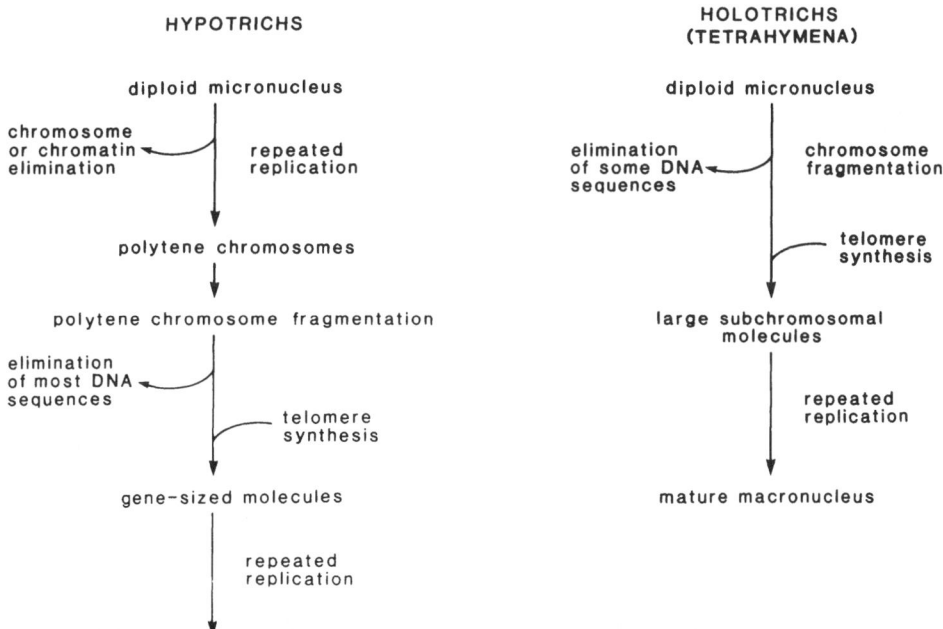

Figure 1. Comparison of the processing of DNA in hypotrichs and holotrichs during macronuclear development.

formed, each containing the material present in a chromosome band plus one half of each adjacent interband. Most of the DNA in each vesicle is then degraded, leaving small molecules ranging in size from a few hundred base pairs to 15 kbp. The number average size is about 2200 bp in *Oxytricha* (Swanton *et al.*, 1980). Each molecule contains a single coding sequence (gene), and a particular gene is usually encoded by one but sometimes several sizes of DNA molecules.

These events are a major difference between hypotrichs and holotrichs. In holotrichs, the elimination of DNA sequences occurs in one step—near the beginning of macronuclear development. This elimination, consisting of 15% of the total sequence complexity, apparently generates subchromosomal DNA molecules. With the exception of addition of telomeric sequences (Section 5.1.3), apparently no further changes in the DNA molecules occur, and they replicate repeatedly to form the mature macronucleus. The size of the molecules in the macronucleus ranges from less than 100 kbp to more than 1500 kbp (Conover and Brunk, 1986; Altschuler and Yao, 1985). There are about 270 different-sized molecules, each containing many genes. A particular gene is always encoded by the same-sized molecule.

Fragmentation of the polytene chromosomes in hypotrichs is immediately followed by massive elimination of sequences and formation of very small molecules (Prescott, 1983; Klobutcher and Prescott, 1986; Kraut *et al.*, 1986). Overall, both holotrichs and hypotrichs break up the DNA in their chromosomes and eliminate sequences. In holotrichs, large subchromosomal molecules, each containing many genes, are formed. The hypotrichs carry the process to an extreme, eliminating about 95% of all sequences and reducing all DNA to small molecules, each containing a single gene.

5.1.3. Addition of Telomeres

Fragmentation of chromosomal DNA and elimination of sequences generates large subchromosomal, linear DNA molecules in holotrichs and short, linear DNA molecules in hypotrichs. Telomeric sequences consisting of many repeats of 5′ CCCCAA 3′ in holotrichs (Blackburn, 1986) and of a fixed number of 5′ CCCCAAAA 3′ repeats in hypotrichs (Oka *et al.*, 1980; Klobutcher *et al.*, 1981) are added to both ends of every molecule. The telomeric sequences differ somewhat in arrangement in the two groups. The telomeres are synthesized on the ends of the DNA molecules by an enzyme called telomere terminal transferase (Greider and Blackburn, 1985, 1987; Zahler and Prescott, 1988). The telomeres are important for the replication of the ends of the DNA molecules (Pluta *et al.*, 1984), hence for the maintenance (stability) of the molecules in vegetative macronuclei. They may also have functions in organization of the DNA into chromatin.

5.1.4. Intricacies of Sequence Elimination

The simplest scheme for the fragmentation of chromosomes and the elimination of sequences in holotrichs would be the recognition and destruction of specific short segments of DNA between longer sequences. The longer sequences would then be the subchromosomal molecules of the macronucleus. In fact, this probably happens, but additional events occur that make the process more complex, for example events in formation of the DNA molecule in the macronucleus encoding rRNA. The molecule is 11 kb long and contains two copies of the transcription unit for rRNA arranged in inverted order with a spacer between them (Yao *et al.*, 1985; Yao, 1986). In addition, noncoding stretches of DNA are present just inside the telomeres at the two ends. This molecule is derived from a single copy of the rRNA transcription unit in micronuclear DNA. Thus, excision of the rDNA is accompanied by replication in a way that creates a palindromic molecule of 11 kb containing two copies of the transcription unit. Other subchromosomal molecules are formed by deletion of sequences between segments, the ends of which are then spliced together to make longer subchromosomal molecules (Austerberry *et al.*, 1984). Although such cutting, deletion, and splicing occurs in a regular way, so that the same-sized molecule always contains the same genes, alternative cutting and splicing can occur (T. C. White and Allen, 1986; Allitto and Karrer, 1986). In addition, some sequences that are eliminated in one strain of *Tetrahymena* may be retained in the macronucleus of another strain (Karrer, 1986). Altogether, it is estimated that there are about 5000 sites in the micronuclear genome at which breakage, deletion, and splicing take place to yield the 270 different subchromosomal molecules.

The amount of sequence eliminated in connection with chromosomal fragmentation in hypotrichs is much greater than that which occurs in holotrichs. The final molecular products are much smaller, and more kinds are formed. In *Oxytricha*, about 24,000 different gene-sized molecules are present in the macronucleus (Lauth *et al.*, 1976). These molecules are not simply excised from chromosomes during macronuclear development but are formed by cutting, deletion, and splicing of DNA segments. Most, if not all, macronuclear molecules, as they are represented within micronuclear chromosomes, contain sequences of a few dozen base pairs up to 1000 bp that are deleted when the macronuclear version of the gene is generated (Klobutcher *et al.*, 1984; Ribas-Aparicio *et al.*, 1987). These sequences, called internal eliminated sequences (IES), interrupt regions that are destined to become a part of the transcribed segments in macronuclear DNA molecules or to become part of the non-transcribed segments (leaders and trailers) in the macronuclear molecules.

Some micronuclear DNA segments in *Oxytricha* undergo alternative cutting, deletion, and splicing, giving rise to several macronuclear DNA molecules of different sizes (Herrick *et al.*, 1987a,b) that share a high degree of homology.

Finally, in the formation of the macronuclear molecule that encodes actin, IES are deleted and fragments spliced, but the fragments are spliced in a different order and polarity than that present in the micronuclear chromosome (A. F. Greslin and D. M. Prescott, unpublished results). This is the only case of a genuine rearrangement in DNA sequences in a ciliate that occurs as a normal part of ciliate development. All other changes in ciliates described as rearrangements are in fact deletions followed by splicing.

Enzymes are presumed to catalyze the various cutting–deletion–splicing–

reordering events that transform micronuclear chromosomes into macronuclear genes, but none has yet been identified, except the telomere terminal transferase.

It remains a possibility that all sequence processing is accomplished through RNA intermediates (transcribed from micronuclear DNA) in a manner analogous to the cutting, deletion, and splicing of RNA in the removal of introns from primary transcripts, followed by reverse transcription into DNA. Such a process is responsible for the generation of pseudogenes in higher eukaryotes. Lipps (1985) detected reverse transcriptase activity in the developing macronuclei of *Stylonychia*. Models designed to accommodate all the events of macronuclear DNA formation in hypotrichs via RNA intermediates requires a set of complex assumptions. In holotrichs such as *Tetrahymena*, it is virtually certain that formation of macronuclear DNA does not involve RNA intermediates.

5.1.5. Differential Amplification of rDNA

During the formation of the macronucleus in *Tetrahymena*, the various subchromosomal fragments are replicated equivalently to form the DNA-rich macronucleus. The final copy number is about 45 for each molecule. However, the gene encoding rRNA is derived from a single copy of the gene in the micronucleus (Yao and Gall, 1977) but is differentially amplified to yield about 9000 copies in the mature macronucleus (Yao and Gall, 1979). Likewise, in *Oxytricha* each of the 24,000 or so different molecules is present in about 1000 copies per macronucleus in the G1 period (Lauth *et al.*, 1976), but the gene for rRNA is present in 100,000 copies. The 100,000 copies are derived from probably two, but no more than a few, copies of the gene (per haploid genome) in the micronucleus of *Oxytricha* (J. C. Chappell and D. M. Prescott, unpublished results). Therefore, during macronuclear development, replication of the single gene for rRNA in *Tetrahymena* and of the two to several genes in *Oxytricha* is differentially regulated.

A few other DNA molecules undergo differential amplification during the clonal aging of *Stylonychia* (Steinbrück, 1983). Different clones show amplification of different molecules. Which molecules amplify is a genetic trait of the clone. The coding functions of these molecules are not known. Similar amplification of a few kinds of molecules has been observed in *Oxytricha*, including amplification of molecules encoding rRNA to the increased level of 1.5×10^6 copies per macronucleus. Attempts to produce cell clones with differential amplification of metallothionein genes by chronic exposure to heavy metal ions have so far not succeeded.

5.1.6. Overall Organization of Genes in Ciliate Chromosomes

In holotrichs, the micronuclear DNA is cut into about 270 large subchromosomal segments, each containing many genes. The only known exception is the gene encoding rRNA; it is present by itself in a separate, relatively small DNA molecule of 11 kb in two palindromically arranged copies. By contrast, in hypotrichs all micronuclear DNA is reduced to small molecules with a number average of 2200 bp (Swanton et al., 1980), with a single gene per molecule. For the few molecules that have been examined in detail, roughly one half of each molecule consists of a coding sequence; the remainder form nontranscribed leaders and trailers. The hypotrichs, in effect, define the maximum length of a gene, including its transcription and replication accessory sequences [each molecule is also a separate replication unit (Murti and Prescott, 1983)]. This circumstance can be used to show how all the genes are arranged in chromosomes relative to one another and without needing to know the coding function of the genes. Hybridization mapping of large segments of cloned, micronuclear DNA has shown that most genes are grouped along the chromosome (Boswell et al., 1983; Klobutcher et al., 1986). Each group contains two to several genes separated by short spacers. One group is separated from the next by spacers many kilobases in length, consisting of both unique and repetitious sequences. Both rRNA genes that have been identified occur singly and are separated from adjacent genes to both sides by many kilobases (J. C. Chappell, D. M. Prescott, and M. T. Swanton, unpublished results). The destruction of spacers between genes accounts for a major part of the loss of DNA sequence complexity during macronuclear development. The presence of spacers between genes (along with IES and introns) is a plausible explanation for the C-value (S-value) paradox in eukaryotes (Prescott, 1983).

What advantages accrue to ciliates by the extensive processing of DNA is not completely clear. Some possible explanations are the following. Amplification of the total genome in the formation of a macronucleus presumably allows for much larger cell size, which is important in the life-style of the organisms. Differentially higher amplification of genes encoding rRNA must occur to provide for adequate ribosome production. The differentially amplified rRNA genes form a band in a gel in which total macronuclear DNA has been electrophoretically distributed by size. Several more bands of varying intensities are present, suggesting that several additional genes may undergo differential amplification. The coding functions of these amplified DNA molecules are not known. The destruction of spacer DNA, eliminating all repetitious sequences and more than 90% of unique sequences in the genome of hypotrichs, eliminates the cost of carrying the large amount of nonessential DNA that would be present in the highly polyploid macronucleus required to support a very large cell size. However, Tetrahymena (and presumably other holotrichs) eliminates only 15% of its genome. The creation of thousands of subchromosomal DNA molecules (holotrichs) or millions of gene-sized molecules (hypotrichs) in the macronucleus may preclude division by mitosis for some reason. On the other hand, amitotic division of the macronucleus is only successful because of the high multiplicity of copies of each DNA molecule in ciliates. Finally, the DNA processing that accompanies macronuclear development may not be advantageous but may simply be necessary for transcriptional activation of genes, since micronuclear genes in general are never expressed.

Whatever the ciliates have gained by the unusual manipulations of their genomes, they have come about after the early separation of ciliates from other eukaryotes. Some events held in common with other eukaryotes, such as differential amplification of rDNA in oocytes, and cutting and splicing of DNA to generate Ig genes, probably arose independently in evolution. Further attestation to the singularity of the ciliates in evolution is the alteration of their genetic code. The codons UAA and UAG, which serve as stop codons in eukaryotes generally, are read as glutamine in ciliates (Helftenbein, 1985; Horowitz and Gorovsky, 1985; Hanyu et al., 1986). UGA is the only stop codon in ciliates.

6. Conclusions

It is remarkable how few examples are known of the alteration of the genome in eukaryotes. These are listed in Table I. Most eukaryotes examined so far show no evidence of genome change in relationship to organismic life cycles. Among the changes that do occur, the loss of germ-line sequence remains without compelling explanation of purpose. Other kinds of normally occurring changes are based on demand for more of a particular gene product (rRNA and chorion proteins) or generation of new genes from gene parts (Ig genes). In all other known situations, increased demand for gene products is met through increase in gene expression without change in gene copy number. By far the most extensive DNA changes known among all eukaryotes are those that take place as a part of macronuclear development in ciliated protozoa. The ciliates provide unique opportunities for studying mechanisms of DNA manipulation, as well as the fine structure of eukaryotic genomes.

ACKNOWLEDGMENTS. This work was supported by research grant R01GM19199 from the National Institute of General Medical Sciences, and by a grant from the National Foundation for Cancer Research to D. M. Prescott.

References

Allitto, B. A., and Karrer, K. M., 1986, A family of DNA sequences is reproducibly rearranged in the somatic nucleus of Tetrahymena, Nucleic Acids Res. **14**:8007–8025.

Altschuler, M. I., and Yao, M.-C., 1985, Macronuclear DNA of Tetrahymena thermophila exists as defined subchromosomal-sized molecules, Nucleic Acids Res. **85**:5817–5831.

Ammermann, D., 1985, Chromatin diminution and chromosome elimination: Mechanisms and adaptive significance, in: The Evolution of Genome Size (T. Cavalier-Smith, ed.), pp. 427–442, Wiley, New York.

Ammermann, D., 1987, Germline specific DNA and chromosomes of the ciliate Stylonychia lemnae, Chromosoma **95**:37–43.

Ammermann, D., Steinbrück, G., von Berger, L., and Hennig, W., 1974, The development of the macronucleus in the ciliated protozoan Stylonychia mytilus, Chromosoma **45**:401–429.

Austerberry, C. F., Allis, C. D., and Yao, M.-C., 1984, Specific DNA rearrangements in synchronously developing nuclei of Tetrahymena, Proc. Natl. Acad. Sci. USA **81**:7383–7387.

Bantock, C., 1961, Chromosome elimination in Cecidomyiidae, Nature (Lond.) **190**:466–467.

Beermann, S., 1977, The diminution of heterochromatic chromosomal segments in Cyclops (Crustacea, Copepoda), *Chromosoma* **60**:297–344.

Beermann, S., 1984, Circular and linear structures in chromatin diminution of Cyclops, *Chromosoma* **89**:321–328.

Blackburn, E. H., 1986, Structure and formation of telomeres in holotrichous ciliates, in: *Int. Rev. Cytol.* **99**:29–47.

Borst, P., 1986, How proteins get into microbodies (peroxisomes, glyoxysomes, glycosomes), *Biochim. Biophys. Acta* **866**:179–203.

Bostock, C. J., and Sumner, A. T., 1978, *The Eukaryotic Chromosome*, North-Holland, Amsterdam.

Boswell, R. E., Jahn, C. L., Greslin, A. F., and Prescott, D. M., 1983, Organization of gene and nongene sequences in micronuclear DNA of *Oxytricha nova*, *Nucleic Acids Res.* **11**:3651–3663.

Boveri, T., 1887, Ueber Differenzierung der Zellkerend waehrend der Furchung des Eies von *Ascaris megalocephala*, *Anat. Anz.* **2**:688–693.

Brunk, C. F., 1986, Genome reorganization in *Tetrahymena*, in: *Int. Rev. Cytol.* **99**:49–83.

Brunk, C. F., and Conover, R. K., 1985, Elimination of micronuclear specific DNA sequences early in anlagen development, *Mol. Cell. Biol.* **5**:93–98.

Conover, R. K., and Brunk, C. F., 1986, Macronuclear DNA molecules of *Tetrahymena thermophila*, *Mol. Cell. Biol.* **86**:900–905.

Cowell, J. K., 1982, Double minutes and homogeneously staining regions: Gene amplification in mammalian cells, *Annu. Rev. Genet.* **16**:21–59.

Davies, A. H., and Carter, C. E., 1980, Chromatin diminution in *Ascaris suum*, *Exp. Cell Res.* **128**:59–62.

De Lange, T., 1986, The molecular biology of antigenic variation in Trypanosomes: Gene rearrangements and discontinuous transcription, in: *Int. Rev. Cytol.* **99**:85–117.

Gall, J. G., 1976, Early studies on gene amplification, *Harvey Lect.* **71**:55–70.

Geyer-Duszynska, J., 1966, Genetic factors in oogenesis and spermatogenesis in *Cecidomyiidae*, in: *Chromosomes Today* (C. D. Darlington and K. R. Lewis, eds.), pp. 174–178, Oliver and Boyd, London.

Goldstein, P., and Straus, N. A., 1978, Molecular characterization of *Ascaris suum* DNA and of chromatin diminution, *Exp. Cell Res.* **116**:462–466.

Gorovsky, M. A., 1980, Genome organization and reorganization in *Tetrahymena*, *Annu. Rev. Genet.* **14**:203–239.

Greider, C. W., and Blackburn, E. H., 1985, Identification of a specific telomere terminal transferase activity in *Tetrahymena* extracts, *Cell* **43**:405–413.

Greider, C. W., and Blackburn, E. H., 1987, The telomere terminal transferase of *Tetrahymena* is a ribonucleoprotein enzyme with two kinds of primer specificity, *Cell* **51**:887–898.

Hanyu, N., Kuchino, Y., and Nishimura, S., 1986, Dramatic events in ciliate evolution: Alteration of UAA and UAG termination codons to glutamine codons due to anticodon mutations in two *Tetrahymena* tRNAs[Gln], *EMBO J.* **5**:1307–1311.

Helftenbein, E., 1985, Nucleotide sequence of a macronuclear DNA molecule coding for α-tubulin from the ciliate *Stylonychia lemnae*. Special codon usage: TAA is not a translation termination codon, *Nucleic Acids Res.* **13**:415–433.

Herrick, G., Cartinhour, S. W., Williams, K. R., and Kotter, K. P., 1987a, Multiple sequence versions of the *Oxytricha fallax* 81-MAC alternate processing family, *J. Protozool.* **34**:429–434.

Herrick, G., Hunter, D., Williams, K., and Kotter, K., 1987b, Alternative processing during development of a macronuclear chromosome family in *Oxytricha fallax*, *Genes Dev.* **1**:1047–1058.

Horowitz, S., and Gorovsky, M. A., 1985, An unusual genetic code in nuclear genes of *Tetrahymena*, *Proc. Natl. Acad. Sci. USA* **82**:2452–2455.

Kafatos, F. C., Mitsialis, S. A., Spoerel, N., Mariani, B., Lingappa, J. R., and Delidakis, C., 1985, Studies on the developmentally regulated expression and amplification of insect chorion genes, *Cold Spring Harbor Symp. Quant. Biol.* **50**:537–547.

Kalfayan, L., Levine, J., Orr-Weaver, T., Parks, S., Wakimoto, B., DeCicco, D., and Spradling, A., 1985, Localization of sequences regulating *Drosophila* chorion gene amplification and expression, *Cold Spring Harbor Symp. Quant. Biol.* **50**:527–535.

Karrer, K. M., 1986, The nuclear DNAs of holotrichous ciliates, in: *The Molecular Biology of Ciliated Protozoa* (J. G. Gall, ed.), pp. 85–110, Academic, Orlando, Florida.

Klar, A. J. S., Strathern, J. N., and Hicks, J. B., 1984, Developmental pathways in yeast, in: *Microbial Development* (R. Losick and L. Shapiro, eds.), pp. 151–196, Cold Spring Harbor, New York.

Klobutcher, L. A., and Prescott, D. M., 1986, The special case of the hypotrichs, in: *The Molecular Biology of Ciliated Protozoa* (J. G. Gall, ed.), pp. 111–154, Academic, Orlando, Florida.

Klobutcher, L. A., Swanton, M. T., Donini, P., and Prescott, D. M., 1981, All gene-sized molecules in four species of hypotrichs have the same terminal sequence and an unusual 3' terminus, *Proc. Natl. Acad. Sci. USA* **78:**3015–3019.

Klobutcher, L. A., Jahn, C. L., and Prescott, D. M., 1984, Internal sequences are eliminated from genes during macronuclear development in the ciliated protozoan *Oxytricha nova*, *Cell* **36:** 1045–1055.

Klobutcher, L. A., Vailonis-Walsh, A. M., Cahill, K., and Ribas-Aparicio, R. M., 1986, Gene-sized macronuclear DNA molecules are clustered in micronuclear chromosomes of the ciliate *Oxytricha nova*, *Mol. Cell. Biol.* **6:**3606–3613.

Kloetzel, J. A., 1970, Compartmentalization of the developing macronucleus following conjugation in *Stylonychia* and *Euplotes*, *J. Cell Biol.* **47:**395–407.

Koropatnick, J., Winning, R., Wiese, E., Heschl, M., Gedamu, L., and Duerksen, J., 1985, Acute treatment of mice with cadmium salts results in amplification of the metallothionein-1 gene in liver, *Nucleic Acids Res.* **13:**5423–5439.

Kraut, H., Lipps, H. J., and Prescott, D. M., 1986, The genome of hypotrichous ciliates, in: *Int. Rev. Cytol.* **99:**1–28.

Lauth, M. R., Spear, B. B., Heumann, J., and Prescott, D. M., 1976, DNA of ciliated protozoa: DNA sequence diminution during macronuclear development of *Oxytricha*, *Cell* **7:**67–74.

Lipps, H. J., 1985, A reverse transcriptase like enzyme in the developing macronucleus of the hypotrichous ciliate *Stylonychia*, *Curr. Genet.* **10:**239–243.

Miller, O. L., 1981, The nucleolus, chromosomes and visualization of genetic activity, *J. Cell Biol.* **91:**15S–27S.

Moritz, K. B., and Roth, G. E., 1976, Complexity of germline and somatic DNA in *Ascaris*, *Nature (Lond.)* **259:**55–57.

Mueller, F., Walker, P., Aeby, P., Neuhaus, H., Back, E., and Tobler, H., 1982, Molecular cloning and sequence analysis of highly repetitive DNA sequences contained in the eliminated genome of *Ascaris lumbricoides*, in: *Prog. Clin. Biol. Res.* **85A:**127–138.

Murti, K. G., and Prescott, D. M., 1983, Replication forms of the gene-sized DNA molecules of hypotrichous ciliates, *Mol. Cell Biol.* **3:**1562–1566.

Oka, Y., Shiota, S., Nakai, S., Nishida, Y., and Okubo, S., 1980, Inverted terminal repeat sequence in the macronuclear DNA of *Stylonychia pustulata*, *Gene* **10:**301–306.

Orr, W., Komitopoulou, K., and Kafatos, F. C., 1984, Mutants suppressing in *trans* chorion gene amplification in *Drosophila*, *Proc. Natl. Acad. Sci. USA* **81:**3773–3777.

Osheim, Y. N., and Miller, O. L., 1983, Novel amplification and transcriptional activity of chorion genes in *Drosophila melanogaster* follicle cells, *Cell* **33:**543–553.

Pasternak, J., and Barrell, R., 1976, Quantitation of nuclear DNA in *Ascaris lumbricoides*: DNA constancy and chromatin diminution, *Genet. Res.* **27:**339–348.

Pavan, C., and Da Cunha, A. B., 1969, Chromosomal activities in *Rhynosciara* and other *Sciaridae*, *Annu. Rev. Genet.* **3:**425–450.

Pluta, A. F., Dani, G. M., Spear, B. B., and Zakian, V. A., 1984, Elaboration of telomeres in yeast: Recognition and modification of termini from *Oxytricha* macronuclear DNA, *Proc. Natl. Acad. Sci. USA* **81:**1475–1479.

Prescott, D. M., 1983, The C-value paradox and genes in ciliated protozoa, in: *Modern Cell Biology*, Vol. 2 (J. R. McIntosh, ed.), pp. 329–352, Liss, New York.

Ribas-Aparicio, R. M., Sparkowski, J. J., Proulx, A. E., Mitchell, J. D., and Klobutcher, L. A., 1987, Nucleic acid splicing events occur frequently during macronuclear development in the protozoan *Oxytricha nova* and involve the elimination of unique DNA, *Genes Dev.* **1:**323–336.

Roth, G. E., and Moritz, K. B., 1981, Restriction enzyme analysis of the germline limited DNA of *Ascaris suum*, *Chromosoma* **83:**169–190.

Schimke, R. T., Kaufman, R. J., Nunberg, J. H., and Dana, S. L., 1979, Studies on the amplification of dihydrofolate reductase genes in methotrexate resistant cultured mouse cells, *Cold Spring Harbor Symp. Quant. Biol.* **43**:1297–1303.

Spear, B. B., and Lauth, M. R., 1976, Polytene chromosomes of *Oxytricha*: Biochemical and morphological changes during macronuclear development in a ciliated protozoan, *Chromosoma* **54**:1–13.

Spradling, A. C., and Mahowald, A. P., 1980, Amplification of genes for chorion proteins during oogenesis in *Drosophila melanogaster*, *Proc. Natl. Acad. Sci. USA* **77**:1096–1100.

Steinbrück, G., 1983, Overamplification of genes in macronuclei of hypotrichous ciliates, *Chromosoma* **88**:156–163.

Swanton, M. T., Heumann, J. M., and Prescott, D. M., 1980, Gene-sized DNA molecules of the macronuclei in three species of hypotrichs: Size distributions and absence of nicks, *Chromosoma* **77**:217–227.

Tobler, H., Smith, K. D., and Ursprung, H., 1972, Molecular aspects of chromatin elimination in *Ascaris lumbricoides*, *Dev. Biol.* **27**:190–203.

White, M. J. C., 1954, *Animal Cytology and Evolution*, Cambridge University Press, Cambridge.

White, T. C., and Allen, S. L., 1986, Alternative processing of sequences during macronuclear development in *Tetrahymena thermophila*, *J. Protozool.* **33**:30–38.

Yao, M.-C., 1986, Amplification of ribosomal RNA genes, in: *The Molecular Biology of Ciliated Protozoa* (J. G. Gall, ed.), pp. 179–201, Academic, Orlando, Florida.

Yao, M.-C., and Gall, J. G., 1977, A single integrated gene for ribosomal RNA in a eucaryote, *Tetrahymena pyriformis*, *Cell* **12**:121–132.

Yao, M.-C., and Gall, J. G., 1979, Alteration of the *Tetrahymena* genome during nuclear differentiation, *J. Protozool.* **26**:10–13.

Yao, M.-C., Zhu, S.-G., and Yao, C.-H., 1985, Gene amplification in *Tetrahymena thermophila*: Formation of extrachromosomal palindromic genes coding for rRNA, *Mol. Cell. Biol.* **5**:1260–1267.

Yokoyama, R. W., and Yao, M.-C., 1982, Elimination of DNA sequences during macronuclear differentiation in *Tetrahymena thermophila*, as detected by in situ hybridization, *Chromosoma*, **85**:11–22.

Zahler, A. M., and Prescott, D. M., 1988, Telomere terminal transferase activity in the hypotrichous ciliate *Oxytricha nova* and a model for replication of the ends of linear DNA molecules, *Nucleic Acids Res.* **16**:6953–6972.

Chapter 3

Fate and Nuclear Localization of Germinal Vesicle Proteins during Embryogenesis

CHRISTINE DREYER

1. Oogenesis

In amphibian oogenesis, essential preconditions are established for a strategy of early development, in which growth and cell division are uncoupled. The fertilized egg cleaves rapidly to form a blastula without a significant expression of the embryonic genes during the first 12 cell cycles and without net intake of food during the first several days. The resources required for fast autonomous embryogenesis are accumulated during the several months of oocyte growth. The axial organization of the embryo is first established in the oocyte, since the direction of the animal–vegetal axis of the egg can be traced back to the nuclear–centriolar axis of the oogonium (Witschi, 1956; Coggins, 1973; Tourte *et al.*, 1981). Also the dorsoventral axis is predetermined in the oocyte of some amphibian species (Wittek, 1952; for a review, see Gerhart, 1980).

The nucleus or germinal vesicle (GV) grows proportionately with the oocyte, although its DNA content, representing four times the haploid genome at diplotene of first meiosis, plus about 3000 extrachromosomal copies of the ribosomal RNA genes, is small as compared with the amount of protein accumulated (4 μg protein in *Xenopus laevis* as compared with 12 pg chromosomal DNA). In some species with extremely large eggs, multinuclear oogenesis is observed (MacGregor and Kezer, 1970; del Pino and Humphries, 1978; MacGregor and del Pino, 1982). There is so far no evidence for functional or structural differences in the multiple nuclei, although only one will finally survive the complete meiosis.

There appears to be a correlation between the quality and quantity of the maternal pool of macromolecules in the oocyte, the size of the egg, and its rate of embryonic development in different species (Woodland, 1982; Davidson, 1986).

CHRISTINE DREYER • Max Planck Institute for Developmental Biology, Department for Cell Biology, D-7400 Tübingen, Federal Republic of Germany.

Whereas in the smaller oocytes of sea urchins, maternal information is mainly stored in the form of untranslated messenger RNA (mRNA) (Brandhorst, 1985), in the larger oocytes of amphibians both protein and maternal mRNA are stored. The establishment of the maternal store of nuclear protein and mRNA has been recently reviewed (Smith and Richter, 1985; Stick and Dreyer, 1989) and is therefore treated concisely here. Unless stated otherwise, *Xenopus laevis* is the subject of the work reported.

1.1. Accumulation of a Store of Nuclear Proteins during Oogenesis

Proteins accumulated in the germinal vesicle consist of enzymes and structural proteins required for DNA synthesis (Fox *et al.*, 1980), chromosome assembly, transcription (Roeder, 1974), processing, RNA storage, and a variety of polypeptides whose function is so far unknown (Dreyer *et al.*, 1981, 1983, 1985; Dreyer and Hausen, 1983). The four core histones are stored in amounts far in excess of the amount of DNA. The oocyte histone pool should suffice to complex the DNA equivalent of 20 000 nuclei (Woodland and Adamson, 1977). In addition to the histone itself, maternal mRNA coding for histones is stored that will be activated during egg maturation, and after fertilization (reviewed by Woodland, 1980; Woodland *et al.*, 1983). Titration of the nucleosome assembly capacity with exogeneous DNA has led to the estimate that DNA equivalent to about 10,000 nuclei can be packaged by the histones present in one oocyte (Laskey *et al.*, 1977, 1978). This implies that other factors aiding nucleosome assembly (e.g., topoisomerases and nucleoplasmin) are also present in excess (Glikin *et al.*, 1984; reviewed by Laskey, 1985).

Stored core histones are most probably complexed to the acidic nuclear proteins N1 and N2, which bind histones H3 and H4 (Kleinschmidt and Franke, 1982) and nucleoplasmin, which mainly binds to H2A and H2B (Kleinschmidt *et al.*, 1985, Dilworth *et al.*, 1987). Both nucleoplasmin and N1 have been shown to promote nucleosome assembly *in vitro* (Laskey *et al.*, 1978; Earnshaw *et al.*, 1980; Dilworth *et al.*, 1987). All three acidic proteins are prevalent in the germinal vesicle.

The transcriptional capacity of the oocyte nucleus has been titrated by injection of exogenous DNA templates (Brown and Gurdon, 1978). The results show that functional RNA polymerases of all three classes are present in excess of actively transcribing templates (reviewed by Gurdon and Melton, 1981; Etkin, 1982). In oogenesis, maximal activity of RNA polymerase III transcription is reached before vitellogenesis and precedes that of polymerase I and II (Gurdon, 1974). One of the transcription factors that stimulates polymerase III genes (Pieler *et al.*, 1987), TFIIIA, specifically regulates transcription of 5S ribosomal RNA. TFIIIA is one of the most prevalent protein species in pre-vitellogenic oocytes. It activates a special set of oocyte-specific 5S RNA genes that are present in 20,000 copies per haploid genome as well as somatic 5S RNA genes present in 400 copies (Brown *et al.*, 1971). TFIIIA binds to the noncoding strand of the internal control region of the 5S RNA genes (Sakonju and Brown, 1982), and to its product 5S RNA (reviewed by D. D. Brown, 1984; Krämer, 1985). Its DNA-binding domain contains Zn-binding cysteine and histidine residues periodically spaced in a highly conserved order, described as a finger structure (Brown *et al.*, 1985; Miller *et al.*, 1985). Homologous structures have been identified in a variety of proteins relevant for gene regulation throughout the animal kingdom (reviewed by Berg, 1986). A gene coding for a protein with 37 finger domains (Xfin) is expressed in oocytes and tadpoles of *X. laevis* (Ruiz i Altaba *et al.*, 1987). At least 14 different genes that have finger domains and that are not identical to TFIIIA or Xfin are expressed in *Xenopus* oocytes. Some of these are also transcribed after the mid-blastula transition (Köster *et al.*, 1988).

RNA-binding proteins other than TFIIIA are accumulated in the oocyte: ribosomal proteins (Baum and Wormington, 1985, reviewed by Wormington, 1988), oocyte-specific mRNA-binding proteins (Darnbrough and Ford, 1981), mRNA-masking proteins (Richter and Smith, 1984) that may prevent maternal mRNA from being assembled into polysomes, and an excess of UsnRNA-binding proteins (reviewed by Mattaj *et al.*, 1985) (see Section 1.2).

The nuclear lamina, a network of proteins belonging to the intermediate filament family underlying the inner nuclear membrane (Fawcett, 1981; Aebi *et al.*, 1986; McKeon *et al.*, 1986), disappears in oocytes during early stages of meiotic prophase. This structure is formed again at diplotene (Stick and Schwarz, 1983). Whereas the lamina of most somatic cell nuclei and most probably that of oogonia contains lamins L_I and L_{II}, the germinal vesicle lamina consists solely of lamin L_{III} (Stick and Krohne, 1982). The germinal vesicle lamina itself provides the pool of lamins used in early embryogenesis (reviewed by Stick, 1987, and Stick and Dreyer, 1989) (see Section 3.2).

The nuclear envelope of the amphibian oocyte contains a high density of pores (about $50/\mu m^2$) as compared with somatic cell nuclei (Franke *et al.*, 1981). Moreover, there are indications that nuclear pores, other than those in the nuclear envelope, are stored in the stacks of annulate lamellae (Kessel, 1983, 1985; Feldherr *et al.*, 1984).

1.2. Compartmentation of Oocyte Nuclear Proteins

Although the phenomenon of reaccumulation of nuclear proteins into the daughter nuclei after mitosis has been studied in other systems, targeting of nuclear proteins into their compartment has been studied most intensely in the amphibian oocyte (reviewed by Bonner, 1978; De Robertis, 1983; Dingwall and Laskey, 1986). In contrast to proteins that are to be directed through membranes of the endoplasmic reticulum and that lose their targeting signal upon passage to their compartment (Blobel, 1980), nuclear proteins retain their capability to enter the nucleus. This must occur after each mitosis and has been proved by injection of isolated oocyte nuclear proteins into the cytoplasm of oocytes. These experiments show that the injected nuclear proteins reaccumulate into the germinal vesicle (Gurdon, 1970; Bonner, 1975). However, different nuclear proteins are concentrated in the germinal vesicle to a different extent, with the consequence that the partition coefficient measured as concentration in the nucleus over cytoplasmic concentration appears to reach a characteristic value for each nuclear protein in the steady state (Paine, 1982).

The intracellular distribution of individual proteins changes during oogenesis. Early in oogenesis, although transcription factor IIIA has a nuclear function, the protein is nevertheless mainly found in the cytoplasm (Mattaj et al., 1983) in the form of a 7S storage particle consisting of the factor and the 5S RNA (Picard and Wegnez, 1979). Later in oogenesis, the 5S RNA is transferred to the nucleoli, where it is assembled into preribosomal subunits. In full-grown oocytes, there is about one tenth the amount of TFIIIA found early in oogenesis, and this is mainly in the nucleus (Pelham et al., 1981), where traces can be found as early as stage III (Johnson et al., 1984). (For stage classification in oogenesis, see Dumont, 1972.)

A pool of protein normally bound to small nuclear UsnRNA is accumulated in the cytoplasm of oocytes. If the corresponding UsnRNA is injected into the cytoplasm, it is bound by these proteins and the complex is transferred to the nucleus (Zeller et al., 1983; Mattaj and De Robertis, 1985; reviewed by Mattaj et al., 1985).

The observation that nuclear proteins can repeatedly be sequestered into their compartment has led to the hypothesis that nuclear proteins contain a signal in their sequence that targets them into the nucleus (De Robertis et al., 1978; De Robertis, 1983). Two prototypes of such nuclear targeting signals were first identified on the yeast Matα2 (Hall et al., 1984) and on the large T antigen of SV40 (Kalderon et al., 1984; Lanford and Butel, 1984). In these two instances, a single short peptide appears to be both necessary and sufficient to translocate a protein into the nucleus, as was shown by deletion analysis and by fusion of the putative signal sequence to an otherwise nonnuclear protein. Similar sequences have been found in the primary structure of several nuclear proteins of X. laevis, as deduced from the corresponding cDNAs (see Table I). Nucleoplasmin contains four different putative signal sequences (Bürglin et al., 1987; Dingwall et al., 1986, 1987), all located on the carboxy-terminal-tail-region of the molecule, which can migrate into the nucleus on its own if proteolytically cleaved from the core molecule (Dingwall et al., 1982). Deletion analysis revealed that sequences A and B, but not C and D, are indispensable for nuclear transport (Bürglin and De Robertis, 1987; Dingwall et al., 1988). Recently, a contiguous sequence of 17 amino acids, comprising both sequences A and B, was shown to be both necessary and sufficient for nuclear location of a pyruvate kinase fusion protein after microinjection into the cytoplasm of Vero cells (Dingwall et al., 1988). Similarly, both putative signals identified in N_1 of Xenopus (Table I) are required for nuclear transport (Kleinschmidt and Seiter, 1988). In conclusion, the sequence requirements for nuclear targeting may in many instances be more complex than in the case of SV40 large T antigen.

Chemical coupling of signal sequences to nonnuclear proteins also leads to their uptake into the nucleus (Goldfarb et al., 1986, Lanford et al., 1986). Coupling of such sequences to proteins of different molecular size has demon-

TABLE I. Amino Acid Sequences Presumed to Act as Signals for Translocation into the Nucleus

Source	Nuclear proteins	Putative nuclear targeting sequence[a]		Homologous to	References
Yeast	Matα2	Lys[3]	ile pro ile lys[b]	—	Hall et al. (1984)
SV40	Large T antigen	Pro[126]	lys lys lys arg lys val[b]	—	Kalderon et al. (1984)
					Lanford and Butel (1984)
Xenopus laevis	N₁	Val[531]	arg lys lys arg lys thr[d]	SV40	Kleinschmidt et al. (1986)
		Ala[544]	lys lys ser lys gln glu[d]		Kleinschmidt and Seiter (1988)
	Lamin L_A	Ser[412]	lys arg arg arg leu glu	SV40	Wolin et al. (1987)
	Lamin L_I	Gly[413]	lys arg lys arg ile glu	SV40	Krohne et al. (1987)
	Lamin L_{III}	Gly[412]	lys arg lys arg lys leu asp	SV40	Stick (1988)
	Nucleoplasmin[c]				Dingwall et al. (1986, 1987, 1988)
	sequence A	Lys[155]	arg pro ala ala thr lys lys[d]	Yeast	
	sequence B	Ala[166]	lys lys lys lys leu asp[d]	SV40	Bürglin et al. (1987)
	sequence C	Pro[183]	thr lys lys gly lys gly	SV40	
	sequence D	Arg[194]	lys pro ala ala lys lys	Yeast	
	Protein7 (37-1A9)	Glu[577]	arg arg lys lys lys thr	SV40	Miller et al. (1989)

[a]Basic amino acids italicized.
[b]Fusion proteins consisting of a non-nuclear protein and the sequence indicated are translocated into the nucleus.
[c]Fusion proteins containing all four sequences (A–D) fused to β-galactosidase are translocated to the nucleus. Deletion of sequence B, but not of sequences C or D, prevents translocation (Bürglin and De Robertis, 1987). To locate pyruvate kinase into the nucleus, a sequence of 17 amino acids, including both sequences A and B, is required (Dingwall et al., 1988).
[d]Both signals required.

strated that ferritin (465 kDa) but not IgM (970 kDa) may be forced through the nuclear pores (Lanford *et al.*, 1986). The speed of accumulation is a function of the number of signals added to the molecule (Lanford *et al.*, 1986, Roberts *et al.*, 1987). The function of artificially inserted signals is dependent on the protein context. They may be inefficient if they are buried in a hydrophobic domain or if they are overruled by another signal that directs the protein to a compartment other than the nucleus (Roberts *et al.*, 1987).

The obvious route for nucleocytoplasmic transport of macromolecules is the nuclear pore (reviewed by Franke *et al.*, 1981). It provides a channel of 9 nm diameter, permitting the diffusion of ions and other molecules up to this size through the nuclear envelope (Bonner, 1978). Thus, proteins smaller than about 60 kDa may diffuse into the nucleus, without necessarily being accumulated there. Larger molecules need a signal sequence to traverse the nuclear pore and their transport requires ATP (Newmeyer *et al.*, 1986a). Feldherr *et al.*, (1984) could visualize gold particles coated with the targeting tails of nucleoplasmin traversing the nuclear pores. Gold particles with more than 20 nm diameter were transported if coated with the targeting protein. This implies that the functional pore diameter for targeted transport of nuclear proteins may exceed the pore diameter open for diffusion and that the nuclear pore complex and not the transported particle has to change conformation to permit passage. Analysis of the transport of protein-coated gold particles of various sizes has revealed that not only the velocity of transport but also the maximal size of particles gated into the nucleus is dependent on the quality of the signal, and on the number of signal sequences per molecule (Dworetzky *et al.*, 1988).

Gold particles coated with nuclear proteins often appear to queue up in front of the nuclear pores on electron micrographs (Feldherr, 1984; Dworetzky *et al.*, 1988, Richardson *et al.*, 1988; Newmeyer and Forbes, 1988). This phenomenon is most likely due to an interaction between the nuclear targeting sequences and fibrils extending from the pore complex (Richardson *et al.*, 1988). Nuclear transport could recently be divided into two steps in tissue culture cells (Richardson *et al.*, 1988) and *in vitro* (Newmeyer and Forbes, 1988): rapid interaction of the protein with the outer face of the nuclear pore, which is dependent on the presence of a nuclear targeting signal; followed by a slower process of translocation through the pore, which is dependent on ATP and is inhibited by the lectin wheat germ agglutinin (WGA; Finlay *et al.*, 1987; Newmeyer and Forbes, 1988).

The pores of rat liver nuclei contain a special class of O-glycosylated proteins that could be localized with application of monoclonal antibodies (Holt *et al.*, 1987; Snow *et al.*, 1987). At least one of the corresponding antigens could also be detected in the nuclear envelope of *Xenopus* oocytes. In this system the antibody inhibits nuclear import of nucleoplasmin and export of RNA (Featherstone *et al.*, 1988). Both processes apparently occur via the same nuclear pore complexes, as was shown with gold particles that were either coated with nuclear protein or with RNA (Dworetzky and Feldherr, 1988).

Although signal sequences that are rich in basic amino acids are now believed to interact with the nuclear pore, a domain required for accumulation in the nucleus rather than for entry has been identified in an influenza virus nuclear protein. This domain has no obvious structural similarity with the sequences shown in Table I (Davey et al., 1985).

A possible role for a domain promoting retention of proteins in the nucleus is that it binds to a hypothetical protein matrix in the nucleus. Evidence favoring this model has come from previous reports on protein accumulation in oocyte nuclei, whose envelopes were punctured in situ (Feldherr and Pomerantz, 1978; Feldherr and Ogburn, 1980). When these experiments were repeated with application of very sensitive immunofluorescent staining techniques, it turned out that puncturing of the nuclear envelope leads to a considerable loss of several nuclear antigens tested, including nucleoplasmin and N_1 (Zimmer et al., 1988). When oocyte nucleoplasm deprived of its envelope is injected into the vegetal region of an oocyte, nuclear antigens diffuse out of the nucleoplasm and are apparently accumulated by the intact host nucleus. This could be demonstrated for N_1 of X. borealis with the aid of a species-specific monoclonal antibody (Fig. 1). Puncturing of the host nucleus inhibits the accumulation process (Fig. 1). The experiment shown in Fig. 1 clearly demonstrates that a hypothetical nuclear matrix in the absence of the envelope is not sufficient to retain the nucleoplasmic proteins studied (N1, N4, nucleoplasmin, and nucleolar antigen b6-6E7). Puncturing of the nuclear envelope will never destroy all the nuclear pores and therefore will still permit transport of nuclear proteins—as has been found by Feldherr and his colleagues. We have concluded that the presence of an intact nuclear envelope is required for both accumulation and retention of large nuclear proteins. Diffusion of small molecules, on the other hand, is not dependent on an interaction between a targeting signal and a nuclear receptor and is not inhibited by WGA (Newmeyer and Forbes, 1988).

2. Oocyte Maturation

2.1. Rearrangement of Nuclear Proteins during Germinal Vesicle Breakdown

The full-grown oocyte can be triggered by a steroid hormone to mature into a fertilizable egg (reviewed by Masui and Clarke, 1979; Maller, 1985; Masui, 1985). A maturation promoting factor (MPF) is activated in the oocyte, the arrest of the chromosomes at diplotene is released, and first meiosis is completed. In analogy to mitosis, nucleoli disappear, transcription ceases, the chromosomes condense, the nuclear envelope dissociates, and the nucleoplasmic proteins are dissipated into the cytoplasm.

The histological changes that occur during oocyte maturation have been studied in great detail in several species of Bufo, Triturus, and Rana, and in Xenopus laevis (Wittek, 1952; Tschou-Su and Wang, 1958; Brachet et al., 1970; Huchon et al., 1981; Hausen et al., 1985). Germinal vesicle breakdown (GVBD) is visibly initiated by the formation of bundles of fibrillar material that appear to originate from a yolk-free zone at the vegetal face of the germinal vesicle (Brachet et al., 1970; Huchon et al., 1981). These fibers can be stained with

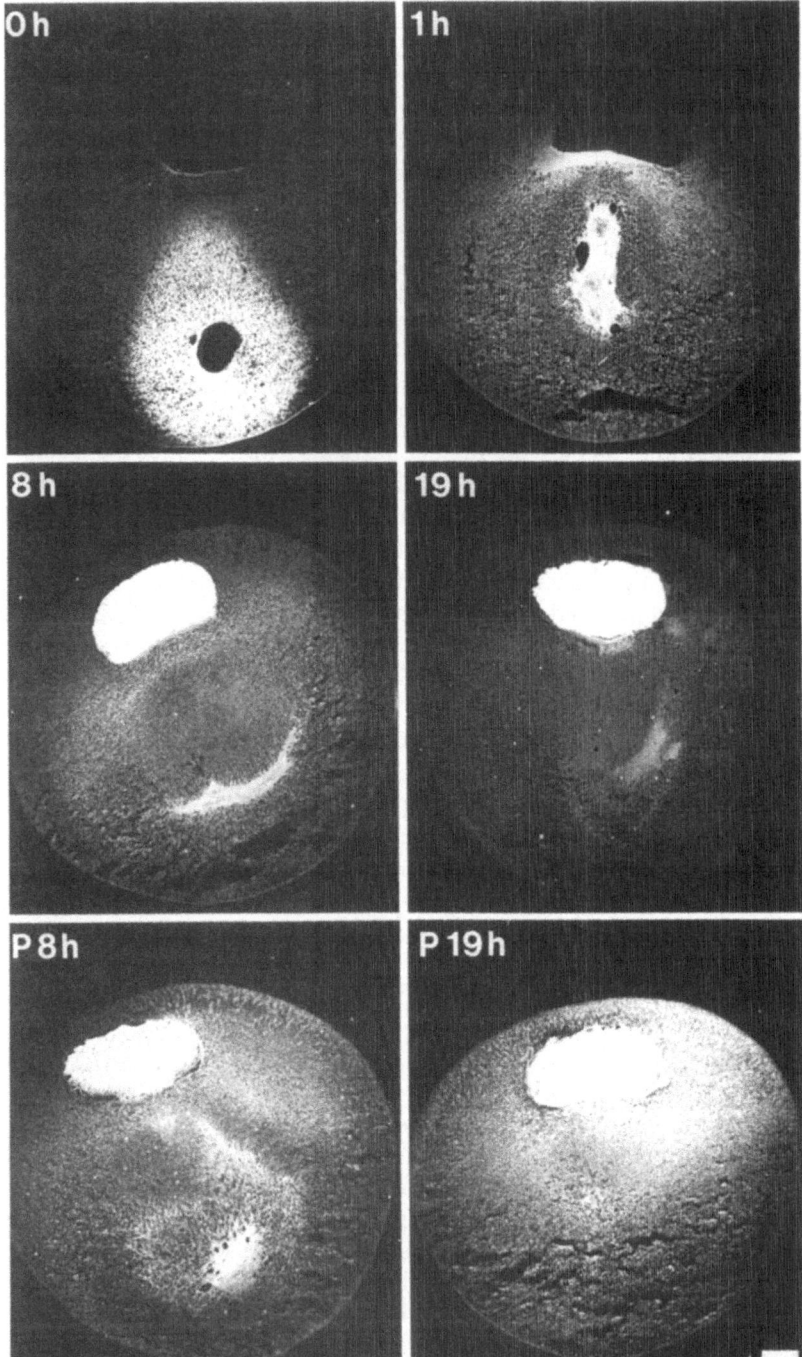

Figure 1. Translocation of N_1 from implanted nucleoplasm into the host nucleus. Germinal vesicles of *Xenopus borealis* were manually isolated, their nuclear envelopes removed, and the nucleoplasm was injected into the vegetal cytoplasm of *X. laevis* oocytes. After the times indicated, injected oocytes were fixed, sectioned, and stained with mAB b6-3B7, specific for N_1 of *X. borealis*. P8h, P19h. The host nucleus was punctured with glass needles after injection of nucleoplasm. Bar = 100 μm. (For details, see Zimmer et al., 1987.)

antitubulin antibodies at the onset of oocyte maturation, (Huchon *et al.*, 1985; P. Hausen and J. Wehland, unpublished observations). Thus, newly formed microtubules appear to invade the germinal vesicle from its basal side and simultaneously push it to the animal pole. During this period, the germinal vesicle changes shape and is finally broken down (Fig. 2a–j). Meanwhile, the condensed chromosomes reach the animal pole, where the first polar body is extruded; the chromosomes are then arrested in the second metaphase (Masui, 1974). The distribution of several nuclear proteins during GVBD has been studied with application of monoclonal antibodies (Dreyer *et al.*, 1983; Hausen *et al.*, 1985; Fig. 2l–o). The nuclear lamina is fragmented, starting at the basal face of the germinal vesicle. It is visible in particulate form, before it appears to be completely dissociated (Hausen *et al.*, 1985). Meiotic disassembly of the lamina is apparently accompanied by hyperphosphorylation of lamin L_{III} (Stick and Hausen, 1985; Krohne and Benavente, 1986; Stick, 1987). Analogous observations have been previously described on somatic cells, where hyperphosphorylation of the lamins occurs during mitosis, and dephosphorylation occurs upon reassembly of the lamina (Gerace and Blobel, 1980; Miake-Lye and Kirschner, 1985).

Most nucleoplasmic proteins appear to be distributed as shown for N1 in Fig. 2n,o. They are first extruded from the germinal vesicle at its basal face. In the mature egg, the major portion of each nuclear protein is localized in the animal hemisphere, where it can be traced up to early gastrula and where later in the embryo most of the nuclei will be formed (Dreyer *et al.*, 1982, 1983). There is a gradual decrease in protein concentration from the animal to the vegetal pole. Contrasting with this pattern, nucleoplasmin appears to be first extruded laterally (Fig. 2l), rather than basally, and finally not only forms a gradient from the animal to the vegetal half, but moreover is found enriched in a peripheral zone in the vegetal hemisphere as well (Fig. 2m).

Nevertheless, the distribution of nucleoplasmin appears to be radially symmetrical, as is that of other nuclear proteins tested. No indications of asymmetry or of other specific localizations of nuclear antigens have been found, but the observation that N1 and nucleoplasmin appear to leave the germinal vesicle via different routes during GVBD implicates a mechanism other than simple diffusion for the distribution of nuclear proteins during oocyte maturation.

2.2. Changes in Metabolism of Nuclear Proteins during Oocyte Maturation

The rate of protein synthesis increases during oocyte maturation by a factor of 2 on average (Wassermann *et al.*, 1982). However, the synthesis rate of specific proteins, e.g. the nucleosomal histones, increases up to 50-fold. Since transcription ceases before GVBD, this burst of translation must be due to selective activation of previously silent mRNA (reviewed by Woodland, 1980, 1982; Smith and Richter, 1985).

Inactivation of stored mRNA is achieved by oocyte-specific mRNA binding

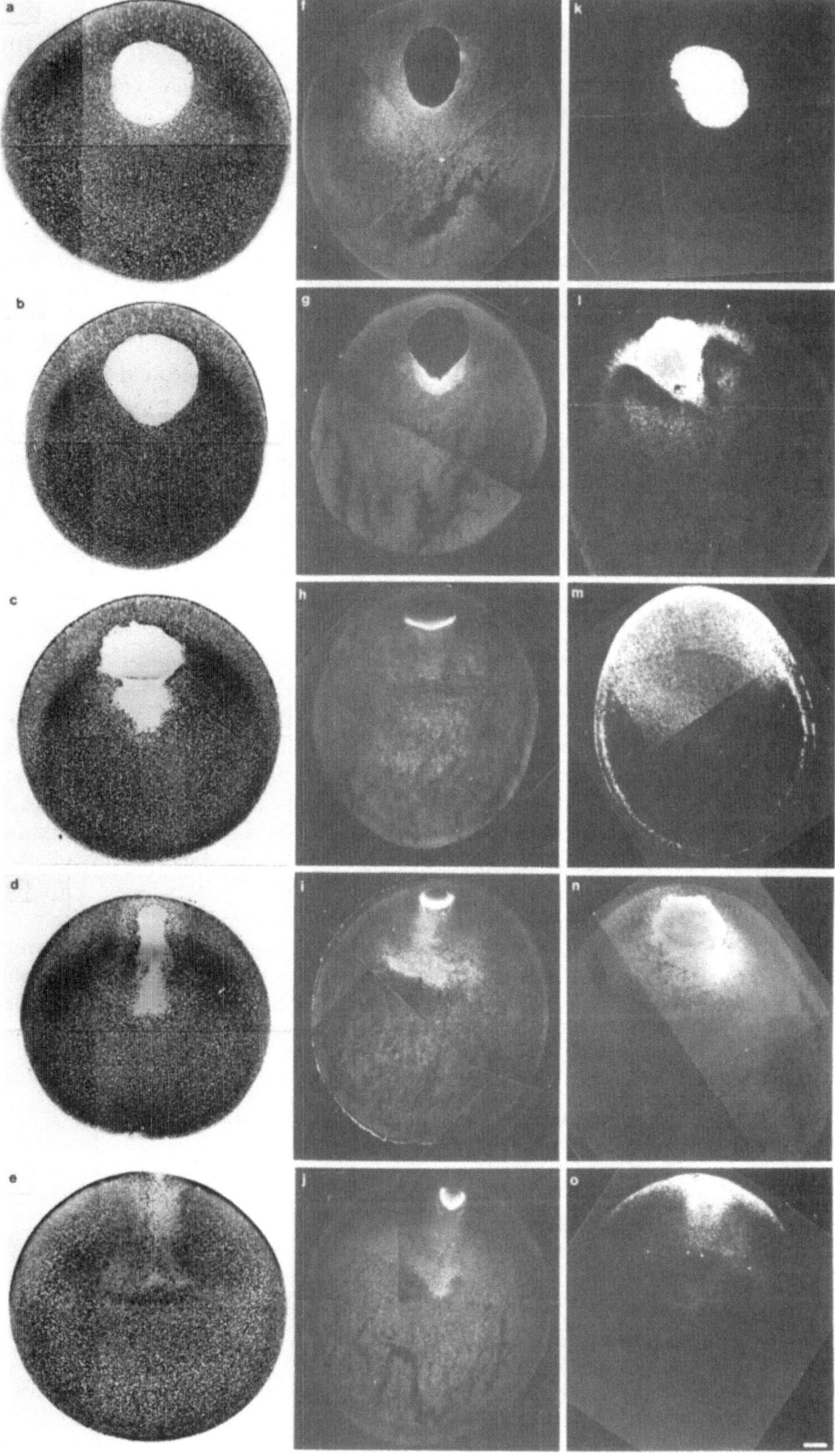

proteins (Richter and Smith, 1984), whereas stimulation of translation is paralleled by phosphorylation of the ribosomal protein S6 (Nielsen *et al.*, 1982). Moreover, there is a general increase in protein phosphorylation upon oocyte maturation (Maller *et al.*, 1977). The abundant nuclear proteins N1, N2, and nucleoplasmin, and lamin L_{III} are among the most highly phosphorylated substrates (Sealy *et al.*, 1986; Dreyer, 1987; Stick, 1987).

The purified maturation promoting factor (MPF), a complex of 200 kDa, has recently been shown to consist of two polypeptides, a 32 kDa protein kinase and a 45 kDa kinase substrate. This kinase also phosphorylates other substrates, e.g. histone H1, but not ribosomal S6 and the nuclear lamins (Lohka *et al.*, 1988). The protein kinase component of MPF is structurally related to the cdc 2 gene product of fission yeast (Dunphy *et al.*, 1988; Gautier *et al.*, 1988), a protein kinase required for the progression from G_1 to S and from G_2 to M in the cell cycle. In Xenopus the MPF promotes the step from G_2 to M not only in meiosis, but is involved in the regulation of mitosis in somatic cells as well (Gerhart *et al.*, 1984).

3. Fate of Oocyte Nuclear Proteins after Fertilization

3.1. Tracing of Germinal Vesicle-Derived Polypeptides in the Embryo

Many of the proteins accumulated in the germinal vesicle play a dual role. They are required for the nuclear architecture and for nuclear activities (e.g., transcription), and they represent a store of nuclear proteins used later in the nuclei of the embryo. All proteins required for chromosome replication are stored and used in the embryo (reviewed by Woodland, 1980; Laskey, 1985; Stick and Dreyer, 1989) (see Section 1.1).

Monoclonal antibodies raised against a variety of prevalent germinal vesicle polypeptides have permitted observation of their fate in the embryo by immunohistological techniques (Dreyer *et al.*, 1981, 1982, 1983). All the antigens studied were found in all nuclei of the embryo between neurula and tadpole stages. Thus, the protein composition of the germinal vesicle does not seem to differ principally from that of nuclei of early embryonic somatic cells.

Figure 2. Distribution of oocyte nuclear antigens during germinal vesicle breakdown (GVBD). Oocytes of *Xenopus laevis* were matured *in vitro*. Oocytes were fixed and sectioned at different times after addition of progesterone. The course of GVBD was monitored after embedding in glycol methacrylate and staining with Orange G, Anilinblue, and azofuchsin (a–e) and after wax embedding and indirect immunohistological staining using mABs YL ½, which binds to microtubules (f–j), b7-1D1, which binds to nucleoplasmin (k–m), and b2-2B10, which binds to N1 (n,o). (a,f,k) Oocyte before addition of progesterone. (b–d.g–i,j,l,n) GVBD in progress. (e) GVBD complete. (m,o) Mature egg. Bar = 100 μm. (For details, see text and Hausen *et al.*, 1985. Courtesy of M. Riebesell, Dr. P. Hausen, and Dr. J. Wehland.)

Since the function of most of these antigens is unknown, it cannot be ascertained whether they are stored for use during embryogenesis or whether they are required in oogenesis as well.

Two-dimensional gel analysis of the total and the newly synthesized polypeptides has shown that prevalent germinal vesicle-derived proteins persist in the embryo. Some proteins are still found in swimming tadpoles, although there is no indication of their *de novo* synthesis in the embryo (Dreyer and Hausen, 1983). Maternal protein N1 is also detected in the nuclei of tadpoles at stage 51, although it is not expressed from the genes of the embryo. This was shown using a species-specific monoclonal antibody for the analysis of interspecies hybrids of *Xenopus* (Dreyer *et al.*, 1983; Wedlich *et al.*, 1985).

3.2. Changes of Protein Composition in Embryonic Nuclei

The complement of nuclear proteins found in the nuclei of the embryo changes with developmental time because (1) compartmentation of maternal proteins changes (see Section 3.3), and (2) the maternal store is supplemented by proteins synthesized in the embryo. If protein species synthesized at different stages of early development are compared by two-dimensional gel analysis, few changes are observed in comparison with the pattern of maternally inherited polypeptides (Ballantine *et al.*, 1979; Dreyer and Hausen, 1983).

Protein species not found in the oocyte are first observed between the late blastula and early gastrula (Ballantine *et al.*, 1979), after transcription of embryonic genes has commenced at the midblastula transition (Signoret and Lefresne, 1971; Newport and Kirschner, 1982). Transcription of exogeneous genes from injected plasmids also starts at mid-blastula (Bendig, 1981; Etkin *et al.*, 1984). One exception to this rule is known: DNA ligase I is expressed from a paternal gene at the 1-cell stage in the axolotl (Signoret *et al.*, 1981; reviewed by Signoret and David, 1986).

Several well-documented examples illustrate how the maternal pool of a nuclear protein is complemented and gradually replaced by protein synthesized *de novo*. The pool of core histones in the oocyte is augmented by a 50-fold increase of their rate of synthesis over maternal mRNA during oocyte maturation; another threefold increase occurs after fertilization. This pool is then sufficient to sustain nucleosome assembly until the gene dose in the embryo is high enough to permit expression of histones to keep pace with DNA replication (Woodland, 1980; Woodland *et al.*, 1983). Species-specific forms of histone H1 permit distinction of paternal and maternal contributions in interspecies hybrids of *X. laevis* and *X. borealis*. By analysis of androgenetic haploid hybrids, Woodland *et al.* (1979) showed that H1 protein during early embryogenesis is synthesized on activated maternal mRNA before it is replaced by H1 expressed by embryonic genes. A comparable sequence of events has been shown for a prevalent nuclear protein, of about 70 kDa of which isoelectric variants exist in the two species of *Xenopus* (Woodland and Ballantine, 1980).

The lamina of the germinal vesicle consists solely of the oocyte lamin L_{III}, whereas the lamins found in adult somatic nuclei are L_I, L_{II}, and L_A (for exceptions, see Benavente et al., 1985). The germinal vesicle lamina is solubilized during oocyte maturation (see Section 2.1) and, after fertilization, lamin L_{III} is reused for the formation of the embryonic nuclei. L_{III} is the only lamin in cleavage nuclei, L_I is added starting at mid-blastula, and L_{II} is added at gastrula (Stick and Hausen, 1985). Concomitantly, the amount of L_{III} decreases to levels that are barely detectable by the swimming tadpole stages (Benavente et al., 1985). The first expression of L_I at mid-blastula (Stick and Hausen, 1985) and of L_{II} at mid-gastrula (Wolin and Kirschner, 1986) occurs independently of transcription and therefore must be attributable to stage-specific activation of maternal mRNA, before both somatic lamins are expressed from the embryonic genes. L_{III} is also transiently translated at mid-blastula from maternal mRNA (reviewed by Stick, 1987; Stick and Dreyer, 1989). Maternal mRNA coding for L_I, L_{II}, and L_{III} can be found in oocytes as early as stage II (Dumont, 1972). This was shown by translocation of RNA from this stage in vitro followed by immunoprecipitation of the products with specific antibodies (Stick, 1988). Recently, a lamin homologous to the human lamin A in its sequence, was detected in Xenopus in the nuclei of all somatic cells tested, except erythrocytes. This lamin A is absent from oocyte nuclei and nuclei of the early embryo at least up to tailbud stage 24 but is found in swimming tadpoles at stage 47 (Wolin et al., 1987).

Among transcripts that are expressed at specific stages of development are several that contain the homeo domain as a consensus sequence. This domain was first identified in the products of homeotic genes of Drosophila (McGinnis et al., 1984; Scott and Weiner, 1984) and these protein products show striking sequence homologies with known DNA-binding proteins of procaryotes and yeast (Laughon and Scott, 1984). In other species, including X. laevis, genes containing the homeo domain have been identified, and their expression in oocytes and embryos has been monitored (Carrasco et al., 1984; Müller et al., 1984; Harvey et al., 1986; Fritz and De Robertis, 1988). The protein product of the gene Xhox 1A has been shown to localize in the nucleus when translated from injected mRNA in oocytes. The zygotic transcript of this gene is found in embryos from mid-gastrula onward at levels 20-fold higher than the maternal transcript found in the oocyte (Harvey et al., 1986). Of the other homeobox genes described so far in Xenopus, only XlHbox 2 (previously called MM3) is transcribed in significant amounts in the oocyte (Müller et al., 1984), where the amount of transcripts is highest at stage II of oogenesis (Wright et al., 1987). The transcripts contain in their 5' leader a sequence that inhibits translation in vitro (Fritz and De Robertis, 1988), and it is not yet known when they are translated in vivo. In Embryos, XlHbox 2 is transcribed from gastrula onward, and the genes XlHbox 3, 5, 6 are transcribed at late neurula, and XlHbox 4 at tailbud stages in a highly stage-specific way (Fritz and De Robertis, 1987; Sharpe et al., 1987; Condie and Harland, 1987).

Region-specific expression of XlHbox 1 (also named Ac1 and Xeb 1) was observed as early as at the end of gastrulation (Carrasco and Malacinski, 1987). whereas at later stages several homeobox-containing gene products are found in

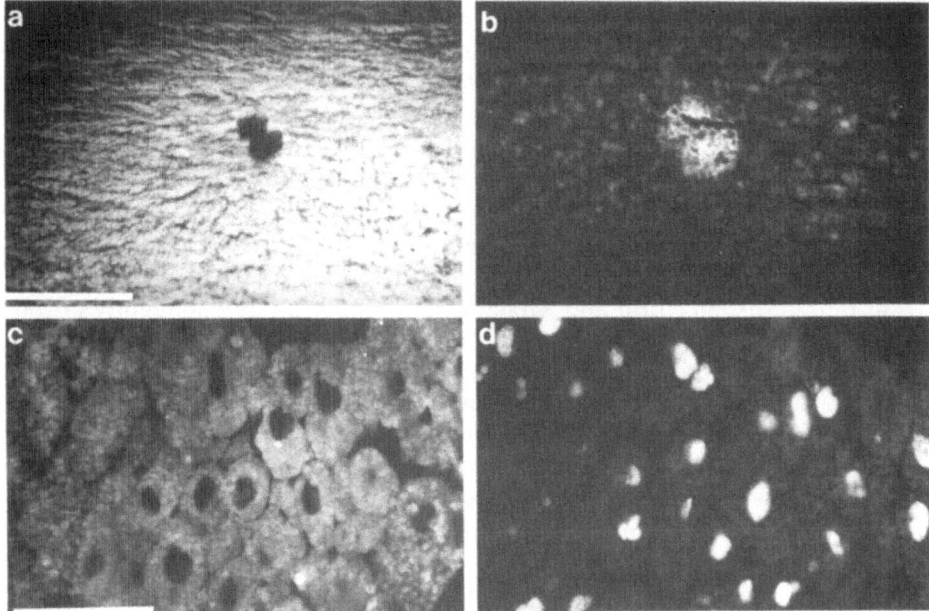

Figure 3. Fate of oocyte nuclear proteins in embryogenesis. Embryos were fixed and embedded at
the 1-cell stage (a,b), at blastula (c,d), gastrula stage 12 (e,f), and tadpole stage 50 (g,h); sectioned
and stained with monoclonal antibodies 32-5B6 (a,c,e,g), 32-4A1 (b), or b2-2B10 (d). Sections e and
g were counterstained with DAPI (f,h) to locate all nuclei. Pronuclei (a) and blastula nuclei (c)
exclude antigen 32-5B6, whereas they accumulate antigens 32-4A1 (b) and b2-2B10 (N1) (d).
Antigen 32-5B6 is gradually translocated to the nuclei during gastrula stages (e) (see opposite page)
and is highly enriched in the nuclei of specific cell types after organogenesis. (g,h) Section through
the pharyngobranchial tract; only the nuclei of secretory epithelial cells are stained by the antibody
(g). Bar = 100 μm. (For details, see Dreyer et al., 1981; Dreyer and Wedlich, 1988a.)

limited regions of the tadpole (Sharpe et al., 1987; Condie and Harland, 1987;
Oliver et al., 1988a). For more details see Section 4.

3.3. Compartmentation of Nuclear Proteins in the Embryo

After germinal vesicle breakdown, nucleoplasmic proteins undergo a
cytoplasmic sojourn before they are reaccumulated by the newly formed nuclei
of the embryo. Since it is not until mid-blastula that the cumulative volume of
all cleavage nuclei comes close to that of one germinal vesicle (Gerhart, 1980),
the nuclei could not be expected to harbor all the oocyte nuclear protein at
early stages.

As was first observed in *Xenopus*, cleavage nuclei do not assemble all the
germinal vesicle-derived antigens proportionately (Dreyer et al., 1982). At least
two classes of proteins can be distinguished, called early shifting and later
shifting proteins for simplicity. The early shifting antigens are accumulated by
both pronuclei and subsequently by all newly formed nuclei (Fig. 3b,d). The

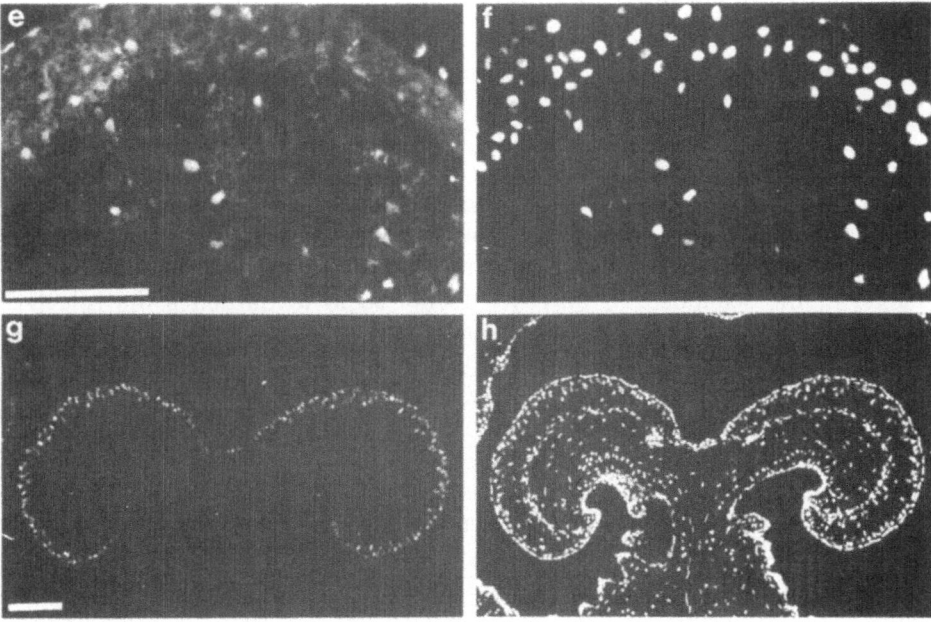

Figure 3. (continued)

later shifting antigens appear to be excluded from embryonic nuclei at early stages (Fig. 3a,c) but are accumulated later, each at a characteristic developmental stage, the latest between gastrula and early neurula (Fig. 3e,f). A similar observation has since been reported on *Drosophila* nuclear proteins (Dequin *et al.*, 1984). There appears to be a correlation between the time of shift of individual proteins into the nucleus and their nuclear function. Among the proteins shifting early are those obviously required for the formation of nuclei (i.e., lamin L_{III} [Stick and Hausen, 1985; Dreyer *et al.*, 1986]; and N1, N2, and nucleoplasmin). The latter are all thought to bind histones and probably aid nucleosome assembly *in vivo* (Laskey *et al.*, 1978; Kleinschmidt *et al.*, 1982, 1985; Dilworth *et al.*, 1987). Less is known about the later shifting proteins. After the mid-blastula transition, they may have a function associated with transcription or posttranscriptional processes. This appears to be the case for antigen b6-6E7, a DNA-binding protein found in the nucleoli (H. Retzbach and C. Dreyer, unpublished observations) and for UsnRNP proteins. The latter are thought to be engaged in RNA processing (Busch *et al.*, 1982). In oogenesis, the UsnRNA proteins are accumulated in stoichiometric excess over the corresponding UsnRNA and are stored in the cytoplasm, since they have to bind to UsnRNA to be transported into the nucleus (Mattaj and De Robertis, 1985). These proteins remain excluded from the nuclei of the embryo before the onset of UsnRNA transcription but are ultimately concentrated in the nuclei at gastrulation (De Robertis, 1983; Zeller *et al.*, 1983; Fritz *et al.*, 1984; Mattaj *et al.*, 1985; Newmeyer and Forbes, 1988).

3.4. Experimental Systems to Study Compartmentation

A multitude of pronuclei can be formed in unfertilized eggs by injection of isolated sperm nuclei (Graham, 1966). Upon swelling they accumulate early- but not late-shifting nuclear proteins. Pronucleus-like bodies are also formed after injection of somatic cell nuclei (Dreyer *et al.*, 1986) and of pure λDNA (Forbes *et al.*, 1983). Their selectivity for protein uptake appears to be the same as that of endogenous pronuclei, indicating that the unfertilized egg contains all components required for the formation of multiple nuclei. The source of the nuclei or the sequence of the DNA do not seem to influence the selective accumulation of early shifting antigens (Dreyer *et al.*, 1986; Dreyer, 1987).

As described by Lohka and Masui (1983), pronuclei can be formed *in vitro* by incubation of sperm nuclei in extracts of activated eggs. These pronuclei also selectively accumulate early- but not late-shifting proteins, indicating that the three-dimensional architecture of the cell is dispensible at least for the accumulation of early shifting proteins (Dreyer *et al.*, 1986). *In vitro* systems consisting of *Xenopus* egg extracts and nuclei of different origins (Newmeyer *et al.*, 1986b) or pure λDNA (Newport, 1987) have been studied in great detail and are of potential use for the analysis of selective translocation of nuclear proteins under controlled conditions (Newmeyer *et al.*, 1986a,b; Finlay *et al.*, 1987).

3.5. Possible Mechanisms for Selective Accumulation of Nuclear Proteins in the Embryo

Nuclear targeting signals homologous to those identified in yeast or virus coded nuclear proteins have been found in the early-shifting nuclear proteins nucleoplasmin (Dingwall *et al.*, 1988; Bürglin and De Robertis, 1987), N1 (Kleinschmidt and Seiter, 1988) and Lamin L_{III} (Stick, 1988). Of the later shifting proteins, antigen 37-1A9 contains a sequence which resembles the large T antigen prototype of nuclear targeting sequence (see Table I; Miller *et al.*, 1989). No such sequence has been found in antigen 32-5B6, which shifts into the nuclei after gastrulation (Etkin, personal communication). It is tempting to speculate that early shifting proteins contain more, or more potent, signal sequences in their structure as compared to late shifting proteins. Since nucleoplasmin is a pentamer consisting of identical subunits (Dingwall *et al.*, 1982), it should contain at least five targeting signals, if only one of the sequences listed in Table I is functional. From a kinetic point of view, multiple targeting sites exposed at the surface of a nuclear protein should greatly facilitate its interaction with the nuclear pore, since the probability of successful collisions between the transport system and its potential substrate should increase with the number of target signals on a given substrate (Lanford *et al.*, 1986). Nuclear targeting signals of different quality or quantity would make the efficiency of transport into nuclei an intrinsic property of each protein. This model might explain why proteins that are most quickly concentrated in the oocyte nucleus after their synthesis are among the early shifting ones in the embryo (Dreyer, 1987).

In the embryo, the very short cell cycle length before the mid-blastula transition (Newport and Kirschner, 1982; Boterenbrood et al., 1983) may not be sufficient for accumulation of slow migrating proteins, which would then be later shifting. Within this context, it is important to realize that all nucleoplasmic proteins are redistributed in the cytoplasm during each mitosis and have to be reaccumulated during each interphase (Dreyer et al., 1982, 1986; Dreyer, 1987). To test this model, the length of the cell cycle at early blastula has been artificially extended by application of inhibitors, and the nuclei have been examined for the presence of later shifting antigens. Their selectivity for early shifting antigens remained unchanged, however (Dreyer, 1987).

The nuclear polypeptides studied do not change their covalent structure with development, as revealed by two-dimensional gel analysis (Dreyer and Hausen, 1983). Post-translational modifications seem not to be necessary for translocation into nuclei, since proteins translated in vitro meet the requirements for being accumulated into nuclei (Dabauvalle and Franke, 1982).

If a later shifting nuclear protein does not contain a nuclear targeting signal, its uptake into the nuclei might be dependent on binding to another protein or RNA that is not synthesized before a certain stage of development. This could be elegantly shown for UsnRNPs, where the protein is only transferred to the nuclei if complexed to UsnRNA (see Section 1.2; reviewed by Mattaj et al., 1985).

In order to investigate whether this model is applicable to other late-shifting nuclear proteins, the in vitro system described in Section 3.4 has been used. Extracts from later embryonic stages were mixed with the egg extract in which pronuclei had formed. This did not lead to a translocation of any of the late-shifting antigens investigated (C. Dreyer, unpublished observations). Whether this is because of the limitations of the in vitro system or whether it means that the model is not generally applicable to all late-shifting proteins remains to be clarified.

Several constituents of the nuclear pore complex have recently been identified at the molecular level (see Finlay et al., 1987, for references). Since antibodies against several of the glycoproteins of the nuclear pore have been raised (Snow et al., 1987), these could be used to study whether there are qualitative or quantitative changes of the nuclear pores in early development.

4. Nuclear Differentiation during Organogenesis

A number of germinal vesicle-derived proteins have been found in all nuclei of the embryo between the time of their reaccumulation into embryonic nuclei and the onset of organogenesis. No significant regional differences in antigen localization have been observed during this period. Only later, during organogenesis, do some of the antigens studied attain a high degree of cell type specificity (Dreyer et al., 1981, 1983; Wedlich et al., 1985; Wedlich and Dreyer,

1988a). Among these is antigen 32-5B6 (protein 21), which is found enriched in the nuclei of secreting cells of the cement gland, the pharyngobranchial tract (see Fig. 3g,h) of the exocrine pancreas, and the gut epithelium, and in certain differentiated kidney cells. Since the maternal transcript of the corresponding gene becomes undetectable in whole embryos by stage 30 (Eastman et al., 1986), the expression of antigen 32-5B6 in specific cell types is most probably due to gene expression in these cells, whereas the protein found earlier in all cells of the embryo may be of maternal origin.

The antigen 37-1A9 (protein 7) is highly enriched in the nuclei of the central nervous system of advanced tadpoles and of adults (Dreyer et al., 1981, 1983; Dreyer and Hausen, 1983; Wedlich and Dreyer, 1988a). Transiently, this antigen is also enriched in nuclei of the intestine during metamorphosis (C. Dreyer, unpublished observation). A cDNA for this protein (xlgv 7) has recently been isolated. Northern blot analysis has revealed that a maternal store of the corresponding mRNA is mainly stored in the animal hemisphere of the fully grown oocyte. In the embryo, more of this mRNA is accumulated at the time of early neurula stage 15–17. In accordance with previous immunohistological data, the mRNA is transiently found in the gut of tadpoles, and in the adult it is highly enriched in the brain, and not or barely detectable in other tissues (Miller et al., submitted for publication). Since there is evidence that the protein binds to DNA in vitro (Dreyer, unpublished observations; Miller et al., 1989), it is intriguing that the amino acid sequence deduced from the cDNA reveals a histidine-rich repeat, that is different from that of known Zn-finger proteins (Miller et al., 1989).

The tissue specificities of several nuclear antigens are highly dynamic during the organogenetic rearrangements during metamorphosis. The cell-type specificities observed correlate with functional differentiation of cells rather than with their descendance from one of the primary germ layers (Wedlich and Dreyer, 1988a). The acidic germinal vesicle proteins N1 and nucleoplasmin are both by far more abundant in cells of the early embryo than in somatic cells of the adult, where traces may exist (Krohne and Franke, 1980). Two transcripts representing N1/N2 have been found in epidermal cells and in cultured Xenopus cell line A6 by Northern blot hybridization. The transcript size that is abundant in oocytes is found only in trace amounts in somatic cells (Kleinschmidt et al., 1986). Although the maternal pool of N1 can be traced up to advanced tadpole stages, it is not expressed to a significant degree from the genes of the embryo, as was shown with application of species-specific antibodies. N1 is, however, expressed in the multiplying gonia of both sexes, where its levels correlate with rapid division of the gonia (Wedlich et al., 1985). Although N1 and nucleoplasmin are both thought to complex histones, only N1 (and not nucleoplasmin) could be detected in the nuclei of dividing gonia. Both proteins are significantly accumulated as the oocytes begin growth at diplotene (Wedlich et al., 1985; Wedlich and Dreyer, 1988b).

The expression of several homeobox-containing genes of Xenopus has been studied by northern blot analysis, in situ hybridization, and immunohistology. The gene XlHbox 6 is transcribed in the middle and posterior neural cord at neurula and tailbud stages (Sharpe *et al.*, 1987). At the same stages, Xhox 36 (XlHbox3) is transcribed in the posterior ectoderm and meso- derm (Condie and Harland, 1987). The gene XlHbox1 (also known as AC1 and Xeb1) is first transcribed mainly in the dorsal lip of the blastopore and later in the neural tube posterior to the brain vesicle, between neurula and swimming tadpole stages, and also in the adult (Carrasco and Malacinski, 1987). Two different promotors give rise to two different transcripts, each with a homeobox (Cho *et al.*, 1988). The two transcripts are translated to proteins of different size, that could be localized in specific nuclei of tadpoles. The spatial limits of expression in the neural cord and in some mesodermal derivatives are distinct for the shorter and the longer protein (Oliver *et al.*, 1988a). In the mesoderm of the forelimb bud, the concentration of this homeobox protein in the nuclei decreases along the antero-posterior and the proximo-distal axis. The protein was not detected in the mesoderm of the hindlimb bud (Oliver *et al.*, 1988b).

5. Summary and Prospects

The germinal vesicle provides a maternal store of proteins that contribute to oogenesis and are moreover required to sustain the rapid development of the egg after fertilization. This maternal pool can be traced in the embryonic nuclei up to feeding tadpole stages.

Maternal nuclear proteins are released into the cytoplasm during oocyte maturation and are reaccumulated by the nuclei of the embryo in a stage- specific manner. The selectivity of this process may be further analyzed in *in vitro* systems. Uptake of nuclear proteins from the cytoplasm occurs via the nuclear pores. An intact nuclear envelope is required for efficient nuclear protein accumulation and retention. A possible functional polarity of the nu- clear pore allowing for import of nuclear protein and export of, for example, ribonucleoprotein should be amenable to experimental test.

Cell type-specific accumulation of nuclear proteins during organogenesis indicates that some of the nuclear proteins found in the oocyte and in all cells of the embryo may serve a nuclear function associated with the differentiated state of the cell. The existence of antibodies against germinal vesicle proteins allows the selection of the corresponding genes from *Xenopus* expression gene libraries. Molecular genetic analysis may demonstrate sequence homologies of these proteins to proteins whose function and sequence are known. Moreover, nuclear targeting sequences of different efficiencies may be identified by this approach.

Nuclear proteins required for the regulation of gene expression might be detected by their homology to developmentally regulated gene products identi- fied by genetic analysis in other species. The search for proteins containing the DNA-binding homeo domain has led to the discovery of a family of proteins that are potentially involved in the regional specification of the vertebrate embryo. Other consensus domains, such as the finger structure, might be exploited in a similar way as the homeobox, with the aim to decipher a network of regulatory interactions that leads to a stepwise cell differentiation in embryogenesis.

ACKNOWLEDGMENTS. The author is grateful to several colleagues for communicating their results before publication, and to Dr. Peter Hausen, Dr. Marc Servetnik, and Dr. Reimer Stick for helpful comments on the manuscript, and to Franz Zimmer, Metta Riebesell, and Peter Hausen for the use of Figs. 1 and 2. I wish to thank Brigitte Gläser and Ulrike Goßweiler for expert technical assistance, Roswitha Groemke-Lutz for photographic work, and last, but not least, Christa Hug and Margot Heller, for carefully typing the manuscript.

References

Aebi, U., Cohn, J., Buhle, L., and Gerace, L., 1986, The nuclear lamina is a meshwork of intermediate-type filaments, *Nature (Lond.)* **323:**560–564.

Ballantine, J. E. M., Woodland, H. R., and Sturgess, E. A., 1979, Changes in protein synthesis during the development of *Xenopus laevis*, *J. Embryol. Exp. Morphol.* **51:**137–153.

Baum, E. Z., and Wormington, W. M., 1985, Coordinate expression of ribosomal protein genes during *Xenopus* development, *Dev. Biol.* **111:**488–498.

Benavente, R., Krohne, G., and Franke, W. W., 1985, Cell type-specific expression of nuclear lamina proteins during development of *Xenopus laevis*, *Cell* **41:**177–190.

Bendig, M. M., 1981, Persistence and expression of histone genes injected into *Xenopus laevis* eggs in early development, *Nature (Lond.)* **292:**65–67.

Berg, J. M., 1986, Potential metal-binding domains in nucleic acid binding proteins, *Science* **232** 485–487.

Blobel, G., 1980, Intracellular protein topogenesis, *Proc. Natl. Acad. Sci. USA* **77:**1496–1500.

Bonner, W. M., 1975, Protein migration into nuclei. II. Frog oocyte nuclei accumulate a class of microinjected oocyte nuclear proteins and exclude a class of microinjected oocyte cytoplasmic proteins, *J. Cell Biol.* **64:**431–437.

Bonner, W. M., 1978, Protein migration and accumulation in nuclei, in: *The Cell Nucleus*, Vol. VI (H. Busch, ed.), pp. 97–148, Academic, New York.

Boterenbrood, E. C., Narraway, J. M., and Hara, K., 1983, Duration of cleavage cycles and asymmetry in the direction of cleavage waves prior to gastrulation in *Xenopus laevis*, *Roux Arch. Dev. Biol.* **192:**216–221.

Brachet, J., Hanocq, F., and Van Gansen, P., 1970. A cytochemical and ultrastructural analysis of *in vitro* maturation in amphibian oocytes, *Dev. Biol.* **21:**157–195.

Brandhorst, B. P., 1985, The information content of the echinoderm egg, in: *Developmental Biology: A Comprehensive Synthesis*, Vol. 1: *Oogenesis* (L. Browder, ed.), pp. 525–576, Plenum, New York.

Brown, D. D., 1984, The role of stable complexes that repress and activate eucaryotic genes, *Cell* **37:** 359–365.

Brown, D. D., Wensink, P. C., and Jordan, E., 1971, Purification and some characteristics of 5S DNA from *Xenopus laevis*, *Proc. Natl. Acad. Sci. USA* **68:**3175–3179.

Brown, D. D., and Gurdon, J. B., 1978, Cloned single repeating units of 5S DNA direct accurate transcription of 5S RNA when injected into *Xenopus* oocytes, *Proc. Natl. Acad. Sci. USA* **75:** 2849–2853.

Brown, R. S., Sander, C., and Argos, P., 1985, The primary structure of transcription factor TFIIIA has 12 consecutive repeats, *FEBS Lett.* **186:**271–274.

Bürglin, T. R., Mattaj, J. W., Newmeyer, D. W., Zeller, R., and De Robertis, E. M., 1987, Cloning of nucleoplasmin from *Xenopus laevis* oocytes and analysis of its developmental expression, *Gene Dev.* **1**:97–107.

Bürglin, T. R., and De Robertis, E. M., 1987, The nuclear migration signal of *Xenopus laevis* nucleoplasmin, *EMBO. J.* **6**:2617–2625.

Busch, H., Reddy, R., Rothblum, L., and Choi, Y. C., 1982, SnRNAs, SnRNPs, and RNA processing, *Annu. Rev. Biochem.* **51**:617–654.

Carrasco, A. E., and Malacinski, G. M., 1987, Localization of *Xenopus* homeobox gene transcripts during embryogenesis and in the adult nervous system, *Dev. Biol.* **121**:69–81.

Carrasco, A. E., McGinnis, W., Gehring, W. J., and De Robertis, E. M., 1984, Cloning of an *X.laevis* gene expressed during early embryogenesis coding for a peptide region homologous to *Drosophila* homeotic genes, *Cell* **37**:409–414.

Cho, K. W. Y., Goetz, J., Wright, C. V. E., Fritz, A., Hardwicke, J., and De Robertis, E. M., 1988, Differential utilization of the same reading frame in a *Xenopus* homoeobox gene encodes two related proteins sharing the same DNA-binding specificity, *EMBO J.* **7**:2139–2149.

Coggins, L. W., 1973, An ultrastructural and radioautographic study of early oogenesis in the toad *Xenopus laevis*, *J. Cell Sci.* **12**:71–93.

Condie, B. G., and Harland, R. M., 1987, Posterior expression of a homeobox gene in early *Xenopus* embryos, *Development* **101**:93–105.

Dabauvalle, M.-C., and Franke, W. W., 1982, Karyophilic proteins: Polypeptides synthesized *in vitro* accumulate in the nucleus on microinjection into the cytoplasm of amphibian oocytes, *Proc. Natl. Acad. Sci. USA* **79**:5302–5306.

Darnbrough, C. H., and Ford, P. J., 1981, Identification in *Xenopus laevis* of a class of oocyte-specific proteins bound to messenger RNA, *Eur. J. Biochem.* **113**:415–424.

Davey, J., Dimmock, N. J., and Colman, A., 1985, Identification of the sequence responsible for the nuclear accumulation of the influenza virus nucleoprotein in *Xenopus* oocytes, *Cell* **40**:667–675.

Davidson, E. H., 1986, *Gene Activity in Early Development*, 3rd ed., Academic, London.

Dequin, R., Saumweber, H., and Sedat, J. W., 1984, Proteins shifting from the cytoplasm into the nuclei during early embryogenesis of *Drosophila melanogaster*, *Dev. Biol.* **104**:37–48.

De Robertis, E. M., 1983, Nucleocytoplasmic segregation of proteins and RNAs, *Cell* **32**:1021–1025.

De Robertis, E. M., Longthorne, R. F., and Gurdon, J. B., 1978, Intracellular migration of nuclear proteins in *Xenopus* oocytes, *Nature (Lond.)* **272**:254–256.

Dilworth, S. M., Black, S. J., and Laskey, R. A., 1987, Two complexes that contain histones are required for nucleosome assembly in vitro: role of nucleoplasmin and N1 in *Xenopus* egg extracts, *Cell* **51**:1009–1028.

Dingwall, C., and Laskey, R. A., 1986, Protein import into the cell nucleus, *Annu. Rev. Cell Biol.* **2**:367–390.

Dingwall, C., Sharnick, S. V., and Laskey, R. A., 1982, A polypeptide domain that specifies migration of nucleoplasmin into the nucleus, *Cell* **30**:449–458.

Dingwall, C., Bürglin, T. R., Kearsey, S. E., Dilworth, S., and Laskey, R. A., 1986, Sequence features of the nucleoplasmin tail region and evidence for a selective entry mechanism for transport into the cell nucleus, in: *Nucleocytoplasmic Transport* (R. Peters and M. Trendelenburg, eds.), pp. 159–169, Springer-Verlag, Berlin.

Dingwall, C., Dilworth, S. M., Black, S. J., Kearsey, S. E., Cox, L. S., and Laskey, R. A., 1987, Nucleoplasmin cDNA sequence reveals polyglutamic acid tracts and a cluster of sequences homologous to putative nuclear localization signals, *EMBO J.* **6**:69–74.

Dingwall, C., Robbins, J., Dilworth, S. M., Roberts, B., and Richardson, W. D., 1988, The nucleoplasmin nuclear location sequence is larger and more complex than that of SV40 large T antigen. *J. Cell Biol.* **107**:841–849.

Dreyer, C., 1987, Differential accumulation of oocyte nuclear proteins by embryonic nuclei of *Xenopus*, *Development* **101**:829–846.

Dreyer, C., and Hausen, P., 1983, Two-dimensional gel analysis of the fate of oocyte nuclear proteins in the development of *Xenopus laevis*, *Dev. Biol.* **100**:412–425.

Dreyer, C., Singer, H., and Hausen, P., 1981, Tissue specific nuclear antigens in the germinal vesicle of *Xenopus laevis* oocytes, *Wilhelm Roux Arch.* **190**:197–207.

Dreyer, C., Scholz, E., and Hausen, P., 1982, The fate of oocyte nuclear proteins during early development of *Xenopus laevis, Wilhelm Roux Arch.* **191**:228–233.

Dreyer, C., Wang, Y. H., Wedlich, D., and Hausen, P., 1983, Oocyte nuclear proteins in the development of *Xenopus*, in: *Current Problems in Germ Cell Differentiation* (A. McLaren and C. C. Wylie, eds.), pp. 329–351, Cambridge University Press, Cambridge, England.

Dreyer, C., Wang, Y. H., and Hausen, P., 1985, Immunological relationship between oocyte nuclear proteins of *Xenopus laevis* and *X. borealis, Dev. Biol.* **108**:210–219.

Dreyer, C., Stick, R., and Hausen, P., 1986, Uptake of oocyte nuclear proteins by nuclei of *Xenopus* embryos, in: *Nucleocytoplasmic Transport* (R. Peters and M. Trendelenburg, eds.), pp. 143–157, Springer-Verlag, Berlin.

Dumont, J. N., 1972, Oogenesis in *Xenopus laevis* (Daudin). I. Stages of oocyte development in laboratory maintained animals, *J. Morphol.* **136**:153–179.

Dunphy, W. G., Brizuela, L., Beach, D., and Newport, J., 1988, The *Xenopus* cdc2 protein is a component of MPF, a cytoplasmic regulator of mitosis, *Cell* **54**:423–431.

Dworetzky, S. I., and Feldherr, C. M., 1988, Translocation of RNA-coated gold particles through the nuclear pores of oocytes, *J. Cell Biol.* **106**:575–584.

Dworetzky, S. I., Lanford, R. E., and Feldherr, C. M., 1988, The effects of variations in the number and sequence of targeting signals on nuclear uptake, *J. Cell Biol.* **107**:1279–1287.

Earnshaw, W. C., Honda, B. M., Laskey, R. A., and Thomas, J. O., 1980, Assembly of nucleosomes: the reaction involving *X. laevis* nucleoplasmin, *Cell* **21**:373–383.

Eastman, E., Dreyer, C., Hausen, P., and Etkin, L., 1986, Cloning of genes coding for GV proteins localized in embryonic nuclei of *Xenopus, J. Cell Biol.* **103**:245a.

Etkin, L. D., 1982, Analysis of the mechanisms involved in gene regulation and cell differentiation by microinjection of purified genes and somatic cell nuclei into amphibian oocytes and eggs, *Differentiation*, **21**:149–159.

Etkin, L. D., Pearman, B., Roberts, M., and Bektesh, S., 1984, Replication, integration and expression of exogenous DNA injected into fertilized eggs of *Xenopus laevis, Differentiation* **26**:194–202.

Fawcett, D. W., 1981, *The Cell*, W. B. Saunders, Philadelphia.

Featherstone, C., Darby, M. K., and Gerace, L., 1988, A monoclonal antibody against the nuclear pore complex inhibits nucleocytoplasmic transport of protein and RNA in vivo, *J. Cell Biol.* **107**:1289–1297.

Feldherr, C. M., Kallenbach, E., and Schultz, N., 1984, Movement of a karyophilic protein through the nuclear pores of oocytes, *J. Cell Biol.* **99**:2216–2222.

Feldherr, C. M., and Ogburn, J. A., 1980, Mechanism for the selection of nuclear polypeptides in *Xenopus* oocytes. II. Two-dimensional gel analysis, *J. Cell Biol.* **87**:589–593.

Feldherr, C. M., and Pomerantz, J., 1978, Mechanism for the selection of nuclear polypeptides in *Xenopus* oocytes, *J. Cell Biol.* **78**:168–175.

Finlay, D. R., Newmeyer, D. D., Price, T. M., and Forbes, D. J., 1987, Inhibition of an in vitro nuclear transport by a lectin that binds to nuclear pores, *J. Cell Biol.* **104**:189–200.

Forbes, D. J., Kirschner, M. W., and Newport, J. W., 1983, Spontaneous formation of nucleus-like structures around bacteriophage DNA microinjected into *Xenopus* eggs, *Cell* **34**:13–23.

Fox, A. M., Breaux, C. B., and Benbow, R. M., 1980, Intracellular localization of DNA polymerase activities within large oocytes of the frog, *Xenopus laevis, Dev. Biol.* **80**:79–95.

Franke, W. W., Scheer, U., Krohne, G., and Jarasch, E.-D., 1981, The nuclear envelope and the architecture of the nuclear periphery, *J. Cell Biol.* **91**:39s–50s.

Fritz, A., and De Robertis, E. M., 1987, *Xenopus* homeobox-containing cDNAs expressed in early development, *Nucl. A. Res.* **16**:1453–1469.

Fritz, A., Parisot, R., Newmeyer, D., and De Robertis, E. M., 1984, Small nuclear U-ribonucleoproteins in *Xenopus laevis* development, *J. Mol. Biol.* **178**:273–285.

Gautier, J., Norbury, C., Lohka, M., Nurse, P., and Maller, J., 1988, Purified maturation-promoting factor contains the product of a *Xenopus* homolog of the fission yeast cell cycle control gene cdc2+, *Cell* **54**:433–439.

Gerace, L., and Blobel, G., 1980, The nuclear envelope lamina is reversibly depolymerized during mitosis, *Cell* **19**:277–287.

Gerhart, J. C., 1980, Mechanisms regulating pattern formation in the amphibian egg and early embryo, in: *Biological Regulation and Development*, Vol. 2 (R. F. Goldberger, ed.), pp. 133–316, Plenum, New York.

Gerhart, J., Wu, M., and Kirschner, M., 1984, Cell cycle dynamics of an M-phase-specific cytoplasmic factor in *Xenopus laevis* oocytes and eggs, *J. Cell Biol.* **98:**1247–1255.

Glikin, G. C., Ruberti, I., and Worcel, A., 1984, Chromatin assembly in *Xenopus* oocytes: *In vitro* studies, *Cell* **37:**33–41.

Goldfarb, D. S., Gariépy, J., Schoolnik, G., and Kornberg, R. D., 1986, Synthetic peptides as nuclear localization signal, *Nature (Lond.)* **322:**641–644.

Graham, C. F., 1966, The regulation of DNA synthesis and mitosis in multinucleate frog eggs, *J. Cell Sci.* **1:**363–374.

Gurdon, J. B., 1970, Nuclear transplantation and the control of gene activity in animal development, *Proc. R. Soc. Lond. [Biol.]* **176:**303–314.

Gurdon, J. B., 1974, *The Control of Gene Expression in Animal Development*, Calderon, Oxford.

Gurdon, J. B., and Melton, D. A., 1981, Gene transfer in amphibian eggs and oocytes, *Annu. Rev. Genet.* **15:**189–218.

Hall, M. N., Hereford, L., and Herskowitz, I., 1984, Targeting of *E. coli* β-galactosidase to the nucleus in yeast, *Cell* **36:**1057–1065.

Harvey, R. P., Tabin, C. J., and Melton, D. A., 1986, Embryonic expression and nuclear localization of *Xenopus* homeobox (Xhox) gene products, *EMBO J.* **5:**1237–1244.

Hausen, P., Wang, Y. H., Dreyer, C., and Stick, R., 1985, Distribution of nuclear proteins during maturation of the *Xenopus* oocyte, *J. Embryol. Exp. Morphol.* **89:**(suppl.):17–34.

Holland, P. W. H., and Hogan, B. L. M., 1988, Expression of homeo box genes during mouse development: a review, *Genes & Development* **2:**773–782.

Holt, G. D., Snow, C. M., Senior, A., Haltiwanger, R. S., Gerace, L., and Hart, G. W., 1987, Nuclear pore complex glycoproteins contain cytoplasmically disposed O-linked N-Acetylglucosamine, *J. Cell Biol.* **104:**1157–1164.

Huchon, D., Crozet, N., Cantenot, N., and Ozon, R., 1981, Germinal vesicle breakdown in the *Xenopus laevis* oocyte: Description of a transient microtubular structure. *Reprod. Nutr. Dev.* **21:**135–148.

Jessus, C., Huchon, D., and Ozon, R., 1986, Distribution of microtubules during the breakdown of the nuclear envelope of the *Xenopus* oocyte: an immunocytochemical study, *Biol. Cell* **56:**113–120.

Kalderon, D., Richardson, W. D., Markham, A. F., and Smith, A. E., 1984, Sequence requirements for nuclear location of simian virus 40 large-T antigen, *Nature (Lond.)* **311:**33–38.

Kessel, R. G., 1983, The structure and function of annulate lamellae: Porous cytoplasmic and intranuclear membranes, *Int. Rev. Cytol.* **82:**181–303.

Kessel, R. G., 1985, Annulate lamellae (porous cytomembranes). With particular emphasis on their possible role in differentiation of the female gamete, in: *Developmental Biology: A Comprehensive Synthesis*, Vol. 1: *Oogenesis* (L. W. Browder, ed.), pp. 179–233, Plenum, New York.

Kleinschmidt, J. A., Dingwall, C., Maier, G., and Franke, W. W., 1986, Molecular characterization of a karyophilic, histone-binding protein: cDNA cloning, amino acid sequence and expression of nuclear protein N1/N2 of *Xenopus laevis*, *EMBO J.* **5:**3547–3552.

Kleinschmidt, J. A., Fortkamp, E., Krohne, G., Zentgraf, H., and Franke, W. W., 1985, Co-existence of two different types of soluble histone complexes in nuclei of *Xenopus laevis* oocytes, *J. Biol. Chem.* **260:**1166–1176.

Kleinschmidt, J. A., and Franke, W. W., 1982, Soluble acidic complexes containing histones H3 and H4 in nuclei of *Xenopus laevis* oocytes, *Cell* **29:**799–809.

Kleinschmidt, J. A., and Seiter, A., 1988, Identification of domains involved in nuclear uptake and histone binding of protein N1 of *Xenopus laevis*, *EMBO J.* **7:**1605–1614.

Köster, M., Pieler, T., Pöting, A., and Knöchel, W., 1988, The finger motif defines a multigene family represented in the maternal mRNA of *Xenopus laevis* oocytes, *EMBO J.* **7:**1735–1741.

Krämer, A., 1985, 5S Ribosomal gene transcription during *Xenopus* oogenesis, in: *Developmental Biology: A Comprehensive Synthesis*, Vol. 1: *Oogenesis* (L. W. Browder, ed.), pp. 431–451, Plenum, New York.

Krohne, G., and Benavente, R., 1986, The nuclear lamins, *Exp. Cell Res.* **162**:1–10.

Lanford, R. E., and Butel, J. S., 1984, Construction and characterization of an SV 40 mutant defective in nuclear transport of T antigen, *Cell* **37**:801–813.

Lanford, R. E., Kanda, P., and Kennedy, R. C., 1986, Induction of nuclear transport with a synthetic peptide homologous to the SV40 T antigen transport signal, *Cell* **46**:575–582.

Laskey, R. A., 1985, Chromosome replication in early development of *Xenopus laevis*, *J. Embryol. Exp. Morphol.* **89**(suppl.):285–296.

Laskey, R. A., Mills, A. D., and Morris, N. R., 1977, Assembly of SV40 chromatin in a cell-free system from *Xenopus* eggs, *Cell* **10**:237–243.

Laskey, R. A., Honda, B. M., Mills, A. D., and Finch, J. T., 1978, Nucleosomes are assembled by an acidic protein which binds histones and transfers them to DNA, *Nature (Lond.)* **275**:416–420.

Laughon, A., and Scott, M. P., 1984, Sequence of a *Drosophila* segmentation gene: protein structure homology with DNA-binding proteins, *Nature* **310**:25–31.

Lohka, M. J., Hayes, M. K., and Maller, J. L., 1988, Purification of maturation-promoting factor, an intracellular regulator of early mitotic events, *Proc. Natl. Acad. Sci. USA* **85**:3009–3013.

MacGregor, H. C., and Kezer, J., 1970, Gene amplification in oocytes with 8 germinal vesicles from the tailed frog *Ascaphus truei* Stejneger, *Chromosoma* **29**:189–206.

MacGregor, H. C., and del Pino, E. M., 1982, Ribosomal gene amplification in multinucleate oocytes of the egg brooding hylid frog *Flectonotus pygmaeus*, *Chromosoma* **85**:475–488.

Maller, J. L., 1985, Oocyte maturation in amphibians, in: *Developmental Biology*, Vol. 1: *Oogenesis* (L. W. Browder, ed.), pp. 289–311, Plenum, New York.

Maller, J. L., Wu, M., and Gerhard, J. C., 1977, Changes in protein phosphorylation accompanying maturation of *Xenopus laevis* oocytes, *Dev. Biol.* **85**:295–312.

Masui, Y., 1985, Meiotic arrest in animal oocytes, in: *Biology of Fertilization*, Vol. 1 (C. B. Metz and A. Monroy, eds.), pp. 189–219, Academic, New York.

Masui, Y., and Clarke, H. J., 1979, Oocyte maturation, *Int. Rev. Cytol.* **57**:185–282.

Mattaj, I. W., and De Robertis, E. M., 1985, Nuclear segregation of U2 snRNA requires binding of specific snRNP proteins, *Cell* **40**:111–118.

Mattaj, I. W., Lienhard, S., Zeller, R., and De Robertis, E. M., 1983, Nuclear exclusion of transcription factor IIIA and the 42S particle transfer RNA-binding protein in *Xenopus* oocytes: A possible mechanism for gene control?, *J. Cell Biol.* **97**:1261–1265.

Mattaj, I. W., Zeller, R., Carrasco, A. E., Jamrich, M., Lienhard, S., and De Robertis, E. M., 1985, U snRNA gene families in *Xenopus laevis*, in: *Oxford Surveys on Eukaryotic Genes*, Vol. 2 (N. Maclean, ed.), pp. 121–140, Oxford University Press,, Oxford, England.

McGinnis, W., Levine, M. S., Hafen, E., Kuroiwa, A., and Gehring, W. J., 1984, A conserved DNA sequence in homeotic genes of the *Drosophila* antennapedia and bithorax complexes, *Nature (Lond.)* **308**:428–433.

McKeon, F. D., Kirschner, M. W., and Caput, D., 1986, Homologies in both primary and secondary structure between nuclear envelope and intermediate filament proteins, *Nature (Lond.)* **319**:463–468.

Miake-Lye, R., and Kirschner, M. W., 1985, Induction of early mitotic events in a cell-free system, *Cell* **41**:165–175.

Miller, M., Kloc, M., Eastman, E., Reddy, B., Dreyer, C., and Etkin, L., 1989, xlgv: a maternal gene product localized in nuclei of the central nervous system *Genes and Develop.*

Miller, J., McLachlan, A. D., and Klug, A., 1985, Repetitive zinc-binding domains in the protein transcription factor IIIA from *Xenopus* oocytes, *EMBO J.* **4**:1609–1614.

Müller, M. M., Carrasco, A. E., and De Robertis, E. M., 1984, A homeo-box-containing gene expressed during oogenesis in *Xenopus*, *Cell* **39**:157–162.

Newmeyer, D. D., Lucocq, J. M., Bürglin, T. R., and De Robertis, E. M., 1986a, Assembly *in vitro* of nuclei active in nuclear protein transport: ATP is required for nucleoplasmin accumulation, *EMBO J.* **5**:501–510.

Newmeyer, D. D., Finlay, D. R., and Forbes, D. J., 1986b, *In vitro* transport of a fluorescent nuclear protein and exclusion of non-nuclear protein, *J. Cell Biol.* **103**:2091–2102.

Newmeyer, D. D., and Forbes, D. J., 1988, Nuclear import can be separated into distinct steps *in vitro*: nuclear pore binding and translocation, *Cell* **52**:641–653.

Newport, J., 1987, Nuclear reconstitution *in vitro*: Stages of assembly around protein-free DNA, *Cell* **48**:205–217.

Newport, J., and Kirschner, M., 1982, A major developmental transition in early *Xenopus* embryos. I. Characterization and timing of cellular changes at the midblastula stage, *Cell* **30**:675–686.

Nielsen, P. J., Thomas, G., and Maller, J. L., 1982, Increased phosphorylation of ribosomal protein S6 during meiotic maturation of *Xenopus* oocytes, *Proc. Natl. Acad. Sci. USA* **79**:2937–2941.

Oliver, G., Wright, C. V. E., Hardwicke, J., and De Robertis, E. M., 1988a, Differential antero-posterior expression of two proteins encoded by a homeobox gene in *Xenopus* and mouse embryos, *EMBO J.* **7**:3199–3209.

Oliver, G., Wright, C. V. E., Hardwicke, J., and De Robertis, E. M., 1988b, A gradient of homeodomain protein in developing forelimbs of *Xenopus* and mouse embryos, *Cell* **55**:1017–1024.

Paine, P. L., 1982, Mechanisms of nuclear protein concentration, in: *The Nuclear Envelope and the Nuclear Matrix* (G. Maul, ed.), pp. 75–83, Liss, New York.

Pelham, H. R. B., Wormington, W. M., and Brown, D. D., 1981, Related 5S RNA transcription factors in *Xenopus* oocytes and somatic cells, *Proc. Natl. Acad. Sci. USA* **78**:1760–1764.

Picard, B., and Wegnez, M., 1979, Isolation of a 7S particle from *Xenopus laevis* oocytes: A 5S RNA–protein complex, *Proc. Natl. Acad. Sci. USA* **76**:241–245.

Pieler, T., Hamm, J., and Roeder, R. G., 1987, The 5S gene internal control region is composed of three distinct sequence elements, organized as two functional domains with variable spacing, *Cell* **48**:91–100.

del Pino, E. M., and Humphries, A. A., jr. 1978, Multiple nuclei during early oogenesis in *Flectonotus pygmaeus* and other marsupial frogs, *Biol. Bull.* **154**:198–212.

Richter, J. D., and Smith, L. D., 1984, Reversible inhibition of translation by *Xenopus* oocyte-specific proteins, *Nature (Lond.)* **309**:378–380.

Richardson, W. D., Mills, A. D., Dilworth, S. M., Laskey, R. A., and Dingwall, C., 1988, Nuclear protein migration involves two steps: rapid binding at the nuclear envelope followed by slower translocation through nuclear pores, *Cell* **52**:655–664.

Roberts, B. L., Richardson, W. D., and Smith, A. E., 1987, The effect of protein context on nuclear location signal function, *Cell* **50**:465–475.

Roberts, B. L., Richardson, W. D., and Smith, A. E., 1987, The effect of protein context on nuclear location signal function, *Cell* **50**:465–475.

Roeder, R. G., 1974, Multiple forms of deoxyribonucleic acid-dependent ribonucleic acid polymerase in *Xenopus laevis*, *J. Biol. Chem.* **249**:249–256.

Ruiz i Altaba, A., Perry-O'Keefe, H., and Melton, D. A., 1987, Xfin: An embryonic gene encoding a multifingered protein in *Xenopus*, *EMBO J.* **6**:3065–3070.

Sakonju, S., and Brown, D. D., 1982, Contact points between a positive transcription factor and the *Xenopus* 5S RNA gene, *Cell* **31**:395–405.

Scott, M. P., 1988, The molecular biology of pattern formation in the early embryonic development of *Drosophila*, in: *Developmental Biology: A Comprehensive Synthesis*, Vol. 5: *The Molecular Biology of Cell Determination and Cell Differentiation* (L. W. Browder, ed.), pp. 151–185, Plenum, New York.

Scott, M. P., and Weiner, A. J., 1984, Structural relationship among genes that control development: Sequence homology between the Antennapedia, Ultrabithorax, and fushi tarazu loci of *Drosophila*, *Proc. Natl. Acad. Sci. USA* **81**:4115–4119.

Sealy, L., Cotten, M., and Chalkley, R., 1986, *Xenopus* nucleoplasmin: Egg vs. oocyte, *Biochemistry* **25**:3064–3072.

Sharpe, C. R., Fritz, A., De Robertis, E. M., and Gurdon, J. B., 1987, A homeobox-containing marker of posterior neural differentiation shows the importance of predetermination in neural induction, *Cell* **50**:749–758.

Signoret, P. J., and David, J. C., 1986, Control of the expression of genes for DNA ligase in eucaryotes, *Int. Rev. Cytol.* **103**:249–279.

Signoret, P. J., and Lefresne, J., 1971, Contribution à l'etude de la segmentation de l'oeuf d'axolotl. I. Définition de la transition blastuléenne, *Ann. Embryol. Morphol.* **4**:113–123.

Signoret, P. J., Lefresne, J., Vinsson, D., and David, J. C., 1981, Enzymes involved in DNA replication in the Axolotl, *Dev. Biol.* **87**:126–132.

Smith, L. D., and Richter, J. D., 1985, Synthesis, accumulation, and utilization of maternal macromolecules during oogenesis and oocyte maturation, in: *Biology of Fertilization* (C. B. Metz and A. Monroy, eds.), pp. 141–188, Academic, Orlando, Florida.

Snow, C. M., Senior, A., and Gerace, L., 1987, Monoclonal antibodies identify a group of nuclear pore complex glycoproteins, *J. Cell Biol.* **104**:1143–1156.

Stick, R., 1987, Dynamics of the nuclear lamina during mitosis and meiosis, in: *Molecular Regulation of Nuclear Events in Mitosis and Meiosis* (R. A. Schlegel, M. S. Halleck, and P. N. Rao, eds.), pp. 43–66, Academic, Orlando, Florida.

Stick, R., 1988, cDNA cloning of the developmentally regulated lamin L_{III} of *Xenopus laevis*, *EMBO J.* **7**:3189–3197.

Stick, R., and Dreyer, C., 1989, Developmental control of nuclear proteins in Amphibia, in: *The Molecular Biology of Fertilization* (H. Schatten and G. Schatten, eds.), pp. 153–188. Academic, Orlando, Florida.

Stick, R., and Hausen, P., 1985, Changes in the nuclear lamina composition during early development of *Xenopus laevis*, *Cell* **41**:191–200.

Stick, R., and Krohne, G., 1982, Immunological localization of the major architectural protein associated with the nuclear envelope of the *Xenopus laevis* oocyte, *Exp. Cell Res.* **138**:319–330.

Stick, R., and Schwarz, H., 1983, Disappearance and reformation of the nuclear lamina structure during specific stages of meiosis in oocytes, *Cell* **33**;949–958.

Tchou-Su, P., and Wang, Y. L., 1985, Etudes comparatives sur l'ovulation et la maturation in vivo et in vitro chez le crapaud asiatique (*Bufo Bufo Asiaticus*), *Acta Biol. Exp. Sin.* **6**:129–180.

Tourte, M., Mignotte, F., and Mounolou, J. C., 1981, Organization and replication activity of the mitochondriae mass of oogonia and previtellogenic oocytes in *Xenopus laevis*, *Dev. Growth Diff.* **23**:9–21.

Wasserman, W. J., Richter, J. D., and Smith, L. D., 1982, Protein synthesis during maturation promoting factor- and progesterone-induced maturation in *Xenopus* oocytes, *Dev. Biol.* **89**: 152–158.

Wedlich, D., Dreyer, C., and Hausen, P., 1985, Occurence of a species-specific nuclear antigen in the germ line of *Xenopus* and its expression from paternal genes in hybrid frogs, *Dev. Biol.* **108**: 220–234.

Wedlich, D., and Dreyer, C., 1988a, Cell specificity of nuclear protein antigens in the development of *Xenopus*, *Cell Tissue Res.* **252**:479–489.

Wedlich, D., and Dreyer, C., 1988b, The distribution of nucleoplasmin in early development and organogenesis of *Xenopus laevis*, *Cell Tissue Res.* **254**:295–300.

Witschi, E., 1956, *Development of Vertebrates*, W. B. Saunders, Philadelphia.

Wittek, M., 1952, La vitellogenèse chez les Amphibiens, *Arch. Biol.* **63**:133–198.

Wolin, S. L., and Kirschner, M. W., 1986, Translational regulation of nuclear lamin proteins during early development in *Xenopus laevis*, *J. Cell. Biol.* **103**:245a.

Wolin, S. L., Krohne, G., and Kirschner, M. W., 1987, A new lamin in *Xenopus* somatic tissues displays strong homology to human lamin A, *EMBO J.* **6**:3809–3818.

Woodland, H. R., 1980, Histone synthesis during the development of *Xenopus*. *FEBS Lett.* **121**:1–7.

Woodland, H. R., 1982, The translational control phase of early development, *Biosci. Rep.* **2**:471–491.

Woodland, H. R., and Adamson, E. D., 1977, Synthesis and storage of histones during oogenesis of *Xenopus laevis*, *Dev. Biol.* **57**:118–135.

Woodland, H. R., and Ballantine, J. E. M., 1980, Paternal gene expression in developing hybrid embryos of*Xenopus laevis* and *Xenopus borealis*, *J. Embryol. Exp. Morphol.* **60**:359–372.

Woodland, H. R., Flynn, J. M., and Wyllie, A. J., 1979, Utilization of stored mRNA in *Xenopus* embryos and its replacement by newly synthesized transcripts: Histone H1 synthesis using interspecies hybrids, *Cell* **18**:165–171.

Woodland, H. R., Old, R. W., Sturgess, E. A., Ballantine, J. E. M., Aldridge, T. C., and Turner, P. C., 1983, The strategy of histone gene expression in the development of *Xenopus*, in: *Current Problems in Germ Cell Differentiation* (A. M. Laren and C. C. Wylie, eds.), pp. 353–376, Cambridge University Press, Cambridge, England.

Wormington, W. M., 1988, Expression of ribosomal protein genes during *Xenopus* development, in: *Developmental Biology: A Comprehensive Synthesis*, Vol. 5 (L. W. Browder, ed.), pp. 227–240, Plenum, New York.

Wright, C. V. E., Cho, K. W. Y., Fritz, A., Bürglin, T. R., and De Robertis, E. M., 1987, A *Xenopus laevis* gene encodes both homeobox-containing and homeobox-less transcripts, *EMBO J.* **6:** 4083–4094.

Zeller, R., Nyffenegger, T., and De Robertis, E. M., 1983, Nucleoplasmic distribution of snRNPs and stockpiled snRNA-binding proteins during oogenesis and early development in *Xenopus laevis*, *Cell* **32:**425–434.

Zimmer, F. J., Dreyer, C., and Hausen, P., 1988, The function of the nuclear envelope in nuclear protein accumulation, *J. Cell. Biol.* **106:**1435–1444.

Chapter 4

Genomic Imprinting in the Mouse

S. K. HOWLETT, W. REIK, S. C. BARTON, M. L. NORRIS, and
M. A. H. SURANI

1. Introduction

In retrospect, perhaps it is not surprising that in mammals the two parental genomes have different influences on the developing embryo. However, it is only recently that the functional nonequivalence of the maternal and paternal genomes has been established conclusively and so has commanded a considerable amount of interest. This chapter reviews experiments that have demonstrated the complementary roles of the two parental genomes in the mouse, thus establishing the existence of genomic imprinting in murine development. Also, we indicate some of the approaches being employed to elucidate the molecular basis of this phenomenon.

Recent advances in micromanipulation techniques have made the early mouse embryo amenable for detailed embryological and molecular investigations. It is now possible to manipulate the genetic constitution of the 1-cell zygote, for example, to replace a maternal pronucleus by a second male pronucleus or to replace endogenous pronuclei with advanced nuclei (McGrath and Solter, 1983, 1984a,b; Surani et al., 1984, 1986a). Furthermore, it is possible to introduce specific cloned genes by microinjection into pronuclei where some exogenous genes may become integrated into the recipient genome as transgenes (see Palmiter and Brinster, 1986). Such techniques have proved invaluable in studies on genomic imprinting.

Clearly, a zygote containing both a male and female pronucleus is totipotential, that is, it can develop into an adult. As development proceeds, cells differentiate and become restricted in potential. We and others have studied the potential of various embryonic nuclei by asking how they behave when they are introduced into the totipotential environment of the egg. In this way we have assayed the developmental capacity of eggs containing 2- to 8-cell nuclei from embryos that contain both or only one parental genome. It has become apparent that a mouse egg requires both a maternal and a paternal

S. K. HOWLETT, W. REIK, S. C. BARTON, M. L. NORRIS, and M. A. H. SURANI • Department of Molecular Embryology, Institute of Animal Physiology and Genetics Research, Babraham, Cambridge CB2 4AT, England.

genome for full development (McGrath and Solter, 1984a; Surani et al., 1984) and furthermore that the functionally totipotent state of embryonic nuclei becomes restricted at a comparatively very early stage (McGrath and Solter, 1984b; Howlett et al., 1987).

There has been speculation in recent years as to whether it is possible to produce mammalian live young that contain only maternal or only paternal genomes. To be able to reproduce without the necessity of any paternal participation might be viewed theoretically as an economical and efficient means of reproduction. Indeed, parthenogenesis is widespread in the animal kingdom. Eggs of many species, including various insects and marine invertebrates (Morgan, 1927), some fish (Hubbs and Hubbs, 1932), certain lizards (Maslin, 1967), and a few birds (Olsen, 1966), are capable of parthenogenetic development. Some species of Whiptail lizards are entirely composed of parthenogenetically reproducing females that compete successfully with other female species whose eggs are fertilized by sperm (Maslin, 1967). Parthenogenetic reproduction occurs commonly in a particular breed of turkeys (Olsen, 1966; Harada and Buss, 1981), producing sterile males (since males are the homogametic sex in birds). Mammals appear, however, to have largely eliminated parthenogenesis as a means of reproduction, relying on more rigid reproductive strategies (Markert, 1982). Presumably the added benefits of greater genetic flexibility conferred by heterozygosity have favored biparental reproduction. Nevertheless, mammalian eggs are easy to activate, and so it is possible to initiate artificial parthenogenetic development (Graham, 1974). Spontaneous parthenogenetic activation occurs frequently in the LTSv strain of mice; however, their development is limited, and they die soon after implantation (Stevens, 1975).

2. Nuclear Totipotency

One of the aims of our studies is to discover whether there is any relationship between nuclear totipotency and genomic imprinting in mammals. Nuclear transfer experiments in amphibia have established that a small proportion of embryonic nuclei from determined regions retain their ability to support normal development to tadpole stages, but that this ability is lost progressively thereafter (King and Briggs, 1956). Furthermore, even a small proportion of adult nuclei from differentiated tissues retain the ability to promote extensive but incomplete development (Laskey and Gurdon, 1970; reviewed by Gurdon, 1986; see also Chapter 8, this volume). Although in the mouse 2-, 4-, and 8-cell blastomeres are considered totipotential in that they can contribute to many different tissues in the adult (Kelly, 1979), a single 4-cell blastomere cannot make a mouse on its own (Tarkowski and Wroblewska, 1967; Rossant, 1976). Thus, even by the 4-cell stage, the ability to realize full developmental potential from an individual blastomere is not possible. Similarly, nuclear totipotency, as judged by the ability to support development after transfer to an enucleated egg, is lost very rapidly (Modlinski, 1978; McGrath and Solter, 1984b). Thus,

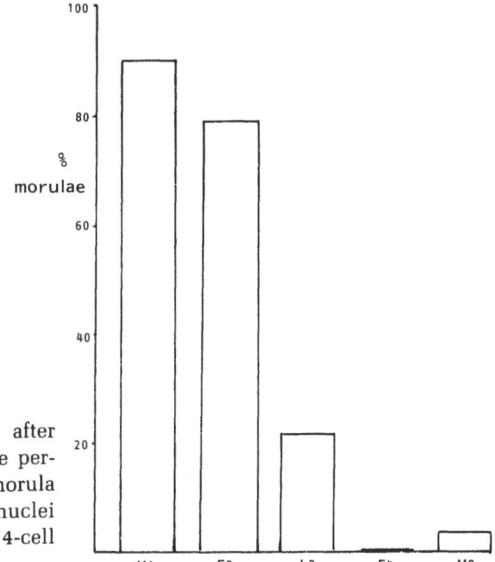

Figure 1. Development of enucleated eggs after transfer of advanced nuclei. Graph shows the percentage of reconstituted eggs that reach the morula stage following enucleation and transfer of pronuclei (M1), early (E2) or late (L2) 2-cell nuclei, early 4-cell (E4) nuclei, or 8-cell (M8) nuclei.

nuclear totipotency is apparently reduced soon after the 2-cell stage (McGrath and Solter, 1984b). We have investigated more precisely at what point the ability to interact successfully with 1-cell cytoplasm is lost by embryonic nuclei (Howlett et al., 1987) (Fig. 1).

Enucleated eggs were used as recipients of 2- to 8-cell stage nuclei. Whereas most early 2-cell nuclei were able to support full in vitro development, this ability was largely lost by nuclei 8–10 hr later. Thus, late 2-cell nuclei produced a few morulae (22%) while 4-cell and 8-cell nuclei were rarely capable of directing more than one cleavage division (Howlett et al., 1987) (Fig. 1). However, if 8-cell nuclei are transferred into enucleated late 2-cell recipients, blastocysts (Robl et al., 1986; Howlett et al., 1987) and even live young (Tsunoda et al., 1987) are produced. Therefore, there appears to be a change in the interaction between nuclei and cytoplasm at the 2-cell stage, which restricts the expression of full totipotency.

The timing of the loss of nuclear totipotency is coincident with the large-scale process of gene activation at the mid 2-cell stage (Flach et al., 1982). Therefore, it is possible that certain information that is used before and during the major transcriptional activation is only available once. Having been used, this particular information may then be altered during the process of major gene activation rendering it unavailable if a transcriptionally active (late 2-cell or beyond) nucleus is returned to 1-cell cytoplasm for a second time. It should be noted that diploid 8-cell nuclei (Modlinski, 1978; Howlett et al., 1987) or nuclei from inner cell mass (ICM) cells (Modlinski, 1981) transferred into eggs that retain one or both pronuclei successfully produce triploid or tetraploid blastocysts. Indeed, haploid advanced nuclei of the appropriate parental genotype when transferred into eggs retaining either of the pronuclei will successfully support full-term development (Surani et al., 1986a) . The inference from these experiments is that a single endogenous pronucleus is sufficient to provide vital information that cannot be extracted by egg cytoplasm from a transferred advanced nucleus. Furthermore, serial transfers of 8-cell nuclei through two exposures to 1-cell cytoplasm in a manner analogous to the serial transfers performed in amphibia (Gurdon and Laskey, 1970) had no beneficial effect on the ability of the advanced nucleus to direct development (Howlett and Barton, unpublished observations).

We could view these combined results as an incompatibility of post-gene-activation nuclei with egg (or early 2-cell) cytoplasm. When looking at the patterns of proteins being synthesized in an egg following transfer of an 8-cell nucleus, we have seen no evidence for synthesis of proteins characteristic of the 8-cell embryo. Furthermore, egg cytoplasm appears to be able to reprogram transcriptionally active nuclei to some extent because expression of the heat-shock 68/70 kDa proteins (hsps) occurs following nuclear transfer at a time that is temporally correct for the recipient cytoplasm (Flach et al., 1982; Bensaude et al., 1983; Howlett et al., 1987). It is not clear whether this is peculiar to the hsps, perhaps caused by an abundance of transcription factors in the egg, or whether this reflects a significant reprogramming of the donor nucleus. Given the appropriate markers, it may prove possible to elucidate fully the molecular mechanisms underlying the interactions between nuclei and cytoplasm in the early embryo. They may also allow us to establish whether there is a link between nuclear totipotency and imprinting.

3. Development of Eggs with Only Maternal Genomes

Removal of the male pronucleus from a fertilized egg and replacement with a second female pronucleus by nuclear transfer produces a gynogenetic embryo. A less laborious means of producing embryos with only a maternal genome is to activate ovulated eggs by exposure to dilute ethanol (Cuthbertson, 1983); these embryos are termed parthenogenetic embryos. Such eggs will normally be haploid, but if the second meiotic cleavage division is suppressed, diploid parthenogenotes will result. Gynogenetic and diploid parthenogenetic embryos differ in that gynogenotes are heterozygous and biparental. Most diploid parthenogenotes develop to the blastocyst stage; if transferred to foster mothers, a small proportion will reach mid-gestation to produce, at best, small 25-somite embryos (Fig. 2). Full-term parthenogenetic fetuses have not been observed (Kaufman et al., 1977). Gynogenetic embryos display similar developmental potential.

Two possible explanations have previously been proposed to explain the failure of parthenogenetic embryos to complete development (Graham, 1974): (1) an extragenetic contribution from sperm is necessary for full embryonic development, and (2) the expression of recessive lethal genes from the homozygous maternal genome results in embryonic death. Several experiments were designed to address this problem. Digynic triploids are produced by suppres-

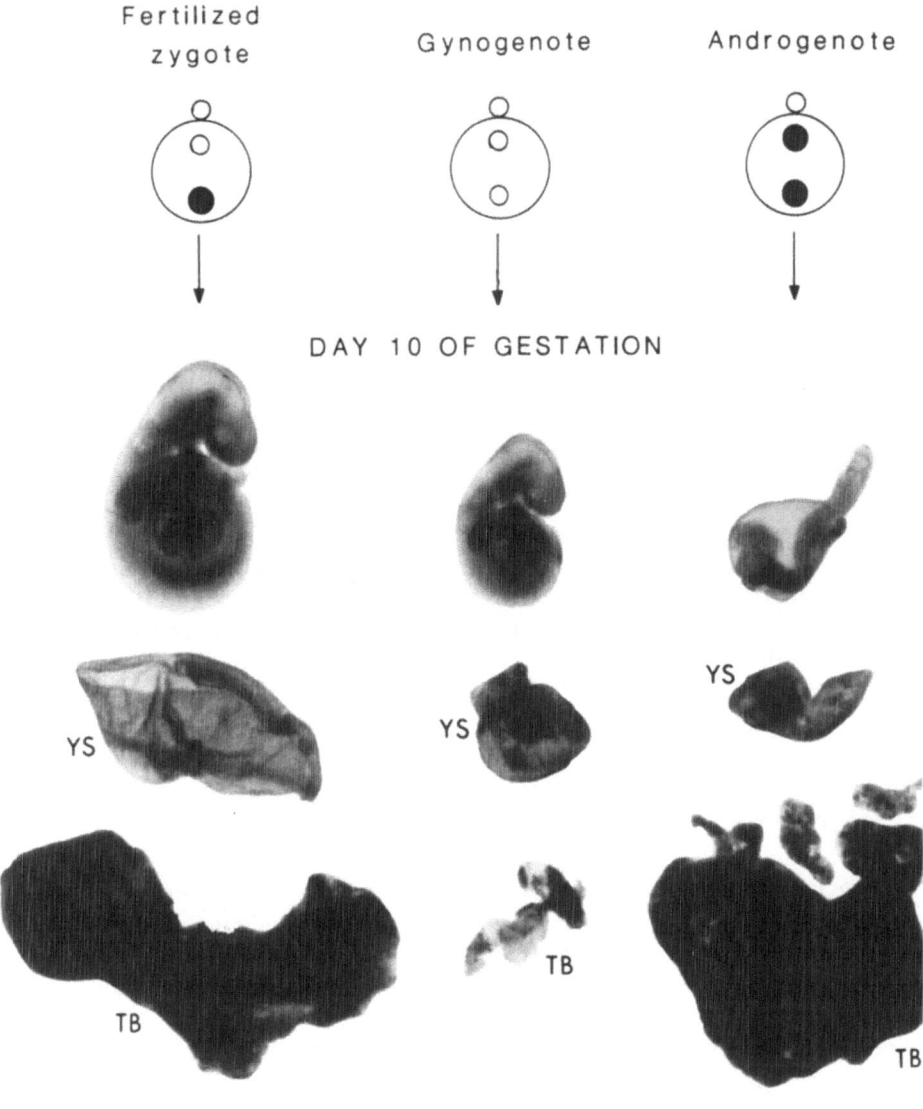

Figure 2. Phenotypes on day 10 of gestation of conceptuses derived from fertilized, parthenogenetic, or androgenetic eggs. Embryo, yolk sac (YS) and trophoblast (TB) have been dissected in each case. (○) Maternal pronucleus. (●) Paternal pronucleus. Bar = 1 mm.

sion of the second meiotic cleavage in fertilized eggs and so contain two female pronuclei and a single male pronucleus (Surani and Barton, 1983). Such digynic triploid embryos exhibit abnormalities by mid-gestation characteristic of all triploids. When one of the female pronuclei was removed, restoring the normal genetic constitution, full embryonic development was possible. However, when the male pronucleus was removed leaving both female pronuclei in the zygote, development closely resembled that of diploid parthenogenetic embryos (Surani and Barton, 1983). A second key series of experiments involved the use of haploid parthenogenetic eggs as recipients for either a male or a female pronucleus. Development to term was only achieved when the recipient egg received a male pronucleus and not when a biparental gynogenetic egg was generated, when development resembled that observed with a parthenogenote (Surani et al., 1984). Similarly, exchange of pronuclei between fertilized eggs (McGrath and Solter, 1984a) or transfer of pronuclei into enucleated parthenogenetic eggs (Mann and Lovell-Badge, 1984) gave the same results. Thus, the experimental evidence suggests that neither the extragenetic contributions from sperm nor the heterozygous state of biparental eggs is sufficient to restore full developmental potential to embryos containing only maternal chromosomes.

Observations of conceptuses lacking paternal chromosomes revealed a striking and consistent phenotype characterized by a chronic lack of extraembryonic tissue (Fig. 2). By mid-gestation, the embryo was normal but small, whereas the yolk sac was slight, the trophoblastic giant cells sparse, and the chorion and ectoplacental cone virtually nonexistent (Surani and Barton, 1983; Surani et al., 1984; Barton, unpublished observations); increasing dependence on placentally derived nutritional support may have prevented development beyond the 25-somite stage. The poor development of the extraembryonic tissue was investigated further in chimeras.

4. Development of Eggs with Only Paternal Genomes

Far fewer studies have been carried out on androgenetic embryos, which contain only paternal genomes. Since there is no equivalent to the convenient method of parthenogenetic activation, androgenotes can only be produced by the removal of the female pronucleus from fertilized eggs, which can then be diploidized, either by preventing the first cleavage division or by the subsequent introduction of a second male pronucleus (Barton et al., 1984; Surani et al., 1984, 1986a). Studies show that androgenetic embryos develop very poorly to the blastocyst stage (Surani et al., 1986a), even allowing for the fact that a quarter of diploid androgenetic eggs will have the genetic constitution YY, limiting development to only a few cleavage divisions (Morris, 1968). Nevertheless, a small percentage of androgenotes showed some postimplantation development. Examination of such mid-gestation conceptuses indicated extensively proliferated extraembryonic tissue with copious giant cells, substantial chorionic tissue, and an expanded yolk sac containing at best a small retarded embryo with 4–6 somites (see Fig. 2); ectoplacental tissue, however, was very poorly developed (Barton et al., 1984; Barton, unpublished observations). Thus, all the evidence from the experiments on reconstituted zygotes showed that development to term requires both parental genomes.

5. Is the Parental Origin of Early Embryonic Nuclei Remembered?

It is important to establish whether the imprinted information present in parental genomes is remembered and propagated through early cleavage divisions or whether it is erased in the 2-cell nucleus during the activation of the embryonic genome. Even during the preimplantation period, androgenetic and parthenogenetic embryos behave differently; whereas parthenogenetic eggs undergo second mitosis at the same time as do fertilized eggs, androgenetic eggs divide to 4-cells about 6 hr earlier, and the few androgenetic cells that divide further compact ahead of their fertilized and parthenogenetic counterparts (Barton, Howlett and Surani, unpublished observations). Even allowing for the fact that 50% of haploid androgenotes will be genetically Y, most (approx. 85%) arrest as 2-cells, and very few divide more than once more; whereas a significant proportion (approx. 30%) of haploid parthenogenetic embryos reach at least the 4-cell stage (Surani et al., 1986a). Intriguingly, we and others have observed that haploid development can be improved by reducing the amount of cytoplasm (McGrath and Solter, 1986; Howlett et al., 1987). By this means, we were able to produce our first haploid androgenetic blastocysts (2 out of 53) and increased the proportion (33%) of haploid parthenogenetic blastocysts (Howlett et al., 1987). By comparison, about 20% of diploid androgenotes reach the blastocyst stage, but diploid parthenogenotes develop as well as normal fertilized embryos.

Advanced haploid embryos were used to test whether the distinction between parental genomes persisted after the activation of the embryonic genome. Live young were obtained from eggs in which haploid parthenogenetic nuclei (2- to 16-cell stage) were returned to fertilized eggs containing only a male pronucleus. This was, however, not the case when the recipient egg contained a female pronucleus; the presence of the two maternal genomes gave rise to mid-gestation conceptuses that phenotypically resembled parthenogenetic fetuses with very poorly developed extraembryonic tissue (Surani et al., 1986a) (see Fig. 2). Conversely, when nuclei (2- to 8-cell stage) from haploid androgenetic embryos were transferred to fertilized eggs containing only a female pronucleus, live young were produced. Correspondingly, the typical mid-gestation androgenetic phenotype was found when the recipient egg retained only a male pronucleus (Surani et al., 1986a) (see Fig. 2).

These experiments demonstrated that functional differences in maternal and paternal genomes are retained during early development. Therefore, presumably the imprint conferred on chromosomes at some point during gametogenesis is faithfully replicated at least through the first four cell cycles. Furthermore, these results suggest that no reversal or removal of the imprint can be achieved simply by exposing the donor nucleus to egg cytoplasm. The limited degree of reprogramming that we observed when advanced fertilized nuclei were transferred back to enucleated eggs, and which resulted in the production of the hsp 68/70 kDa proteins, is apparently insufficient either to interfere with the chromosomal modification reflecting their parental origin or to restore an advanced nucleus to a fully functional totipotential state. We might suppose that it is only during passage through the germline that an imprint can be erased and reestablished anew and that totipotency can be restored.

6. Embryo Reconstruction

Two different experimental approaches have been used to determine how long the differential parental information persists through development and the precise developmental potential of cells containing only chromosomes of one parental origin. Both techniques of blastocyst reconstruction and the production of aggregation chimeras have yielded important information.

6.1. Blastocyst Reconstruction

Techniques for microsurgical and immunosurgical dissection of the blastocyst into the two definitive tissues of the inner cell mass (ICM) and trophectoderm (TE) (Gardner, 1968, 1978; Solter and Knowles, 1975) have permitted reconstruction of blastocysts with a desired genotypic makeup. The embryo and yolk sac are derived from the ICM, while the ectoplacenta, trophoblastic cells, and most of the chorion are derived from TE (Gardner, 1982, and references cited therein).

In the absence of paternal chromosomes, parthenogenotes are able to produce small mid-gestation fetuses that may simply fail to develop to term because of the chronic lack of extraembryonic tissue. Parthenogenetic ICM were therefore placed within a TE vesicle from a blastocyst derived from a fertilized egg to test for their developmental potential. Such reconstituted blastocysts did indeed give rise to substantially improved fetuses, reaching the 40-somite stage on day 12 (Barton *et al.*, 1985). This observation supported the notion that parthenogenetic fetal development is limited to an extent by poor parthenogenetic trophoblast development. The reciprocal blastocyst reconstruction with a normal ICM introduced into parthenogenetic TE vesicle resulted in development resembling that of an unoperated parthenogenetic blastocyst (Barton *et al.*, 1985). Therefore, parthenogenetic TE could not be induced to proliferate even in the presence of ICM cells from a normal blastocyst. Indeed, the converse was true, that is, the normal fertilized ICM developed rather poorly, resembling a parthenogenote.

The most advanced 40-somite parthenogenetic fetuses produced with the help of normal trophoblast were smaller than their normal counterparts and showed evidence of abnormalities and rather poor yolk sacs. It would be of interest to determine the developmental potential of reconstructed ICM containing parthenogenetic ectoderm and normal endoderm placed inside normal TE vesicles.

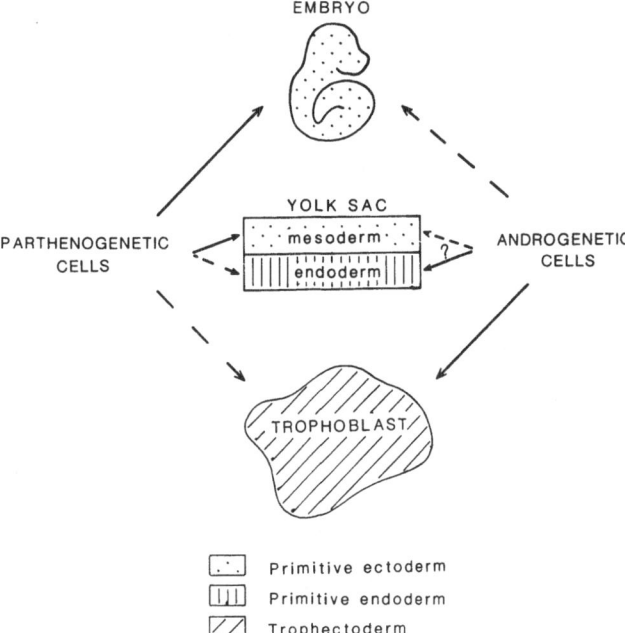

Figure 3. Developmental potential of parthenogenetic and androgenetic cells in chimeras. Substantial contribution of cells to the day-10 embryo, yolk sac, or trophoblast is indicated by a solid line (———). Very low contribution by a broken line (– – –). The contribution of cells to the yolk sac has been subdivided into the mesodermal and endodermal layers. (The relative contributions of parthenogenetic but not androgenetic cells to these two layers has been established.)

6.2. Aggregation Chimeras

Aggregation of two preimplantation embryos provides an opportunity for all cells to contribute to the tissues of the conceptus (Tarkowski, 1961; Mintz, 1964; McLaren, 1976). However, aggregation of asynchronous embryos will bias the contribution by the more advanced embryo to the ICM, hence to the embryo itself (Kelly *et al.*, 1978; Surani and Barton, 1984). Thus, aggregation chimeras between embryos of chosen parental genotypes can test the developmental potential of cells containing either maternal or paternal chromosomes. In all cases, the spatial distribution of cells in mid-gestation conceptuses was analyzed. In the first series of experiments, parthenogenetic and androgenetic 2- and 4-cell embryos were aggregated (Surani *et al.*, 1987). In a second and more detailed study, parthenogenetic or androgenetic embryos were aggregated with normal fertilized embryos (Surani *et al.*, 1988).

Aggregation of 4-cell parthenogenetic and androgenetic embryos produced rather poor fetuses that at best resembled parthenogenotes with trophoblast which was variable but often similar to that found in androgenotes. Analysis of the spatial distribution of cells revealed that the embryos were composed almost entirely of parthenogenetic cells and the trophoblast almost entirely of androgenetic cells; the yolk sac contained both cell types (Fig. 3). Even when an attempt was made to bias the potential contribution of androgenetic cells to the embryo (by aggregating 4-cell androgenetic with 2-cell parthenogenetic embryos) there was no alteration to the overall spatial segregation observed (Surani *et al.*, 1987).

Analysis of mid-gestation conceptuses derived from aggregations of normal fertilized embryos and parthenogenetic or androgenetic embryos demonstrated the same tendency for spatial segregation. Parthenogenetic cells were found mostly in the embryo and yolk sac, whereas androgenetic cells were confined to the trophoblast and yolk sac (Fig. 3). In the parthenogenetic ↔ fertilized chimeras, more detailed analysis of the yolk sac demonstrated that the parthenogenetic contribution was greater to the mesoderm layer than to the endoderm layer (Surani *et al.*, 1988) (Fig. 3). Furthermore, there appears to be a selection against parthenogenetic cells in chimeric embryos, but in rather a more protracted fashion (Nagy *et al.*, 1987). It is curious that the androgenetic cells survive in the yolk sac; we suspect that they are restricted to the yolk sac endoderm layer, which derives from the outer layer of primitive endoderm cells.

These analyses show a striking and reciprocal arrangement of cells in the mid-gestation conceptus according to their parental genotype. Thus, there appears to be an inherent tendency of cells containing only maternal chromosomes to contribute to the embryo itself and to the yolk sac mesoderm; that is, they are better able to survive in the primitive ectoderm lineages (Fig. 3). Conversely, cells containing only paternal chromosomes contribute to the trophoblast and yolk sac possibly only to the endoderm layer, although this has yet to be determined (Fig. 3). Furthermore, despite the presence of chromosomes of both parental origins in aggregation chimeras between androgenetic and parthenogenetic embryos, there was no functional complementation to produce development to term. This finding suggests strongly that both parental sets of chromosomes must be present within the same cells at some point during development in order to obtain full development.

We do not know at what level or exactly at what time this spatial segregation of cells occurs in chimeras. In view of the inability of parthenogenetic cells to survive and proliferate in the trophoblast it is possible that in chimeras their failure to proliferate may be due to their inability to respond to diffusible substances such as growth factors form normal cells. Therefore, even if there are normal cells present to produce the requisite growth factors, if parthenogenetic cells cannot respond they will be selected against. Conversely, the extensive proliferation of androgenetic cells in the trophoblast could be explained by an overproduction of or overresponse to growth factors. Indeed, in human cytotrophoblast autocrine factors are implicated in the control of cell proliferation where high levels of expression of c-*myc* and c-*fos* have been seen, possibly in response to the PDGF-like activity (Waterfield *et al.*, 1983) of c-*sis* (Goustin *et al.*, 1985). These observations could explain the unrestricted growth of the human syncytiotrophoblast in hydatidiform moles where only paternal chromosomes are found (Bagshawe and Lawler, 1982; Szulman and Surti, 1984).

7. Chromosomal Imprinting

The picture that emerges so far suggests that chromosomes of maternal origin are relatively more important for development of the embryo while the paternal chromosomes seem unable to permit normal embryonic development even to mid-term; however, the paternal chromosomes are very effective at directing proliferation of the trophoblast. This functional complementarity is reflected in, and in some way explained by, the spatial arrangement and restriction of cells of the two parental genotypes (see Fig. 3). These observations imply that the parental chromosomes are recognized as being different and that their behavior reflects these differences. Thus, we have speculated that there are DNA modifications that tag chromosomes as being of maternal or of paternal origin and that these imprints are responsible for causing differential gene expression that result in the observed embryonic phenotypes.

The work of Cattanach and Searle involving the judicious use of Robertsonian translocations has shown that otherwise chromosomally balanced mice that are monoparental for certain chromosomal regions show anomalous phenotypes (reviewed Cattanach, 1986). These studies demonstrate that some chromosomal regions must be of maternal origin (e.g., chromosomes 7 and 17) and others of paternal origin (e.g., chromosomes 6 and 8) (Fig. 4). Differential activity of regions of chromosomes 2 and 11 are also implicated by the observed complementary phenotypes. Maternal duplication of a region of chromosome 2 results in hypokinetic and flat-sided young; the corresponding paternal duplication produces hyperkinetic, short, square-bodied offspring. Similarly, a region of chromosome 11 when maternally duplicated gives rise to small young

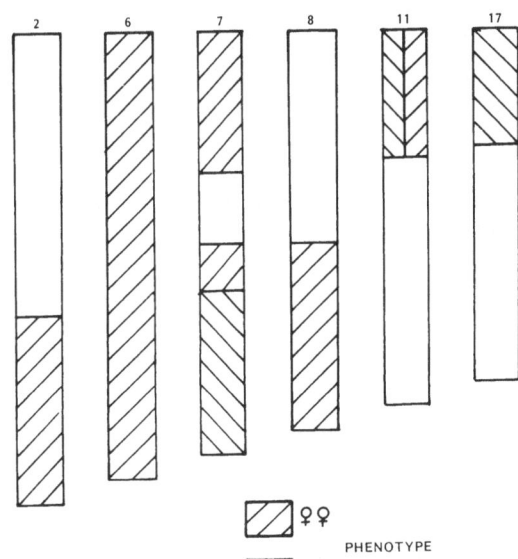

Figure 4. Noncomplementation of the mouse genome. The chromosomal domains that show parental origin effects are illustrated by cross lines and the parental origin of these regions that exhibit phenotypes is indicated.

PHENOTYPE

and, when paternally duplicated, to larger young. Thus, such genetic analyses have identified the chromosomal domains that exhibit parental origin effects. Regions of chromosomes 2, 6, 7, 8, 11, and 17 fit into this category (Searle and Beechey, 1985) suggesting that up to one eighth of the mouse genome is imprinted (Fig. 4).

8. Probing Imprinted Domains

The large chromosomal domains identified as showing parental origin effects define boundaries within which imprinted information presumably resides (see Fig. 4), but we have no idea how many genes are involved. As yet no genes have been identified that show differential timing of activation of parental alleles in the embryo (see McGrath and Solter, 1986). We have postulated that parental differences arise from germline-specific modifications of specific chromosomal regions (Surani et al., 1986b). If this is indeed the case, then such modifications should be faithfully replicated through embryonic development and probably into adulthood. Moreover, these modifications must be reversed during passage through gametes of the opposite sex. DNA methylation is a possible candidate modification, since it is heritable, reversible, and implicated in gene expression (Jahner and Jaenisch, 1984). As an approach toward an understanding of the modifications involved in imprinting, we have made use of random inserts of newly introduced genes (transgenes) as molecular probes. This approach assumes that a transgene that becomes integrated into an imprinted region will respect its position and behave as an imprinted gene. It is then possible to compare the behavior of the transgene in heterozygotes that inherit the transgene either maternally or paternally.

We began our analysis with a study of the methylation status of day 10 embryos carrying a transgene composed of a chloramphenicol acetyl transferase (CAT) gene linked to an immunoglobulin heavy-chain enhancer (IgH-E) (Reik et al., 1987). Conceptuses were dissected into extraembryonic membranes and embryos and the methylation status of the transgene in each was determined and compared following maternal or paternal inheritance. Seven such strains were analyzed. In six strains, the transgene was found to be highly methylated in the embryo and undermethylated in the trophoblast and yolk sac, irrespective of the parental origin. However, in embryos of one strain (CAT 17), a striking difference in male and female-derived alleles was observed. When maternally derived the transgene was highly methylated, however when it was derived from the father it was virtually unmethylated. Moreover, when the transgene was passed from grandfather to mother to son, it went from being unmethylated to methylated and back to unmethylated again (Fig. 5). Therefore, the methylation pattern was faithfully reversed as the transgene was shuttled between the male and female germlines and always reflected that of its immediate parental origin (Reik et al., 1987). We have since found one out of five strains of mice containing a κ-light-chain transgene (OX1-5) that shows the same lack of methylation when paternally derived and high methylation when

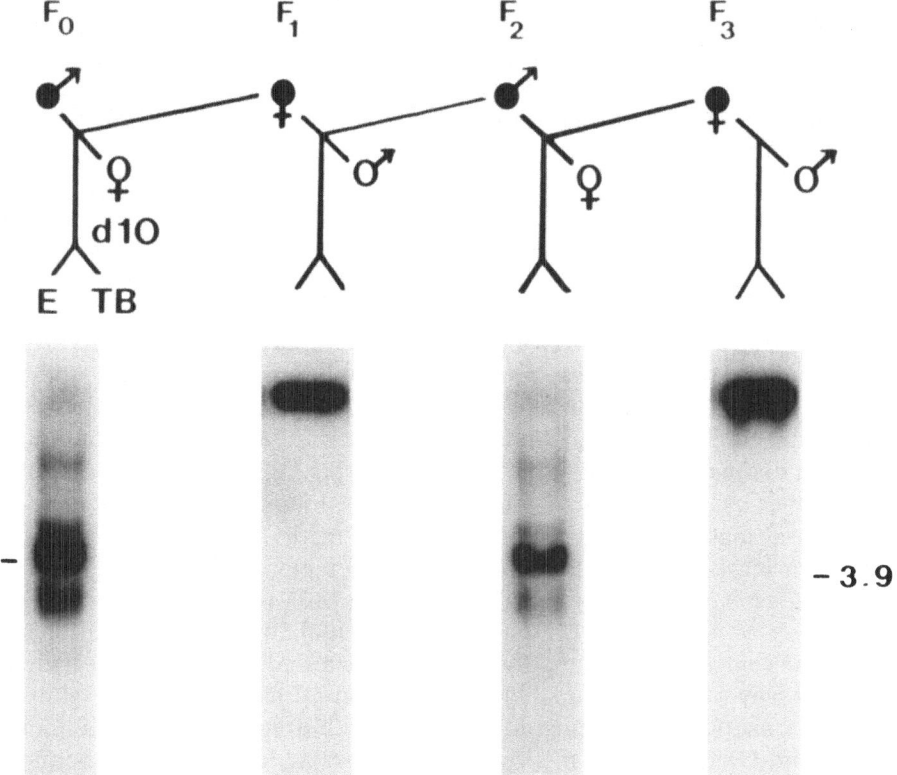

Figure 5. Alternate patterns of methylation of CAT 17 through male and female germlines. Beginning with the heterozygous male founder ($F_0 \male$), animals were bred with wild type mice ($\male$, $\female$) to derive heterozygous animals of the next generation. Some fetuses were taken on day 10 for methylation analysis of the embryo. *Hpa* II digestion revealed a CAT-specific fragment that was characteristically undermethylated when paternally derived (i.e., from F_0 and $F_2 \male$) and methylated when maternally derived (i.e., from F_1 and $F_3 \female$). E, embryonic tissue; TB, trophoblast tissue.

maternally derived (Reik, unpublished observations). Two other laboratories have reported similar observations with other transgenic strains containing a Troponin 1 transgene (Sapienza *et al.*, 1987) and an RSV-*myc* construct (Swain *et al.*, 1987).

It is not clear to what extent these parental differences in methylation affect or reflect gene expression. For the RSV-*myc* transgenic strain undermethylation of the paternal allele correlated with expression and methylation of the maternal allele with repression (Swain *et al.*, 1987). However, neither CAT 17 parental alleles (Reik *et al.*, 1987), but both OX1-5 alleles (Reik, unpublished observations) were expressed. In the case of a Hepatitis B surface antigen (HBs Ag) transgene, passage through the male germline resulted in undermethylation and expression, but on passage through the female germline the transgene became irreversibly methylated and repressed (Hadchouel *et al.*,

1987). Although it is quite possible that methylation is not, in fact, the primary imprinting signal, these observations of differential methylation clearly demonstrate a means by which parental alleles may be distinguished (Figs. 5 and 6). Therefore, the methylation status of imprinted genes may only reflect a potential for activity, with additional *trans*-acting factors being ultimately responsible for the regulation of expression.

It is not clear when such a pattern is imposed on the DNA. It may occur during gametogenesis when parental alleles are spatially segregated. Our preliminary investigations suggest that even as early as in spermatogonia a maternally derived CAT 17 is undermethylated, and so it remains throughout the rest of spermatogenesis and embryogenesis. We do not know whether CAT 17 is already methylated in oocytes, but if so it probably remains so during embryogenesis. Alternatively, if the transgene is undermethylated in the oocyte then it would imply that *de novo* methylation occurs in the embryo and that methylases can distinguish between parental alleles (Fig. 6). This mechanism could be explained by the presence of DNA-binding proteins that enhance the accessibility of maternal chromosomal regions to methylases or conversely that interfere or mask certain paternal chromosomal regions preventing the action of methylases. It is intriguing that overall sperm DNA is more methylated than oocyte DNA (Monk et al., 1987; Sanford et al., 1987); thus, the observation that several transgenes are under methylated when paternally derived appears to be the wrong way round. Presumably methylases are present and active globally during spermatogenesis but certain regions, including the transgenic loci of CAT 17 and OX1-5, escape methylation. Conversely, during oogenesis methylases are not globally active but the loci protected in sperm are accessible to methylases in the oocyte or at some point during embryogenesis (Fig. 6).

The frequency of transgenic strains that show differential methylation, which in our hands is 1 in 5 and 1 in 7 with respect to the two different constructs, correlates quite well with expectations from genetic data of Cattanach and Searle (reviewed by Cattanach, 1986) suggesting that up to one eighth of the genome may be imprinted (see Fig. 4). This may mean that the loci of CAT 17 and OX1-5 lie within the genetically-defined imprinted domains. Indeed, from linkage studies OX1-5 has been localized to chromosome 7, although it is not clear whether it falls within an imprinted domain; the location of CAT 17 remains to be determined. For certain other constructs, e.g., the troponin transgene (Sapienza et al., 1987), germline-dependent methylation was observed in four out of five strains, which is too high a frequency for integration into imprinted domains. It is possible that some transgenes contain sequences that are more susceptible to methylases; indeed, methylation at the site of integration may be the cellular response to mutation. If introduced genes insert preferentially into open chromatin, which would be expected to be undermethylated, and because microinjection is usually into the male pronucleus, this could explain why all reported imprinted transgenes have so far shown undermethylation following paternal transmission. However, we have preliminary evidence which suggests that before germline passage integrated genes do not take on a paternal-type transmission pattern of undermethylation following injection into

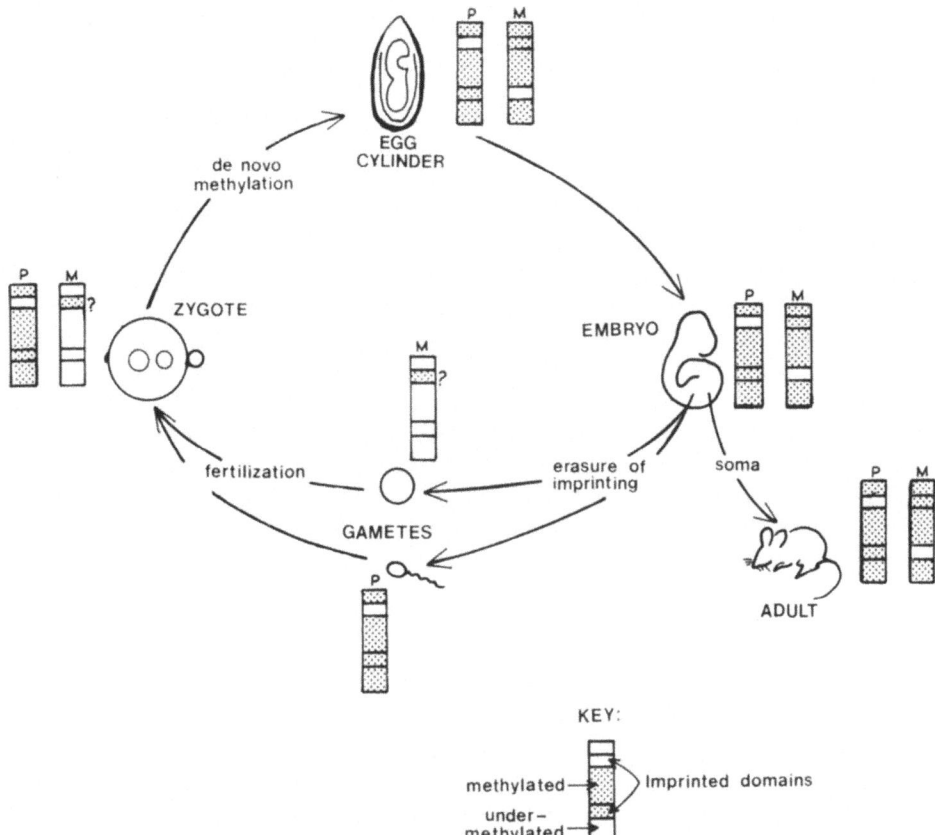

Figure 6. Proposed scheme for genomic imprinting. Diagram shows various stages of gameto-genesis and embryogenesis and indicates the methylation status of homologous chromosomes (paternal, P; maternal, M) at each stage. A hypothetical chromosome is illustrated that contains two imprinted domains and methylation is indicated by the presence of shading. It is not known whether the primary imprinting signal involves methylation, but differential parental methylation has been observed for several transgenic loci (as discussed in the text) and is diagrammatically represented here as a means of distinguishing between parental domains. The upper imprinted domain is shown as remaining undermethylated during passage through the male germline with the paternal allele escaping *de novo* methylation in the embryo, whereas the maternal allele becomes selectively methylated during oogenesis or early embryogenesis. This behavior most accurately represents that of the imprinted transgenes described in the text (although the point at which the maternal allele becomes methylated remains to be determined). The lower domain is illustrated as behaving in a complementary manner. The pattern established in the embryo by the egg cylinder stage probably remains through to adulthood in somatic tissues. The methylation status of extraembryonic tissue, where no differences in methylation levels have been detected, is not included.

the male pronucleus. Instead, preliminary results suggest that embryos derived directly from injected eggs show a random 50:50 methylation pattern (Reik, unpublished observations). This supports the notion that imprinting can only be imposed during passage through the germline (Fig. 6).

In each case studied of a locus showing parental origin effects, the paternal allele is undermethylated. If this undermethylation indeed reflects the status of the paternal domains, this may suggest that paternally derived genes residing within imprinted domains are relatively more active than the corresponding maternal alleles. Under normal conditions higher expression of a paternal allele may balance the lower maternal expression. This possible explanation is consistent with the effect on growth, as demonstrated by the duplication of chromosome 11 (Cattanach, 1986). If both regions are paternally derived, an overdose of certain gene products may result, which will probably be as detrimental as an underdose when both regions are of maternal origin. In this regard, it is perhaps pertinent that we have identified a transgenic strain (OX1-2) which causes embryonic lethality when paternally derived but not when of maternal origin. A simple explanation may be that an excess of certain paternal gene products is the cause of the embryonic lethality (Reik, unpublished observations). It is entirely possible that, although the transgenic loci identified so far suggest that paternal alleles may be more active, other imprinted domains may exist where maternal expression is higher. This view may be supported by the studies on the proximal region of chromosome 17 involving the hairpin tail deletion, which is lethal when the mutation is maternally derived (Johnson, 1975; McGrath and Solter, 1984c). If indeed certain chromosomal domains are subject to different levels of methylation depending on their experiences during gametogenesis and that this is reflected in expression levels, there may be a link between imprinting and dosage compensation. This is a particularly attractive idea in view of the recent findings that the testis-determining factor (TDF) gene on the Y chromosome has a homologue on the X. If both alleles are capable of producing TDF, then females, in which one X is inactivated, will only have one dose of TDF; hence, maleness may result from two doses of TDF (Page *et al.*, 1987). Therefore, sex determination may be a function of dosage compensation arising from random X inactivation (Chandra, 1985).

9. Conclusions

Maternal and paternal chromosomes of mice, while carrying essentially the same information, have evolved to express some of this information so as to play differential and complementary roles during embryogenesis (and perhaps through into adults). Essential information may be imprinted during gametogenesis and not erased and re-established until it is next passed through gametogenesis. This could explain the rapid restriction in nuclear totipotency in the mouse. It has yet to be established exactly when during gametogenesis imprinting occurs or indeed whether there really is a link between imprinting and nuclear totipotency.

Despite the observations that certain transgenes show a parental origin effect, we feel that it is important to identify imprinted endogenous genes. It is possible that transgenes do not faithfully mirror the behavior of endogenous genes and that the methylation patterns that we see in fact reflect the cellular response to mutation. Nevertheless, the patterns and reversibility of methylation of certain transgenic loci provide us with a workable model to explore genomic imprinting. The identification of particular genes that are imprinted may help toward an explanation of the failure of parthenogenotes and androgenotes. Our hope is that careful studies of genes that map within the known imprinted chromosomal domains will reveal the answers to this parental paradox.

References

Bagshawe, K. D., and Lawler, D. S., 1982, Unmasking moles. *Br. J. Obstet. Symp.* **89:**255–259.

Barton, S. C., Surani, M. A. H., and Norris, M., 1984, Role of paternal and maternal genomes in mouse development, *Nature (Lond.)* **311:**374–376.

Barton, S. C., Adams, C. A., Norris, M. L., and Surani, M. A. H., 1985, Development of gynogenetic and parthenogenetic inner cell mass and trophectoderm tissues in reconstituted blastocysts in the mouse, *J. Embryol. Exp. Morphol.* **90:**267–285.

Bensaude, O., Babinet, C., Morange, M., and Jacob, F., 1983, Heat shock proteins, first major products of zygotic gene activity in mouse embryos, *Nature (Lond.)* **305:**331–332.

Cattanach, B. M., 1986, Parental origin effects in mice. *J. Embryol. Exp. Morphol.* **97**(suppl.):137–150.

Chandra, H. S., 1985, Is human X chromosome inactivation a sex-determining device? *Proc. Natl. Acad. Sci. USA* **82:**6947–6949.

Cuthbertson, K. S. R., 1983, Parthenogenetic activation of mouse oocytes *in vitro* with ethanol and benzyl alcohol, *J. Exp. Zool.* **226:**311–314.

Flach, G., Johnson, M. H., Braude, P. R., Taylor, R. S., and Bolton, V. N., 1982, The transition from maternal to embryonic control in the 2-cell mouse embryo, *EMBO J.* **1:**681–686.

Gardner, R. L., 1968, Mouse chimeras obtained by the injection of cells into the blastocyst, *Nature (Lond.)* **220:**596–597.

Gardner, R. L., 1978, Production of chimeras by injecting cells or tissue into the blastocyst, in: *Methods in Mammalian Reproduction* (J. C. Daniel, Jr., ed.), pp. 137–165, Academic, New York.

Gardner, R. L., 1982, Investigation of cell lineage and differentiation in the extraembryonic endoderm of the mouse embryo, *J. Embryol. Exp. Morphol.* **68:**175–198.

Graham, C. F., 1974, The production of parthenogenetic mammalian embryos and their use in biological research, *Biol. Rev.* **49:**399–422.

Goustin, A. S., Betsholtz, C., Pfeifer-Ohlson, S., Persson, H., Rydnert, J., Bywater, M., Holmgren, G., Heldin, C.-H., Westermark, B., and Ohlsson, R., 1985, Coexpression of the *sis* and *myc* proto-oncogenes in developing human placenta suggests autocrine control of trophoblast growth, *Cell* **41:**301–312.

Gurdon, J. B., 1986, Nuclear transplantation in eggs and oocytes, *J. Cell Sci.* **4**(suppl.):287–318.

Gurdon, J. B., and Laskey, R. A., 1970, The transplantation of nuclei from single cultured cells into enucleate frog's eggs, *J. Embryol. Exp. Morphol.* **24:**227–248.

Hadchouel, M., Farza, H., Simon, D., Tollias, P., and Pourcel, C., 1987, Maternal inhibition of hepatitis B surface antigen gene expression in transgenic mice correlates with de novo methylation, *Nature (Lond.)* **329:**454–456.

Harada, K., and Buss, E. G., 1981, Cytogenetic studies of embryos developing parthenogenetically in turkeys, *Poultry Sci.* **60:**1362–1364.

Howlett, S. K., Barton, S. C., and Surani, M. A. H., 1987, Nuclear cytoplasmic interactions following nuclear transplantation in mouse embryos, *Development* **101**:915–923.

Hubbs, C. L., and Hubbs, L. C., 1932, Apparent parthenogenesis in nature, in a form of fish of hybrid origin, *Science* **76**:628–630.

Jahner, D., and Jaenish, R., 1984, DNA methylation in early mammalian development, in: *DNA Methylation, Biochemistry and Biological Significance* (R. Razin, H. Cedar, and A. D. Riggs, eds.), pp. 189–219, Springer-Verlag, New York.

Johnson, D. R., 1975, Further observations on the hairpin tail (T^{hp}) mutation in the mouse, *Genet. Res.* **24**:207–213.

Kaufman, M. H., Barton, S. C., and Surani, M. A. H., 1977, Normal post-implantation development of mouse parthenogenetic embryos to the forelimb bud stage, *Nature (Lond.)* **265**:53–55.

Kelly, S. J., 1979, Studies of the developmental potential of 4- and 8-cell stage mouse blastomeres, *J. Exp. Zool.* **200**:365–370.

Kelly, S. J., Mulnard, J. G., and Graham, C. F., 1978, Cell division and cell allocation in early mouse development, *J. Embryol. Exp. Morphol.* **48**:37–51.

King, T. J., and Briggs, R., 1956, Serial transplantation of embryonic nuclei, *Cold Spring Harbor Symp. Quant. Biol.* **21**:271–290.

Laskey, R. A. and Gurdon, J. B., 1970, Genetic content of adult somatic cells tested by nuclear transplantation from cultured cells, *Nature* **228**:1332–1333.

Mann, J. R., and Lovell-Badge, R. H., 1984, Inviability of parthenogenones is determined by pronuclei, not egg cytoplasm, *Nature (Lond.)* **310**:66–67.

Markert, C. L., 1982, Parthenogenesis, homozygosity and cloning in mammals, *J. Hered.* **73**:390–397.

Maslin, T., 1967, Skin grafting in the bisexual Teiid lizard *C. nemidophorous sexlineatus* and in the unisexual *C. tesselatus*, *J. Exp. Zool.* **166**:137–150.

McGrath, J., and Solter, D., 1983, Nuclear transplantation in the mouse embryo by microsurgery and cell fusion, *Science* **220**:1300–1303.

McGrath, J., and Solter, D., 1984a, Completion of mouse embryogenesis requires both the maternal and paternal genomes, *Cell* **37**:179–183.

McGrath, J., and Solter, D., 1984b, Inability of mouse blastomere nuclei transferred to enucleated zygotes to support development *in vitro*, *Science* **226**:1317–1319.

McGrath, J., and Solter, D., 1984c, Maternal T^{hp} lethality in the mouse is a nuclear not cytoplasmic defect, *Nature(Lond.)* **308**:550–551.

McGrath, J., and Solter, D., 1986, Nucleocytoplasmic interactions in the mouse embryo, *J. Embryol. Exp. Morphol.* **97**(suppl.):277–289.

McLaren, A., 1976, *Mammalian Chimaeras*, Cambridge University Press, Cambridge.

Mintz, B., 1964, Formation of genetically mosaic mouse embryos and early development of "lethal (t^{12}/t^{12})-normal" mosaics, *J. Exp. Zool.* **157**:273–292.

Modlinski, J. A., 1978, Transfer of embryonic nuclei to fertilized mouse eggs and development of tetraploid blastocysts, *Nature (Lond.)* **273**:466–467.

Modlinski, J. A., 1981, The fate of inner cell mass and trophectoderm nuclei transplanted to fertilized mouse eggs, *Nature (Lond.)* **292**:342–343.

Monk, M., Boubelik, M., and Lehnert, S., 1987, Temporal and regional changes in DNA methylation in the embryonic, extraembryonic and germ cell lineages during mouse development, *Development* **99**:371–382.

Morgan, T., 1927, *Experimental Embryology*, Columbia University Press, New York.

Morris, J., 1968, The XO and OY chromosome constitution in the mouse, *Genet. Res.* **12**:125–137.

Nagy, A., Paldi, A., Dezso, L., Varga, L., and Magyar, A., 1987, Prenatal fate of parthenogenetic cells in mouse aggregation chimaeras, *Development* **101**:67–71.

Olsen, M. W., 1966, Segregation and replication of chromosomes in turkey parthenogenesis, *Nature (Lond.)* **212**:435–436.

Page, D. C., Mosher, R., Simpson, E. M., Fischer, E. M. C., Mardon, G., Pollack, J., McGillivray, B., de la Chapelle, A., and Brown, L. A., 1987, The sex-determining region of the human Y chromosome encodes a finger protein, *Cell* **51**:1091–1104.

Palmiter, R. D., and Brinster, R. L., 1986, Germ line transformation of mice, *Annu. Rev. Genet.* **20**:465–499.

Reik, W., Collick, A., Norris, M. L., Barton, S. C., and Surani, M. A. H., 1987, Genomic imprinting determines DNA methylation of parental alleles in transgenic mice, *Nature (Lond.)* **328**:248–251.

Robl, J. M., Gilligan, B., Critser, E. S., and First, N. L., 1986, Nuclear transplantation in mouse embryos: assessment of recipient cell stage, *Biol. Reprod.* **34**:733–739.

Rossant, J., 1976, Post-implantation development of blastomeres from 4- and 8-cell mouse eggs, *J. Embryol. Exp. Morphol.* **36**:283–290.

Sanford, J. P., Clark, H. J., Chapman, V. M., and Rossant, J., 1987, Differences in DNA methylation during oogenesis and spermatogenesis and their persistence during early embryogenesis in the mouse, *Genes Dev.* **1**:1039–1046.

Sapienza, C., Peterson, A. C., Rossant, J., and Balling, R., 1987, Degree of methylation of transgenes is dependent on gamete of origin, *Nature (Lond.)* **328**:251–254.

Searle, A. G., and Beechey, C. V., 1985, Noncomplementation phenomena and their bearing on nondisjunctional events, in: *Aneuploidy* (V. L. Dellarco, P. E. Voytek, and A. Hollaender, eds.), pp. 363–376, Plenum, New York.

Solter, D., and Knowles, B. B., 1975, Immunosurgery of mouse blastocysts, *Proc. Natl. Acad. Sci. USA* **72**:5099–5102.

Stevens, L. C., 1975, Teratocarcinogenesis and spontaneous parthenogenesis in mice, in: *Developmental Biology of Reproduction* (C. L. Markert and J. Papaconstantinou, eds.), pp. 13–106, Academic, New York.

Surani, M. A. H., and Barton, S. C., 1983, Development of gynogenetic eggs in the mouse: implications for parthenogenetic embryos, *Science* **222**:1034–1036.

Surani, M. A. H., and Barton, S. C., 1984, Spatial distribution of blastomeres is dependent on cell division order and cell interactions in mouse morulae, *Dev. Biol.* **102**:335–343.

Surani, M. A. H., Barton, S. C., and Norris, M. L., 1984, Development of reconstituted mouse eggs suggests imprinting of the genome during gametogenesis, *Nature (Lond.)* **308**:548–550.

Surani, M. A. H., Barton, S. C., and Norris, M. L., 1986a, Nuclear transplantation in the mouse: heritable differences between parental genomes after activation of the embryonic genome, *Cell* **45**:127–136.

Surani, M. A. H., Reik, W., Norris, M. L., and Barton, S. C., 1986b, Influence of germline modifications of homologous chromosomes on mouse development, *J. Embryol. Exp. Morphol.* **97** (suppl.):123–136.

Surani, M. A. H., Barton, S. C., and Norris, M. L., 1987, Influence of parental chromosomes on spatial specificity in androgenetic↔parthenogenetic chimeras in the mouse, *Nature (Lond.)* **326**:395–397.

Surani, M. A. H., Barton, S. C., Howlett, S. K., and Norris, M. L., 1988, Influence of chromosomal determinants on development of androgenetic and parthenogenetic cells, *Development* **103**:171–178.

Swain, J. L., Stewart, T. A., and Leder, P., 1987, Parental legacy determines methylation and expression of an autosomal transgene: a molecular mechanism for parental imprinting, *Cell* **50**:719–727.

Szulman, A. E., and Surti, U., 1984, Complete and partial moles: cytogenetic and morphological aspects, in: *Human Trophoblast Neoplasma* (R. A. Pattillo and P. O. Hussa, eds.), pp. 135–145, Plenum, New York.

Tarkowski, A. K., 1961, Mouse chimaeras developed from fused eggs, *Nature (Lond.)* **190**:857–860.

Tarkowski, A. K., and Wroblewska, J., 1967, Development of blastomeres of mouse eggs isolated at the 4- and 8-cell stage, *J. Embryol. Exp. Morphol.* **18**:155–180.

Tsunoda, Y., Yasui, T., Shioda, Y., Nakamura, K., Uchida, T., and Sugie, T., 1987, Full-term development of mouse blastomere nuclei transplanted into enucleated two-cell embryos, *J. Exp. Zool.* **242**:147–151.

Waterfield, M. D., Scrace, G. T., Whittle, N., Stroobant, P., Johnsson, A., Wasteson, A., Westermark, B., Heldin, C.-H., Huang, J. S., and Devel, T. F., 1983, Platelet-derived growth factor is structurally related to the putative transforming p28[sis] of simian sarcoma virus, *Nature (Lond.)* **304**:35–39.

Chapter 5
Phenotypic Changes in Cell Culture

MORGAN HARRIS

1. Introduction

The history and evolution of cell populations *in vitro* stem initially from characteristics of the tissues or organs used for explantation. As developmental products, each of these carries a spectrum of identifying markers: distinctive patterns of enzyme activities or unique metabolic products, diagnostic cytoarchitecture or elements of fine structure, and other specificities recognizable microscopically or biochemically. Phenotypic changes in early cultures of animal cells thus commonly derive from forward or retrograde shifts in state of differentiation, rather than from alterations at the genetic level.

Progressive differentiation may in some instances continue to occur *in vitro*, particularly when intact embryonic rudiments are isolated from the body as a whole. In a classic study, Fell and Robison (1929) showed that the precursor of the chick embryonic femur is capable of progressive development if explanted to the surface of a plasma clot. Histotypically normal bone arose from cells of the initial primordium; while in external configuration, the head, trochanter, and condyles of the normal femur appeared in clearly recognizable form. Conditions in culture rarely permit progressive morphogenesis to this degree, but it is noteworthy that even when cells from embryonic rudiments are physically dissociated by tryptic digestion the isolated cells can still reassociate and reconstitute normal tissue fabrics (see Moscona, 1962). The potential for re-establishment of tissue fabrics is type specific, as was shown by constructing aggregates form cell suspensions of two different tissue types, e.g., chick mesonephros and cartilage cells from wing bud mesenchyme (Trinkhaus and Groves, 1955; Moscona, 1956). Within such aggregates, type-specific reassortment occurs, with epithelial tubules formed from kidney cells and cartilage masses from chondrogenic cells.

Phenotypic shifts are more often regressive when populations of fully differentiated cells are placed in culture. Disappearance or reduction in expression of cell-type specific markers is in fact commonly observed for most cells in conventional monolayer cultures (Harris, 1964). Distinctions between

MORGAN HARRIS • Department of Zoology, University of California–Berkeley, Berkeley, California 94720.

related cell types become difficult to make after one or more serial passages, and the cells in culture assume more generalized patterns. Thus, patches of a simple pavement epithelium appear in cultures of kidney or other epithelial organs, in lieu of the distinctly different subtypes apparent *in vivo*. Similarly, stellate or spindlelike fibroblasts arise from diverse sources, such as cartilage, muscle, linings of blood vessels, and nerve sheaths, as well as from connective tissue per se. To early investigators (Champy, 1912), this convergence to simpler growth forms *in vitro* suggested reversion to a more generalized embryonic state. But true dedifferentiation in the sense of regaining broad developmental potentials has never been demonstrated in cultures of animal cells, although limited conversion between cell types (transdifferentiation) does occur (see Chapter 7, *this volume*). In other cases, there is a firm commitment to differentiation, even though the phenotypic expression of cell type-specific features may decline. Despite the apparent loss of differentiated properties as a result of these regressive shifts, it is often potentially reversible, and re-expression of specialized features may occur under altered culture conditions. Reversible transitions of this sort in cellular phenotype, without change in basic commitment, come under the heading of modulation as defined by Weiss (1939, 1950) and may occur in cultures of both normal and tumor cells (see Chapter 9, *this volume*).

Phenotypic change in animal cells takes on a new dimension when populations are maintained for prolonged periods in serial culture. Early investigators hoped to isolate pure strains of each cell type, which could then be subcultivated indefinitely *in vitro* without change, like cultures of microorganisms. Initially, such efforts did meet with some success, e.g., pure cultures of chick iris epithelium (Fischer, 1922), chick monocytes (Carrel and Ebeling, 1922), and thyroid epithelium (Ebeling, 1925). These primary isolates from fresh tissue could not, however, be maintained indefinitely in their original form. Eventually it became clear that primary populations of cells taken from the intact organism will eventually die out in serial culture (i.e., show limited viability), or else give rise to variants that deviate from the normal precursor cell types (transformation).

These two patterns can be exemplified by the fate of cells in cultures isolated from human fetal tissues and mouse embryos respectively. The classic studies of Hayflick and Moorhead (1961) showed, for example, that fibroblasts from human tissues proliferate for only about 50 serial passages when divided at each transfer on a 1 : 1 basis. Limited viability in this sense becomes apparent as population growth in such cultures slows to a standstill, debris accumulates, and further cell propagation cannot be obtained. By contrast, cells derived from mouse tissues do not show limited viability but instead undergo transformation at the population level to form established or permanent cell lines. In this process, freshly explanted mouse cells proliferate luxuriantly for a few weeks and then go into a sharp growth decline. This static state, however, is temporary and can be reversed if the population is maintained with periodic fluid changes. Proliferation begins again and, after a few more weeks, the population of mouse cells is multiplying as vigorously as before. The cells can then be subcultivated indefinitely by serial transfer as an established or permanent line. The emergence of such lines can be viewed as a mass progression or transformation, in which the original diploid cells become aneuploid, marked changes in growth properties occur, and cells may also alter in morphology and other features, including the ability to produce tumors in many cases if injected into the hosts of origin (see Rothfels et al., 1963; Todaro and Green, 1963). Cells from a number of other species tend to follow the sequence described for mouse cells; cultures from chick embryos show limited viability, as do those from human tissues. Combinations of these patterns may also be seen, but in no case do cells taken directly from the intact organism proliferate indefinitely in unmodified form.

Evolutionary and morphogenetic changes may thus condition the properties of animal cells in vitro, but heritable changes in phenotype can also stem from discrete events within single cells. A classic problem here, faced earlier with microorganisms, is whether a given change in cellular phenotype results from adaptation within the population as a whole, or by spontaneous variation in one or a few cells, which then spreads by selection to characterize the entire system. In the case of bacteria, the fluctuation test, a statistical approach devised by Luria and Delbrück (1943), provided an experimental answer. Acquired resistance in Escherichia coli to an infecting virus (bacteriophage, or simply phage) was used as a model system. From a mass culture of E. coli containing a low frequency of variant cells resistant to phage, a series of sublines were initiated. The inocula used were too small to contain any resistant cells initially. When each of the subline cultures had grown to a large size, samples from all were tested for the frequency of variants resistant to bacteriophage, along with a similar number of replicate samples obtained from the original undivided mass population. Statistically, variance among the replicate samples from the undivided culture was small and essentially identical to the mean frequency of resistant variants in these cultures, as would be expected for a Poisson distribution. However, variance among samples taken from different sublines was very large, suggesting that variation had occurred independently, at random, and at different times in the separate subline cultures. The simplest explanation for this disparity was that genetic mutation was the random event; and that such mutants were then selected for by subsequent exposure to phage. This inference for bacteria was later borne out by a varied array of direct genetic tests which show that most, if not all, bacterial variants do indeed arise from mutational changes in the DNA coding system. Fluctuation tests have also been adapted for use with cultures of animal cells, and have aided in the recognition of genetic mutants in these populations as well.

It is important to recognize, however, that random and spontaneous changes in cell populations do not necessarily mean as such that genetic mutation has occurred. A positive result in fluctuation tests does not indicate the nature of the event concerned; this information must be obtained by other means. Specifically, fluctuation tests do not reveal where in the cell spontaneous and random changes have occurred, or even if a given variation is genetic or nongenetic in nature. These caveats need to be emphasized because in recent years it has been found that in addition to genetic mutants, phenotypically stable variants, based on nongenetic changes, can arise in cultures of animal cells. These so-called epigenetic variants, like genetic mutants, give positive results with a fluctuation test, and their frequency of appearance can likewise be calculated by this means. As later discussion will make clear, the essential difference is that epigenetic variants result from a stable shift in gene expression rather than from a structural change (mutation) at the nucleotide level. The spontaneous change picked up in fluctuation tests with these variants appears to be a random but stable shift in DNA methylation pattern, which is then perpetuated in subsequent cell replications. Because DNA methylation is known to affect gene expression, a stable shift in methylation pattern may accordingly lead to an equally stable change in cellular phenotype, without altering the basic DNA coding system. Epigenetic as well as genetic variants thus exist in cultures of animal cells, and when present in low frequency may masquerade as gene mutations (Riggs and Jones, 1983).

2. Modulative Shifts in Phenotype

One of the earliest examples of modulation to be studied *in vitro* was the formation of cartilage by chondrocytes in serial culture. (Holtzer *et al.*, 1960; Stockdale, Abbott *et al.*, 1963). The advantages of this experimental system include the ready availability of chondrocytes in nearly pure form, when obtained from such sources as the vertebral rudiments of 10-day-old chick embryos. The freshly isolated cells can be dissociated and cultivated as monolayers, reaggregated masses, or in gels, and numerous cell type-specific markers are available for study at morphological and biochemical levels. In their initial experiments, Holtzer and colleagues examined the properties of chondrocytes when grown serially in monolayer form. Although at explantation the cells retained the typical morphology of chondroblasts or chondrocytes, formed matrix, and incorporated ^{35}S in this process, these cell type-specific features disappeared in subsequent subcultures. The population as a whole continued to grow vigorously but became fibroblast-like in appearance, with cells no longer surrounded by matrix as detected with appropriate staining procedures. Synthesis of chondroitin sulfate in these cultures, as determined by incorporation of ^{35}S, also progressively declined. Attempts by Holtzer and co-workers to reverse the transition by altering culture protocols were unavailing, and the conclusion reached was that progressive and irreversible dedifferentiation had occurred.

Later work by Coon (1966) showed that the loss of specialized function of chondrocytes in monolayer culture was more apparent than real. With improved media and altered cultural conditions the differentiated state of freshly isolated chondrocytes could be stabilized in serial culture, without progressive decline in cartilage-making function. Similarly, if dedifferentiated fibroblast-like populations were produced by subcultivation or chondrocytes in mass populations, it was possible to restore cartilage formation in these cells by suitable shifts to more optimal conditions. In this process, nonexpressing cells regained the polygonal morphology of chondrocytes, formed typical cartilage capsules, and showed increased uptake of ^{35}S as synthesis of chondroitin sulfate was regained. In addition to demonstrating the reality of modulation for chondrocytes *in vitro*, these findings demonstrate that specialized function of cells is not necessary for heritable retention of the differentiated state. Success in restoring cartilage-making function was due in part to the use of improved synthetic media, combined with fetal calf serum, in lieu of the undefined horse serum-chick embryo extract employed for earlier work. But a more important factor was the innovative cultivation by Coon of chondrocytes at low population densities instead of passaging the cells with high cell numbers in serial culture. Coon noted that cartilage-making function declined progressively if mass populations were used, even in optimal media, although the cells continued to grow vigorously. But if similar cultures were initiated at low cell densities, isolated colonies developed from single cells. In the improved media used for this work the cells of these colonies took on the typical morphology of chondrocytes, which they retained even if passed serially for four or more transfers in the same way. While such observations do not define the underlying mechanisms of modulation or stabilization of the differentiated state, they do indicate that regulation is based on both extrinsic and intrinsic factors.

In more recent experiments, the emphasis has shifted to molecular patterns of synthesis, as cartilage cells undergo modulative shifts between specialized and unspecialized states. The studies of Benya and Shaffer (1982) with rabbit auricular chondrocytes are representative of the data obtained. For this work, dedifferentiated chondrocytes were harvested after five passages at high density in monolayer culture and were placed in nutrient medium containing 0.5% agarose, a procedure that immobilizes individual cells and allows expression of cartilage-making function even when relatively high inocula are used. As others have found, Benya and Shaffer observed that differentiated chondrocytes are distinguished at the molecular level by the synthesis of type II collagen and cartilage-specific proteoglycans. This biochemical phenotype disappears in serial monolayer culture and is replaced by a complex collagen phenotype consisting primarily of type I collagen and a low level of proteoglycans synthesis. When dedifferentiated chondrocytes are transferred to agarose, type II collagen reappears and the level of proteoglycans synthesis rises to a normal level. These changes coincide with a shift in morphology of the cells from a flattened form to the spherical morphology of anchorage-independent cells in suspension culture. Such observations suggest that cell shape may play a primary role in the initiation or maintenance of specialized function in differentiated cells.

Stabilization of differentiation in other culture systems seems also to depend on an interplay between explanted populations and microenvironmental conditions. Mammary epithelial cells, with their distinctive structural and functional markers, offer a second illustration of these relationships. As with other types of freshly isolated epithelia, primary cultures from mammary gland on a solid substrate soon give rise to a population of flattened epithelial cells, in which specialized functional activities decline or are lost. The kinetics of this process were demonstrated with particular clarity in the experiments of Ebner et al. (1961) with bovine mammary cells. Following the relative activity of several functional components in these cultures over a 15-day period, they were able to show that the rates of decrease are nonparallel and distinctively different for individual markers. Thus, lactose production was initially high in freshly isolated cells but fell rapidly and could not be detected after 24 hr in culture. UDPGal-4-epimerase activity decreased more slowly, disappearing at 7–10 days after explanation. Synthesis of β-lactoglobulin declined even more gradually and stabilized at a plateau level after 15 days. These data are of particular interest because they indicate that declines in functional activity are not controlled by a single cell-wide mechanism at a quantitative level, although the potential for modulative return to a more active state was not directly demonstrated here.

Cultures prepared directly from mouse mammary gland have been used extensively in recent years for studies on morphogenesis and modulation *in vitro* (reviewed by Levay-Young et al., 1987). Freshly isolated cells in monolayers were used initially to determine the requirements of epithelial cells for differentiation. In such experiments, Emerman and Pitelka (1977) showed that cells if grown directly on a plastic surface give rise to a thin sheet with the surface polarity of a transporting epithelium and a poorly developed internal secretory apparatus. On a surface coated with collagen there was a modest morphological increase for secretion of rough endoplasmic reticulum and Golgi complexes. Striking differentiation of mammary cells occurred, however, if the collagen membrane was freed physically from the substrate, so that it contracted sharply and floated freely in the medium with the explanted cells on the upper surface. As Michalopoulos and Pitot (1975) had found previously with liver cells, this leads to a change in shape of the cultured cells and onset of a more differentiated state. Mammary cells on floating collagen gels become cuboidal to columnar in shape, with long microvilli on the apical surface, and formation of a basal lamina separating epithelial cells from the gel below. Internally, the ultrastructure of these cells shows rough endoplasmic reticulum filling most of the cytoplasm and stacks of cisternae typical of actively secreting cells.

Definition of conditions for stabilizing the differentiated state in cultures of mammary epithelium enabled Emerman and Pitelka (1977) to demonstrate that the flattened nonsecretory epithelia grown directly on plastic had not been irreversibly altered. When these dedifferentiated cells were subcultured to floating collagen gels, the cells regained the normal appearance of secretory cells in 2 days, and were then indistinguishable from populations explanted directly by this procedure. Using floating collagen gels, Emerman et al. (1977) were further able to define hormonal requirements for maintenance of the differentiated state. In media containing insulin alone, the cells had few secretory organelles and only a small amount of casein as detected by radioimmunoassay. When the nutrient was supplanted with insulin, cortisol, and prolactin, prominent development of the internal secretory apparatus occurred, with large increases in the amount of casein intracellularly and in the culture medium. These experiments show that mouse mammary cells on floating collagen membranes differentiate in response to hormones as do the same cells *in vivo*.

Modulation in the production of secreted proteins by mouse mammary cells was examined in more detail in experiments by Lee *et al.* (1984). These studies showed that cells on floating collagen gels secrete γ-casein and also secrete α_1-, α_2-, and β-caseins into the culture medium. None of these were detectable in media from cells grown on plastic, and only to a slight extent for populations maintained on attached collagen gels. Contraction of gels on detachment does seem significant, since cells on glutaraldehyde-treated gels, which do not contract on removal from plastic, did not secrete any of the caseins. Two other proteins in milk, probably transferrin and butyrophilin, were only partially modulated in secretion by changes in the surface substrate, while acidic protein was synthesized abundantly by freshly isolated mammary cells, but appeared in only very small amounts in cultured cells regardless of the substrate used. These observations again suggest that while maintenance of mammary epithelial cells on different substrates can profoundly alter secretory patterns, regulation is distinct for individual proteins and is not governed by any single mechanism.

Since casein secretion is a direct indicator of differentiation in mammary cells, the question arises whether sensitivity of this marker to the extracellular matrix in floating collagen gels is mediated by secretion, synthesis, or stability of the proteins produced. In their original experiments, Lee *et al.* (1984) measured the amounts of intracellular and media caseins for mouse mammary cells maintained on different substrates. For cells on floating gels, considerable quantities of casein accumulated in the media, and only a small amount remained in the cells. Although cells on plastic did not secrete caseins into the medium, only the same low level was found within the cells. These data show that casein found in the culture medium of floating gels had not simply accumulated without release within cells on plastic. In a later study, Lee *et al.* (1985) showed that cells on plastic rapidly lose the ability to secrete caseins. mRNA is produced and synthesis of milk proteins continues at a low level in these cells, but the newly synthesized caseins are degraded intracellularly. In comparison, large increases in mRNA for casein are observed for cells on floating gels, and the caseins synthesized are not degraded but secreted instead into the extracellular medium. These investigations illuminate factors at the molecular level that underlie differentiation of mammary cells as well as modulative shifts to alternative functional states.

3. Stable Changes in Gene Expression

Differentiation in embryos and stem cell systems implies a predetermined pattern of gene action for each of the tissues or organs concerned. While modulation may occur within these populations *in vitro*, the final morphogenetic state is unchanged. This leads to an interesting question: Can developmental changes continue within individual cells once the overall process of differentiation is complete? Answers to this question in experimental terms are not obvious. One might expect that if morphogenetic change continued, any further differentiation would take on a new dimension. In this context, gene expression might be visualized as random, unprogrammed events in single cells, or in response to an external inducer. However, these same phenomena are often described as characteristics of a mutational process. This ambiguity means that it is not possible at the present time to say definitively whether such complex processes as the emergence of tumors *in vivo*, or appearance of variants in cell cultures depend on genetic changes, epigenetic shifts, or more probably to a combination of the two. In the present climate of opinion, variation in animal cells, as in microorganisms, is thought to follow predominantly a genetic model. Much evidence for this view can be found in numerous reviews and monographs currently available as sources of information (Littlefield, 1976; Morrow, 1983; Gottesman, 1985). However, less attention has been given to the possibility that epigenetic shifts may participate as nongenetic mechanisms for variation in animal cells, and the discussion that follows will emphasize this topic with data that have become available in recent years.

Spontaneous phenotypic changes in cultures of neoplastic and normal cells on occasion can serve as a basis for epigenetic studies, but alterations in gene expression induced by direct exposure to external inductors have led to more useful experimental systems. The early observations of Charlotte Friend that virus-transformed mouse erythroleukemic cells were blocked in differentiation (Friend *et al.*, 1971) provided the impetus for a broad spectrum of experiments on gene expression in these cells, which continue to the present time (see summaries by Reuben *et al.*, 1980; Marks *et al.*, 1985). Friend cells will grow indefinitely in culture as unspecialized paraproerythroblasts, and they can be induced at any time to differentiate to normoblasts, with the production of hemoglobin, by brief exposure to a disparate series of chemical inducers, e.g., sodium butyrate, dimethylsulfoxide (DMSO), hexamethylenebisacetamide (HMBA), hormones, and inhibitors of ion transport. Substances such as butyrate that are naturally occurring compounds without mutagenic potential are particularly interesting, because they indicate clearly that induction is based on a change in gene expression rather than genetic alteration in the reaction system. Moreover, butyrate-treated Friend cells respond by synthesizing many new species of proteins that are not detectable in control cells (Reeves and Cserjesi, 1979). These *de novo* changes in protein and RNA synthesis are not seen when Friend cells are induced with other agents such as DMSO and are rapidly reversible when butyrate is removed. The mode of action of butyrate as an inducer is unknown. The coordinate induction of histone hyperacetylation may be significant, but it is not clear how the latter could provide a mechanism or working model for the induction process. Similarly, much has been made of molecular configurations in polar–planar inducers such as DMSO or HMBA (Reuben *et al.*, 1980), but the role of these unique properties for the induction of gene expression remains to be determined.

Epigenetic changes can be examined in a broader format with another inducer of gene expression, the nucleoside analogue 5-azacytidine (5-aza-CR). Experimental analysis of mechanisms in this case is facilitated by the known effects of this agent on DNA methylation, which provide a molecular model for induced cell variation (see recent summaries by Riggs and Jones, 1983; Taylor et al., 1984; Chandler and Jones, 1988). The first indication that 5-aza-CR affects gene action came from a study on transformation of mouse C3H/10T1/2 cells by cancer chemotherapeutic agents (Constantinides et al., 1977). Fully functional striated muscle fibers appeared in large numbers after brief exposure of this fibroblast-like line to micromolar concentrations of 5-aza-CR. These multinucleated myotubes were found to have elevated levels of myosin ATPase and showed a twitching response to added acetylcholine (ACh). Myogenic activity had never been observed in previous studies of 10T1/2 cells exposed to many different agents, although the effect was highly reproducible after treatment with 5-aza-CR. Moreover, the potential for altered differentiation extended to other end products as well. Taylor and Jones (1979, 1982) showed that phenotypic conversion by 5-aza-CR included the appearance of biochemically differentiated chondrocytes and adipocytes, which appeared later and in somewhat smaller numbers than induced muscle fibers. Cloning experiments indicated that conversion proceeded by induction rather than by selection of previously existing variants. Five subclones of 10T1/2 cells all expressed both muscle and fat cell phenotypes after exposure to 5-aza-CR, and in later studies even single cells, plated at limiting dilution in isolated wells, show a high frequency of conversion if treated with 5-aza-CR. More surprising was the discovery that cells from the 10T1/2 line that had undergone neoplastic transformation were still capable of forming typical striated muscle fibers on exposure to 5-aza-CR. Thus, phenotypic conversion can occur in cells with tumorigenic properties as well as in normal cells. The implication of these studies seems to be that 5-aza-CR induces a reversion in target cells to a more pluripotent state, from which redifferentiation can proceed via alternative pathways (Taylor and Jones, 1979).

Although most of these experiments were performed with 10T1/2 mouse cells, other established murine lines and even a diploid Chinese hamster line (Sagar and Kovac, 1982) respond to 5-aza-CR with the formation of muscle, cartilage, and adipocytes. In searching for a common mechanism of action, early observations showed that 5-aza-CR is incorporated into both RNA and DNA (see review by Vesely and Cihak, 1978). However, since 5-aza-3-deoxycytidine (5-aza-CdR) is an even more potent inducer of myogenesis and is incorporated into DNA only, evidence pointed to a process associated with DNA synthesis. In a classic study, Jones and Taylor (1980) linked incorporation of 5-aza-CR and induction of muscle with a clear-cut inhibition of methylation in newly formed DNA. As background, it should be pointed out that many investigations have shown that DNA methylation patterns affect gene expression (see recent reviews by Burdon and Adams, 1980; Doerffler, 1983; Taylor et al., 1984; Taylor, 1984). Since 5-aza-CR also has little or no mutagenic activity (Landolph and Jones, 1982), a nongenetic mechanism mediated by methylation changes is an attractive explanation for the effects of 5-aza-CR. In the experiments mentioned, Jones and Taylor showed that newly formed strands of DNA containing 5-aza-CR remained unmethylated at CpG sequences, the primary site for methylation in mammalian DNA. The maximum effect was observed at 48 hr after treatment with the inhibitor, a time point which coincided with the peak for induction of myotubes as well. This correlation of phenotypic conversion in 10T1/2 cells with systematic undermethylation of newly formed DNA agrees with many other investigations, in which hypomethylation of DNA can be linked directly with increased gene activity, e.g., synthesis of chicken β-globin (McGhee and Ginder, 1979).

A more direct indication that morphogenetic induction by 5-aza-CR stems from structural modifications in DNA has now been provided by studies with 10T1/2 cells at the molecular level (Lassar et al., 1986). These investigators reasoned that myoblasts arising by normal differentiation, as well as those originating from 10T1/2 cells by exposure to 5-aza-CR, must contain a putative regulator locus, since activation triggers a developmental program leading to the coordinated appearance of a series of enzymes and other muscle-specific markers. Such a regulative locus should be active (unmethylated) in myogenic cells, while methylated and inactive in nonmyogenic tissues. High-molecular-weight DNA was accordingly isolated from unmodified 10T1/2 cells, 10T1/2 cells that had been induced by 5-aza-CR to form myoblasts and embryonic mouse myoblasts. Clear-cut conversion to muscle fibers was observed after transfection with genomic DNA from both types of myogenic cells, but not with DNA from untreated 10T1/2 cells. Specificity in these inductions was documented by assays using monoclonal antibody produced against myosin heavy chain, as well as with the sensitive immunoperoxidase technique used to detect this protein in single cells.

In further experiments, Davis et al. (1987) showed that expression of a single transfected cDNA is sufficient to convert 10T1/2 cells to myoblasts. To obtain the specific cDNA sequences active in this process, cDNAs were prepared from a normal mouse myogenic line and from a line or 10T1/2 cells treated with 5-aza-CR. Subtracted probes were then obtained by hybridizing the cDNAs extensively with poly(A)$^+$ RNA isolated from unmodified 10T1/2 cells. These probes were then used to screen a myoblast/myotube cDNA library. Of numerous sequences obtained, one designated MyoD1 induced efficient formation of muscle fibers when transfected into 10T1/2 cells.

It is not yet clear whether the MyoD1 sequence is identical to the azacytidine gene (i.e., the locus activated in situ by treatment of 10T1/2 cells with 5-aza-CR), and/or the active gene in myoblast DNA which induces muscle formation when transfected into 10T1/2 cells. That more than one gene can serve as activator is suggested by the studies of Pinney, Pearson-White, Konieczney, Latham, and Emerson (1988), who induced myogenesis in 10T1/2 cells with a genomic cosmid sequence obtained from a nonmyogenic cell type (human leucocytes). Southern blot analyses showed that the sequence in question, which has been designated myd, is not identical with the MyoD1 gene. One or more genes may therefore serve as regulatory loci in myogenesis but since

MyoD1, at least, does not promote adipogenesis it seems unlikely that these are general differentiation agents.

More recent studies have concentrated on the expression product of MyoD1, which has been shown to be a phosphoprotein present in the nuclei of proliferating myoblasts and differentiated myotubes, but not detectable in 10T1/2 or other non-muscle cell types (Tapscott, Davis, Thayer, Cheng, Weintraub, and Lasser, 1988). As a precursor, MyoD1 cDNA contains an open reading frame of 318 amino acid residues, with a sequence of 22 residues which resemble a region that is highly conserved in the family of c-Myc proteins. Site-directed mutagenesis of MyoD1 cDNA was employed to analyze functional domains in MyoD1 protein. Comparison of deletion mutants showed that ex-tensive regions could be removed at either end of the MyoD1 protein without interference with biological activity. Expression of a continuous sequence of only 68 amino acids from MyoD1 was sufficient to activate myogenesis in 10T1/2 cells. The active portion includes a very basic region (residues 102–135), without which both nuclear localization and myogenesis were blocked. The c-Myc related region (residues 143–162) was likewise essential for myo-genic induction, but in the absence of this sequence nuclear localization of the rest of the probe continued to occur. The exact chromosomal location of MyoD1 has not yet been determined, but the genetic analyses performed map this gene to mouse chromosome 7 and human chromosome 11.

Much progress has thus been made in linking normal and 5-aza-CR-in-duced myogenesis with structural modifications in a regulatory locus at the DNA level. The MyoD1 model serves to explain the coordinated program of changes which appear on 5-aza-CR treatment, but the role of regulator and final effector genes in this integrated pattern remain to be determined. Less complex systems for analysis of single gene changes can be found in the analysis of stable variations which arise spontaneously or by induction in cell cultures. At an earlier time it was believed that essentially all such variants were the result of gene mutations (Siminovitch, 1975). It is now clear, however, that a number of these are in fact epigenetic in origin, particularly those obtained by treatment of cells with 5-aza-CR. DNA methylation changes are particularly useful as an explanation for such variants, which seem to be based on stable shifts in gene expression rather than on genetic alteration in the DNA coding system. The characteristics which permit this interpretation include the following pertinent features (reviewed in the general references cited above) (1) the presence or absence of methyl groups on specific CpG sequences throughout the genome which forms a heritable and reproducible pattern that persists through suc-cessive cell division. This is so because of the semiconservative character of methylase activity, in which a methyl group is added to cytosines in newly formed DNA only if a corresponding methyl group is present on the comple-mentary CpG of the pre-existing partner strand. Second, if a methyl group at a particular site is gained or lost by spontaneous or induced change, the new pattern is perpetuated for the same reason through successive replications, although fidelity is not absolute (Wigler et al., 1981). In combination, these characteristics provide a working model for at least one class of heritable epigenetic changes. DNA methylation patterns here play the role of a second information system, parallel to, but independent from the primary DNA coding sequence (Riggs and Jones, 1983).

A good example of epigenetic variation in these terms is provided by the experiments of Compere and Palmiter (1981) on the metallothionein-1 (MT-1) gene in W7 mouse thymoma cells. Metallothioneins are specialized proteins which bind heavy metals, and can be induced in certain cell lines on exposure to cadmium or glucocorticoid hormones. Although this inductive response does not ordinarily occur in untreated W7 populations, it can be stimulated to take place if W7 cells are previously exposed to 5-aza-CR. The molecular basis for this shift was analyzed by Compere and Palmiter, using the restriction enzymes HpaII and MspI, which in combination show the degree of methylation at CpG sequences. In untreated W7 cells, all the available HpaII sites in the vicinity of the MT-1 gene were fully methylated. After exposure to 5-aza-CR, the same sites were unmethylated, along with associated flanking sequences. Thus, DNA methylation, which exists at control sites in the W7 cells, seems to block inducibility of the MT-1 gene, a locking mechanism which can be removed by demethylation of these sites with 5-aza-CR.

Cells deficient in thymidine kinase (TK) and other specific enzyme activities offer additional study systems for epigenetic variation induced by 5-aza-CR. Variant lines lacking TK activity have been produced by many investigators through prolonged exposure to 5-bromodeoxyuridine (BUdR). In V79 Chinese hamster cells, TK^- variations were found to emerge in this way by a two-stage process (Harris and Collier, 1980). In the first stage, there was an abrupt mutation-like event, later found to be a stochastic drop in copy number at the TK locus (Wise, 1985; Wise and Harris, 1988). This conferred partial resistance to BUdR, along with decline of TK activity to a lower level. The second phase was distinctively different. During this interval, partially resistant first-stage cells were propagated serially for months in media containing BUdR. Cells in these populations continued to lose TK activity on a logarithmic basis for 28 weeks or more. The terminal populations were stably and completely resistant to BUdR, lacking entirely in TK activity and unable to grow in HAT medium. Studies by Wise (1985) and Wise and Harris (1988), subsequently showed that transition to complete BUdR resistance and the TK^- state were associated with gradual hypermethylation in the 5' region of the TK gene, without further change in copy number. Repression by methylation is again seen here as a mechanism for silencing gene activity.

The interpretation given above is strongly supported by experiments designed to induce re-expression of activity in the V79 TK^- cells that had been obtained (Harris, 1982). Brief exposure of such populations to 5-aza-CR resulted in mass conversion to the HAT^+ state, with one-step increases of 10^6 in frequency of TK^+ cells over background levels. Although the incidence of HAT^+ cells arose slightly in cultures treated with the chemical mutagen ethyl methanesulfonate, the increase obtained was orders of magnitude less than effects from treatment with 5-aza-CR. Large increases in frequency of HAT$^+$ cells were also obtained with n-butyrate and L-ethionine, both of which affect gene expression but have no mutagenic potential. Re-expression of TK activity in this system seems clearly to be an epigenetic event, which reverses the gradual extinction in TK expression imposed by BUdR-associated methylation. In all cases, however, TK activity in these revertants was significantly lower than levels in wild-type V79 cells, indicating that there had been no reversal of the initial drop in copy number at the TK locus.

Other examples of epigenetic variation have been found in cell lines that require L-asparagine for growth and lack detectable levels of asparagine synthetase (Sugiyama et al., 1983; Harris, 1986). In the earlier study, made with Jensen rat sarcoma cells, asparagine-independent variants were obtained in high frequency from the parental auxotrophs after treatment with 5-aza-CR. All revertants were phenotypically stable, whether maintained in the presence or absence of asparagine. Asparagine synthetase preparations isolated from revertants were indistinguishable in biochemical properties from those of the normal enzyme obtained from rat liver or other rat cell lines. These comparisons included measurements of thermolability and several kinetic parameters, as well as immunological properties. Other studies showed that the parental Jensen rat sarcoma cells contained no immunologically cross-reacting material. Taken as a whole, these data suggest that asparagine synthetase gene is repressed but intact in asparagine auxotrophs, and can be restored to a normal functional state by spontaneous or induced epigenetic change.

One further instance of epigenetic change is worth mentioning because the marker in question, the well-known requirement of CHO-K1 Chinese hamster cells for proline, has long been regarded as a prototype for auxotrophic genetic mutations in mammalian cells. Such an interpretation was put forth by Kao and Puck (1967) when this naturally occurring line was first characterized, and is consistent with several observed properties: (1) the proline requirement is phenotypically stable, (2) spontaneous reversion to the pro$^+$ state occur with a frequency of less than 10^{-7} as measured by fluctuation tests, (3) there is a small increase in incidence of revertants after treatment of CHO-K1 cells with mutagens, and (4) biochemical differences between pro$^+$ and pro$^-$ cells are readily apparent when proline synthesis is measured in cell extracts. However, no direct proof of a mutational origin was forthcoming, and all the characteristics mentioned above are equally compatible with the hypothesis that lack of proline synthesis in CHO-K1 cells stems from suppression of proline synthesis, perhaps by hypermethylation, rather than from gene mutation. Recent experiments (Harris, 1984) suggest that in fact the epigenetic explanation is the correct one. Thus, exposure of CHO-K1 cells to 5-aza-CR induces a 10^5 to 10^6 increase over background conversion to the proline-independent state. The revertants are stable phenotypically, with enzymatic activities reexpressed for pyrroline-5-carboxylate synthase and ornithine aminotransferase, as in other proline-synthesizing lines. These observations deserve careful consideration, since the view often expressed (e.g., Siminovitch, 1976) has been that stable variants in culture are in general genetic rather than epigenetic in origin. A more salutary approach may be to require direct evidence in each system for the mechanisms involved, rather than simple analogy to a preconceived model.

4. Concluding Remarks

Morphogenesis, differentiation, and modulation occur against a constant and unchanging genome in eukaryotic cells (the few exceptions are discussed in Chapter 2, *this volume*), and provide a unique dimension of variability that is essentially lacking in microbial systems to which they have often been compared (Puck, 1972). In addition, the recent studies discussed here show that stable epigenetic changes can continue to occur even in animal cells that have completed a normal developmental program. Epigenetic shifts of this type deserve careful study, since the results may be of broad significance for understanding the stability of differentiated states, as well as neoplastic transformation and tumor progression (Foulds, 1954, 1958).

DNA methylation changes provide one means for visualizing epigenetic variation in molecular terms, but the significance of this concept should not be overdrawn, nor is it an exclusive one. Many variables are known to affect gene activity, and it is reasonable to expect that other control mechanisms for epigenetic change may operate with, or function independently from, shifts in DNA methylation patterns. This is consistent with the fact that when 5-aza-CR is used as an inductive agent, gene activation is highly selective. Thus, 5-aza-CR affects some genes but not others (Riggs and Jones, 1983; Jones, 1985). A further indication of the need to avoid oversimplification lies in observations that certain genes can be fully expressed even when totally methylated (Gerber-Huber *et al.*, 1983), while others remain transcriptionally inactive even when demethylated extensively (Hsiao *et al.*, 1984). These varying results with 5-aza-CR are understandable if DNA demethylation is seldom the only factor required for gene expression. Thus, DNA demethylation alone might frequently be "necessary but not sufficient" (Riggs and Jones, 1983).

Combinations of genetic and epigenetic change in an overall pattern of phenotypic variation may be a field of promise for future study. A simple composite sequence of this kind has been described for the disappearance of thymidine kinase activity during long-term culture of V79 Chinese hamster cells in BUdR (Harris and Collier, 1980). More important study systems may be found in transitions from normal to the neoplastic state. The gradual long-term character of such shifts, especially *in vivo*, has been generally viewed as a multistage sequence. An initial genetic component is suggested by the correlation between properties of mutagens and carcinogens (McCann *et al.*, 1975). However, it is difficult to understand why one-step mutations, if crucial to initiation of carcinogenesis, are typically followed by a long latent period before neoplastic characteristics are expressed. Long drawn-out latent periods suggest a gradual evolution in cell properties. Thus, carcinogens such as X-rays may create chronic cell damage, with consequences which only slowly emerge at epigenetic and genetic levels. Conceivably, epigenetic shifts may facilitate the occurrence of genetic changes, or be stabilized by them on a selective basis. New model systems in culture may provide the means to dissect the events concerned experimentally.

References

Benya, P.D., and Shaffer, J. D., 1982, Dedifferentiated chondrocytes reexpress the differentiated collagen phenotype when cultured in agarose gels, *Cell* **30**:215–224.

Burdon, R. H., and Adams, R. L. P., 1980, Eukaryotic DNA methylation, *Trends Biochem. Sci.* **5**: 294–297.

Carrel, A., and Ebeling, A. H., 1922, Pure cultures of large mononuclear leucocytes, *J. Exp. Med.* **36**: 365–378.

Champy, C., 1912, Sur les phénomènes cytologique qui s'observent dans les tissus cultivés en dehors de l'organisme, *C.R. Soc. Biol.* **72**:987–988.

Chandler, L. A., and Jones, P. A., 1988, Hypomethylation of DNA in the regulation of gene expression, in: *Developmental Biology: A Comprehensive Synthesis*, Vol. 5: *The Molecular Biology of Cell Determination and Cell Differentiation* (L. W. Browder, ed.), pp. 335–349, Plenum, New York.

Compere, S. J., and Palmiter, R. D., 1981, DNA methylation controls the inducibility of the mouse metallothionein-1 gene in lymphoid cells, *Cell* **25**:233–240.

Constantinides, P. G., Jones, P. A., and Gevers, W., 1977, Functional striated muscle cells from non-myoblast precursors following 5-azacytidine treatment, *Nature (Lond.)* **267**:364–366.

Coon, H. G., 1966, Clonal stability and phenotypic expression of chick cartilage cells *in vitro, Proc. Natl. Acad. Sci. USA* **55**:66–73.

Davis, R. L., Weintraub, H., and Lassar, A. B., 1987, Expression of a single transfected cDNA converts fibroblasts to myoblasts, *Cell* **51**:987–1000.

Doerfler, W., 1983, DNA methylation and gene activity, *Annu. Rev. Biochem.* **52**:93–124.

Ebeling, A. H., 1925, A pure strain of thyroid cells and its characteristics, *J. Exp. Med.* **41**:337–346.

Ebner, K. E., Hageman, E. C., and Larson, B. L., 1961, Functional biochemical changes in bovine mammary cell cultures, *Exp. Cell Res.* **25**:555–570.

Emerman, J. T., Enami, J., Pitelka, D. R., and Nandi, S., 1977, Hormonal effects on intracellular and secreted casein in cultures of mouse mammary epithelial cells on floating collagen membranes, *Proc. Natl. Acad. Sci. USA* **74**:4466–4470.

Emerman, J. T., and Pitelka, D. R., 1977, Maintenance and induction of morphological differentiation in dissociated mammary epithelium on floating collagen membranes, *In Vitro* **13**:316–328.

Fell, H. B., and Robison, R., 1929, The growth, development and phosphatase activity of embryonic avian femora and limb-buds cultivated *in vitro, Biochem. J.* **23**:767–787.

Fisher, A., 1922, A three months old strain of epithelium, *J. Exp. Med.* **35**:367–372.

Foulds, L., 1954, The experimental study of tumor progression: A review, *Cancer Res.* **14**:327–339.

Foulds, L., 1958, The natural history of cancer, *J. Chron. Dis.* **8**:2–37.

Friend, C., Scher, W., Holland, J. G., and Sato, T., 1971, Hemoglobin synthesis in murine virus induced leukemia cells in vitro: Stimulation of erythroid differentiation by dimethyl-sulfoxide, *Proc. Natl. Acad. Sci. USA* **68**:378–382.

Gerber-Huber, S., May, F. E. B., Westley, B. R., Felber, B. K., Hosbach, H. A., Andres, A., and Ryffell, G. U., 1983, In contrast to other *Xenopus* genes the estrogen-inducible vitellogenin genes are expressed when totally methylated, *Cell* **33**:43–51.

Gottesman, M. M. (ed.), 1985, *Molecular Cell Genetics*, Wiley, New York.

Harris, M., 1964, *Cell Culture and Somatic Variation*, Holt, Rinehart, and Winston, New York.

Harris, M., 1982, Induction of thymidine kinase in enzyme-deficient Chinese hamster cells, *Cell* **29**: 483–492.

Harris, M., 1984, High-frequency induction by 5-azacytidine of proline independence in CHO-K1 cells, *Somatic Cell Mol. Genet.* **10**:615–624.

Harris, M., 1986, Induction and reversion of asparagine auxotrophs in CHO-K1 and V79 cells, *Somatic Cell Mol. Genet.* **12**:459–466.

Harris, M., and Collier, K., 1980, Phenotypic evolution of cells resistant to bromodeoxyuridine, *Proc. Natl. Acad. Sci, USA* **77**:4206–4210.

Hayflick, L., and Moorhead, P. S., 1961, The serial cultivation of human diploid strains, *Exp. Cell Res.* **25**:585–621.

Holtzer, H., Abbott, J., Lash, J., and Holtzer, S., 1960, The loss of phenotypic traits by differentiated cells *in vitro*. I. Dedifferentiation of cartilage cells, *Proc. Natl. Acad. Sci. USA* **46**:1533–1542.

Hsaio, W. W., Gattoni-Celli, S., Kirschmeier, P., and Weinstein, I. B., 1984, Effects of 5-azacytidine on methylation and expression of specific DNA sequences in C3H 10T1/2 cells, *Mol. Cell. Biol.* **4**:634–641.

Jones, P. A., 1985, Altering gene expression with 5-azacytidine, *Cell* **40**:485–486.

Jones, P. A., and Taylor, S. M., 1980, Cellular differentiation, cytidine analogs and DNA methylation, *Cell* **20**:85–93.

Kao, F., and Puck, T. T., 1967, Genetics of somatic mammalian cells. IV. Properties of Chinese hamster cell mutants with respect to the requirement for proline, *Genetics* **55**:513–524.

Landolph, J. R., and Jones, P. A., 1982, Mutagenicity of 5-azacytidine and related nucleosides in C3H/10T1/2 clone 8 and V79 cells, *Cancer Res.* **42**:817–823.

Lassar, A. B., Paterson, B. M., and Weintraub, H., 1986, Transfection of a DNA locus that mediates the conversion of 10T1/2 fibroblasts to myoblasts, *Cell* **47**:649–656.

Lee, E. Y. P., Lee, W., Kaetzel, C. S., Parry, G., and Bissell, M. J. 1985, Interaction of mouse mammary epithelial cells with collagen substrate: Regulation of casein gene expression and secretion, *Proc. Natl. Acad. Sci. USA* **82**:1419–1423.

Lee, E. Y., Parry, G., and Bissell, M. J., 1984, Modulation of secreted proteins of mouse mammary epithelial cells by the collagenous substrata, *J. Cell Biol.* **98**:146–155.

Levay-Young, B. K., Imagawa, W., Yang, J., Richards, J. E., Guzman, R. C., and Nandi, S., 1987, Primary culture systems for mammary biology studies, in: *Cellular and Molecular Biology of Experimental Mammary Cancer* (C. Ip, D. Medina, G. Heppner, and E. Anderson, eds.), pp. 181–203, Plenum, New York.

Littlefield, J. W., 1976, *Variation, Senescence, and Neoplasia in Cultured Somatic Cells*, Harvard University Press, Cambridge.

Luria, S. E., and Delbrück, M., 1943, Mutations of bacteria from virus sensitivity to virus resistance, *Genetics* **28**:491–511.

Marks, P. A., Murate, T., Kaneda, T., Ravetch, J., and Rifkind, R. A., 1985, Modulation of gene expression during terminal cell differentiation: Murine erythroleukemia, *M. D. Anderson Symp. Fund. Cancer Res.* **37**:327–340.

McCann, J., Choi, E., Yamasaki, E., and Ames, B. N., 1975, Detection of carcinogens in the Salmonella/microsome test: Assay of 300 chemicals, *Proc. Natl. Acad. Sci. USA* **72**:5135–5139.

McGhee, J. D., and Ginder, G. D., 1979, Specific DNA methylation sites in the vicinity of the chicken β-globin genes, *Nature (Lond.)* **280**:419–420.

Michalopoulos, G., and Pitot, H. C., 1975, Primary culture of parenchymal liver cells on collagen membranes. Morphological and biochemical observations, *Exp. Cell Res.* **94**:70–78.

Morrow, J., 1983, *Eukaryotic Cell Genetics*, Academic, Orlando, Florida.

Mascona, A., 1956, Development of heterotypic combinations of dissociated embryonic check cells, *Proc. Soc. Exp. Biol. Med.* **92**:410–416.

Moscona, A., 1962, Cellular interactions in experimental histogenesis, *Int. Rev. Pathol.* **1**:371–428.

Puck, T. T., 1972, *The Mammalian Cell as a Microorganism*, Holden-Day, San Francisco.

Reeves, R., and Cserjesi, P., 1979, Sodium butyrate induces new gene expression in Friend erythroleukemic cells, *J. Biol. Chem.* **254**:4283–4290.

Pinney, D. F., Pearson-White, S. H., Konieczney, S. F., Latham, K. E., and Emerson, C. F., Jr., 1988, Myogenic linkage determination and differentiation: Evidence for a regulatory gene pathway, *Cell* **53**:781–793.

Reuben, R. C., Rifkind, R. A., and Marks, P. A., 1980, Chemically induced murine erythroleukemic differentiation, *Biochim. Biophys. Acta* **605**:325–346.

Riggs, A. D., and Jones, P. A., 1983, 5-Methycytosine, gene regulation, and cancer, *Adv. Cancer Res.* **40**:1–30.

Rothfels, K. H., Kupelwieser, E. B., and Parker, R. C., 1963, Effects of X-irradiated feeder layers on mitotic activity and development of aneuploidy in mouse-embryo cells *in vitro, Can. Cancer Conf.* **5**:191–223.

Sager, R., and Kovac, P., 1982, Pre-adipocyte determination either by insulin or by 5-azacytidine, *Proc. Natl. Acad. Sci. USA* **79**:480–484.

Siminovitch, L., 1976, On the nature of hereditable variation in cultured somatic cells, *Cell* **7**:1–11.

Stockdale, F. E., Abbott, J., Holtzer, S., and Holtzer, H., 1963, The loss of phenotypic traits by differentiated cells. II. Behavior of chondrocytes and their progeny *in vitro, Dev. Biol.* **7**:293–302.

Sugiyama, R. H., Arfin, S. M., and Harris, M., 1983, Properties of asparagine synthetase in asparagine-independent variants of Jensen rat sarcoma cells induced by 5-azacytidine, *Mol. Cell. Biol.* **3**:1937–1942.

Taylor, J. H., 1984, *DNA Methylation and Cellular Differentiation*, Springer-Verlag, New York.

Taylor, S. M., constantinides, P. A., and Jones, P. A., 1984, 5-Azacytidine, DNA methylation and differentiation, *Curr. Top. Microbiol. Immunol.* **108**:115–127.

Tapscott, S. J., Davis, R. L., Thayer, M. J., Cheng, P., Weintraub, H., and Lassar, A. B., 1988, MyoD1: A nuclear phosphoprotein requiring a Myc homology region to convert fibroblasts to myoblasts. *Science* **242**:405–411.

Taylor, S. M., and Jones, P. A., 1979, Multiple new phenotypes induced in 10T1/2 and 3T3 cells treated with 5-azacytidine, *Cell* **17**:771–779.

Taylor, S. M., and Jones, P. A., 1982, Changes in phenotypic expression in embryonic and adult cells treated with 5-azacytidine, *J. Cell. Physiol.* **111**:187–194.

Todaro, C. J., and Green, H., 1963, Quantitative studies of the growth of mouse embryo cells in culture and their development into established lines, *J. Cell Biol.* **17**:299–313.

Trinkaus, J. P., and Groves, P. W., 1955, Differentiation in culture of mixed aggregates of dissociated tissue cells, *Proc. Natl. Acad. Sci. USA* **41**:787–495.

Vesely, J., and Cihak, A., 1978, 5-azacytidine: Mechanism of action and biological effects in mammalian cells, *Pharmacol. Ther. A* **2**:813–840.

Weiss, P., 1939, *Principles of Development*, Holt, Rinehart, and Winston, New York.

Weiss, P., 1950, Perspectives in the field of morphogenesis, *Q. Rev. Biol.* **25**:177–198.

Wigler, M., Levy, D., and Perucho, M., 1981, The somatic replication of DNA methylation. *Cell* **24**:33–40.

Wise, T. L., 1985, The molecular basis for the loss of thymidine kinase activity in BUdR resistant Chinese hamster cells and for reactivation of thymidine kinase in revertants, Doctoral thesis, University of California, Berkeley.

Wise, T. L., and Harris, M., 1988, Deletion and hypermethylation of the thymidine kinase gene in V79 Chinese hamster cells resistant to bromodeoxyuridine, *Somatic Cell Mol. Genet.* **14**:567–581.

Chapter 6

Developmental Regulation of the Heat-Shock Response

LEON W. BROWDER, MICHAEL POLLOCK,
ROBERT W. NICKELLS, JOHN J. HEIKKILA, and
ROBERT S. WINNING

1. Introduction

Cell differentiation in the development of multicellular organisms occurs as a consequence of the generation of chronologically and spatially distinct patterns of protein synthesis. These unique constellations of proteins confer on cells the functional and structural characteristics that enable them to perform their specialized roles in the organism. However, in addition to the overt protein synthetic profile, cells may retain the potential to produce proteins that they would not normally produce in significant amounts. During periods of stress, this potential is realized and is evidenced by the synthesis of a set of stress proteins. As discussed in this chapter, the pattern of stress protein synthesis is also subject to developmental regulation. Thus, cells are simultaneously engaged in two parallel developmental processes: regulation of the patterns of overt and stress-inducible protein synthesis. Stress-inducible protein synthesis can be evoked by regulation at either the transcriptional or translational levels, or both. Thus, regulation at both levels must be subject to developmental modulation.

In addition to being induced by stress, some of the so-called stress proteins are apparently synthesized at particular times during development and in specific cells, suggesting that they may also have developmental functions under nonstressed conditions. The genes encoding such proteins must therefore be sensitive to developmental cues as well as cues that signal stress. Overt synthesis of heat-shock proteins during development was recently reviewed by Bond and Schlesinger (1987). This chapter emphasizes developmental changes in inducible gene expression.

LEON W. BROWDER, MICHAEL POLLOCK, and ROBERT W. NICKELLS • University of Calgary, Calgary, Alberta, Canada T2N 1N4.   JOHN J. HEIKKILA and ROBERT S. WINNING • University of Waterloo, Waterloo, Ontario, Canada N2L 3G1.

2. The Heat-Shock Response

2.1. *Drosophila* Heat-Shock Response

The first indication that gene expression can be modified by heat shock was the observation by Ritossa (1962) that a distinct set of puffs appears in larval salivary gland polytene chromosomes in response to heat shock. This result, confirmed by Ashburner (1970), led to considerable research activity documenting the heat-shock response in *Drosophila* larvae and cultured cells. The protein products, the heat-shock proteins (or hsps), of the heat-inducible puffs were identified by Tissières *et al.* (1974). Spradling *et al.* (1975) demonstrated by *in situ* hybridization that [^{3}H]uridine-labeled polyadenylated RNA synthesized in cultured cells during heat shock hybridizes almost exclusively to the heat-shock-inducible loci on polytene chromosomes of salivary gland cells, indicating that transcription in heat-shocked cells is restricted to these loci. RNA synthesized by the tissue culture cells at ambient temperature hybridizes to numerous additional loci (but not the loci that form heat-shock puffs). Thus, genes expressed at ambient temperature become transcriptionally silent during heat shock, whereas a small number of genes (corresponding to loci that puff in heat-shocked salivary glands) that are transcriptionally inert at ambient temperature are induced to transcribe by heat shock. These conclusions have been supported by numerous investigations. One of the most graphic demonstrations of selective transcription of heat shock genes is illustrated in Fig. 1. This micrograph shows immunohistological localization of RNA polymerase II in heat-shocked salivary gland chromosomes to the heat-shock loci.

In *Drosophila* salivary glands and tissue culture cells, the pattern of protein synthesis becomes very simple during heat shock with synthesis of constitutive proteins being drastically reduced. This simplification of the protein synthetic profile is not because of the loss of constitutive messengers, as these messengers are still present and can be efficiently translated *in vitro* using a heterologous cell-free translation system (Mirault *et al.*, 1978). Thus, a translational control mechanism must discriminate in favor of translation of heat-shock messengers during heat shock. This selectivity has been demonstrated by the use of cell-free translation lysates from ambient and heat-shocked *Drosophila* cells. When messengers from ambient and heat-shocked cells are added to these lysates, ambient lysate translates both kinds of messengers with high efficiency, whereas the heat-shock lysates produce predominantly hsps (Krüger and Benecke, 1981; Scott and Pardue, 1981). Thus, the homologous system has the selectivity that the heterologous system lacks.

The recognition of hsp mRNA is apparently attributable to some inherent property of these messengers, since deproteinized hsp messengers are correctly selected for translation in *in vitro* assay systems derived from heat-shocked *Drosophila* cells (Storti *et al.*, 1980). In *Drosophila* lysates, the ability to discriminate hsp messengers is at least partially associated with the ribosomes (Scott and Pardue, 1981), although the precise nature of the controlling component is unknown. The return to the normal pattern of protein synthesis after

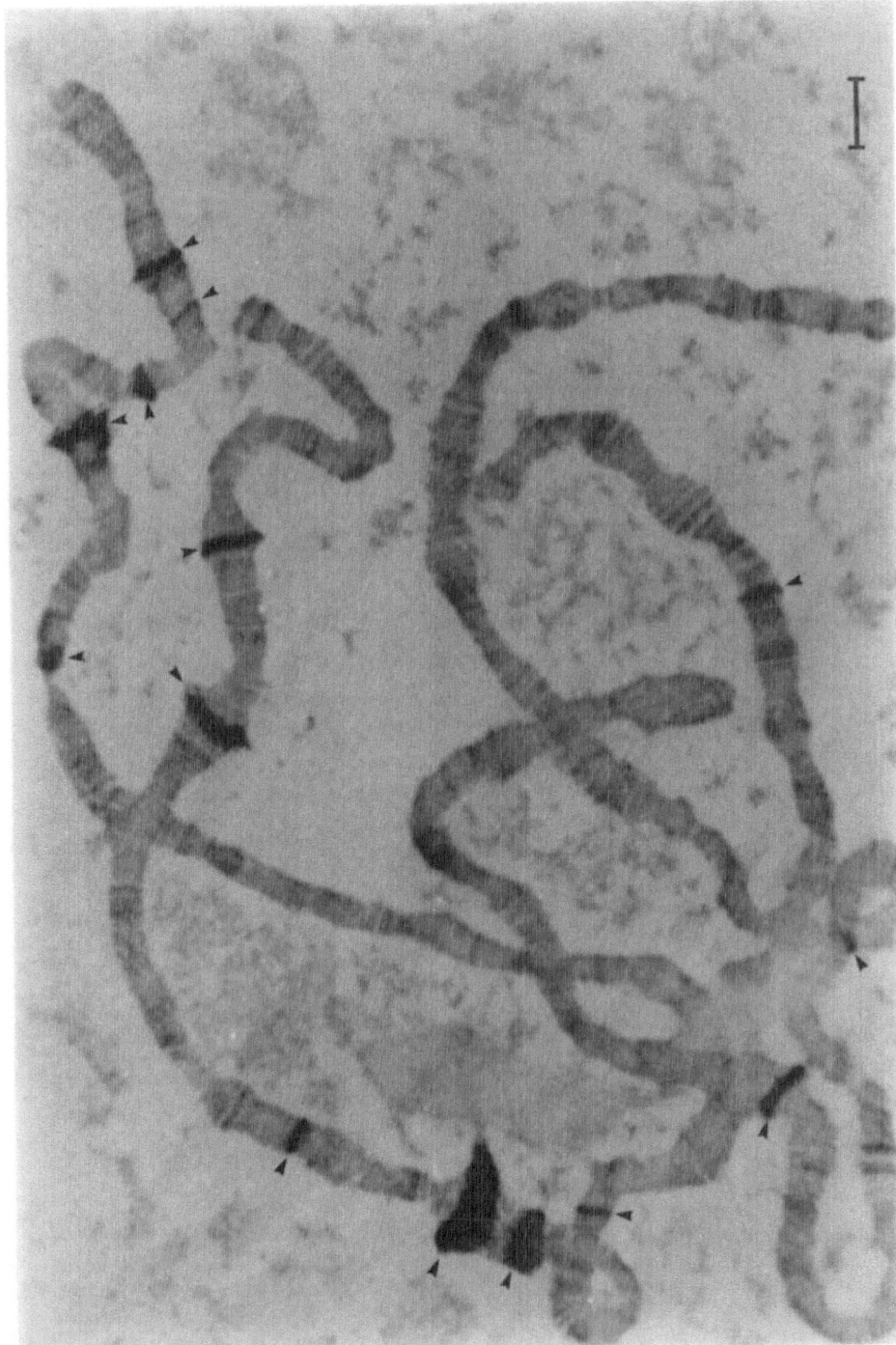

Figure 1. RNA polymerase II localization after heat shock of *Drosophila* salivary gland. Chromosomes were stained immunochemically using antibody to *Drosophila* RNA polymerase II and horseradish peroxidase. Arrowheads indicate heat-shock loci. Scale bar = 10 μm. (From Bonner and Kerby, 1982.)

heat shock is gradual and is dependent on the synthesis of hsps, particularly hsp 70 (Petersen and Mitchell, 1981; DiDomenico *et al.*, 1982; Lindquist and DiDomenico, 1985). The extreme lability of constitutive protein synthesis in certain cells (e.g., *Drosophila* salivary gland cells) might reflect lower levels of hsp accumulation in those cells than in cells with resistant constitutive protein synthesis. Careful determinations of hsp accumulation would be necessary to confirm this hypothesis.

The pattern of hsp synthesis in heat-shocked *Drosophila* salivary glands is shown in Fig. 2a. The major hsps have molecular weights of 83, 70, 68, 27, 26, 23, and 22 kDa. There is also a minor hsp of approximately 37-kDa molecular mass. Figure 2b,c shows that the latter hsp is made more strongly in certain other larval *Drosophila* tissues. Hsp 37 may be homologous to hsp 35 in

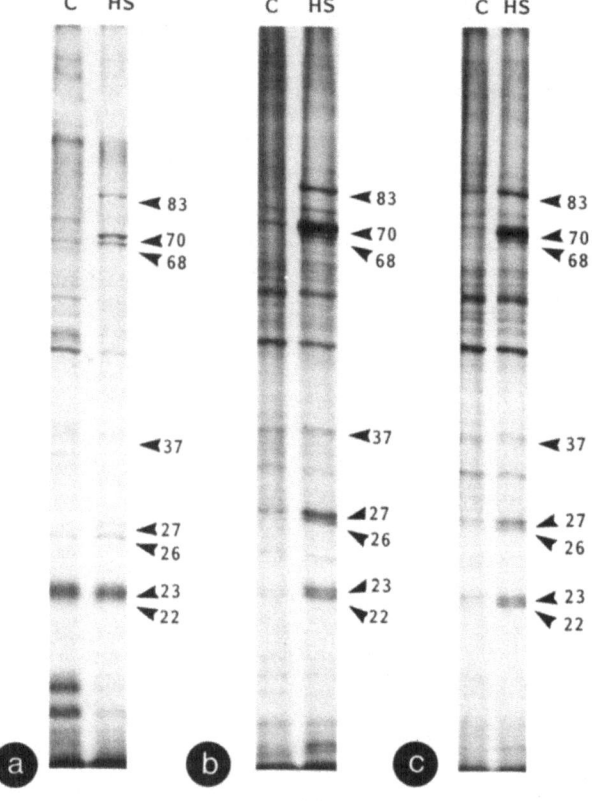

Figure 2. Fluorographs of the proteins synthesized by control (C) and heat-shocked (HS) organs of *Drosophila* third-instar larvae. In the experiment shown here, the salivary glands (a), imaginal wing discs (b), and brains (c) were dissected from larvae and placed in respective drops of *Drosophila* Ringer's solution. These organs were then heat-shocked at 35°C for 1 hr in the presence of [35S]methionine. Controls were maintained at 22°C. The synthesis of hsps (arrowheads, with their relative molecular weights shown in kDa) is detected in all organs tested. An hsp37 is detected in each organ but appears strongest in the brains. (Modified from Nickells, 1987.)

Xenopus embryos (see Sections 2.2 and 2.4). Comparison of the patterns of protein synthesis in heat-shocked salivary glands, wing discs, and brain shown in Fig. 2 reveals that the salivary gland is exceptional in the simplification of its pattern of protein synthesis during heat shock. In the two other tissues, constitutively synthesized proteins clearly continue to be synthesized during heat shock. This is also the case in *Xenopus* (see Section 3.2). Thus, there is considerable variability in the extent to which various cells discriminate against translation of non-heat-shock messengers during heat shock.

In recent years, the *Drosophila* heat-shock genes have been isolated and cloned, providing probes to analyze the transcriptional induction mechanism. The use of these probes has led to the identification of the transcription regulation elements upstream of the hsp coding sequences (Pelham, 1982) and of the heat-shock-activated *trans*-acting factors that bind these elements and initiate transcription (Parker and Topol, 1984; Wu, 1985; Shuey and Parker, 1986; Wiederrecht *et al.*, 1987; Zimarino and Wu, 1987). The transcripts of these genes have nucleotide sequences in the 5′ noncoding leader regions that control their preferential translation at heat-shock temperatures (Klemenz *et al.*, 1985; McGarry and Lindquist, 1985). Thus, the *Drosophila* heat-shock genes contain both the transcriptional and translational response elements.

2.2. Universality of the Response

Since its discovery in *Drosophila*, the heat-shock response has also been observed in a wide variety of organisms covering the spectrum from *Escherichia coli* to higher animals (including humans) and plants (Schlesinger *et al.*, 1982; Atkinson and Walden, 1985). The heat-shock genes have been conserved with high fidelity through evolution, producing proteins with high levels of homology. The eukaryotic hsps that are homologous to the major *Drosophila* hsps have subunit molecular weights of 80–100 kDa, 65–75 kDa, and 15–30 kDa. These genes are all characterized by the presence of the upstream transcriptional regulatory element identified in the *Drosophila* hsp70 gene (Pelham, 1982). An additional small (8-kDa) protein, ubiquitin, has also been identified as an hsp in at least chickens and mammals (Bond and Schlesinger, 1985; Schlesinger, 1986). Regulation of heat-shock gene induction appears to have been conserved through evolution, since heat-shock promoters, when introduced into cells of diverse species, will be activated by temperatures appropriate to the host species (reviewed by Lindquist, 1986). Thus, the *trans*-acting regulators and transcriptional regulatory elements are mutually compatible across phylogenetic boundaries.

2.3. Thermotolerance

The potential to synthesize a specific set of proteins under heat-shock conditions implies that these proteins have adaptive significance under stress

conditions. In this section, we examine the putative functions of hsps in confer-
ring thermotolerance.

Thermotolerance can be defined as the ability of a cell or organism to
survive a potentially lethal change in temperature. The classic experimental
demonstration of thermotolerance is the ability of a cell to survive a lethal heat
shock if it has first been conditioned by a shorter, sublethal heat treatment.
Mammalian tissue culture cells provide good examples of this phenomenon.
For example, Morris hepatoma cells (MH-7777) do not survive prolonged exposure to temperatures of 43°C; however, if given a shorter exposure to this
temperature (30 min), they acquire the ability to survive a second, normally
lethal, extended treatment at 43°C (reviewed by Landry, 1986). This effect is
also observed in Chinese hamster HA-1 cells (Li and Hahn, 1980; reviewed by
Li and Mivechi, 1986), Chinese hamster ovary (CHO) cells (Henle and Leeper,
1976), and HeLa cells (Gerner and Schneider, 1975). The term thermotolerance
can also be applied to the developmentally acquired ability of an organism to
survive heat shock at a temperature that would be lethal at earlier stages (see
Section 3). Thermotolerance, in this instance, is due to developmental events at
ambient temperature rather than previous exposure to a sublethal heat shock.

2.3.1. Relationship between hsps and Thermotolerance

A great deal of correlative evidence has been accumulated to suggest that
hsps confer thermotolerance on heat-shocked cells. These kinds of data have
been well documented in the acquisition of thermotolerance by mammalian
cells (Li and Werb, 1982; Li and Laszlo, 1985). Mammalian tissue culture cells
will survive an otherwise lethal heat shock if they are first conditioned with a
brief sublethal heat treatment. Such conditioning is sufficient to activate the
synthesis of the mammalian hsps, which accumulate with kinetics that mirror
the acquisition of thermotolerance by the conditioned cells. Consequently, the
conditioned cells that have accumulated the largest amount of hsps will show
the highest degree of thermotolerance. Conversely, if the conditioned cells are
allowed to remain at ambient temperature for an extended period of time, the
synthesis and accumulation of hsps decay with the same kinetics as the loss of
thermotolerance (Landry et al., 1982; Landry, 1986; Li and Laszlo, 1985). A
similar correlation exists between cell types exhibiting different sensitivities to
heat shock. Cells that are more thermotolerant have a higher rate of hsp synthesis than do less thermotolerant cell lines (Tomasovic et al., 1984). Hsp
synthesis can also be induced in many mammalian cells by other environmental insults such as ethanol or sodium arsenite (reviewed by Nover et al., 1984;
Lindquist, 1986). These other stimuli are also sufficient to condition some cell
types to withstand lethal temperatures, presumably through the induction of
hsp synthesis (reviewed by Li and Laszlo, 1985). Similar correlative evidence
between hsp synthesis and the acquisition of thermotolerance during development has also been documented (see Section 3). Recently, direct evidence has
confirmed a role for hsp70 in enhancing survival of mammalian cells during
and after stress (Riabowol et al., 1988; Johnston and Kucey, 1988).

Although there are many examples showing a strong correlation between the acquisition of thermotolerance and hsp synthesis, there are several exceptions. These exceptions range from cells that only require the synthesis of some hsps to those that do not require the synthesis of any hsps in order to develop thermotolerance (reviewed by Landry, 1986). *Drosophila* salivary gland cells, for example, can be conditioned to develop thermotolerance by first treating them with the moulting hormone ecdysone. Ecdysone induces the synthesis of the low-molecular-weight hsps of 27, 26, 24, and 23 kDa, but not the large molecular weight hsps of 70 and 83 kDa (Berger and Woodward, 1983). Alternatively, in *Drosophila* Schneider's line 2 tissue culture cells, the expression of hsp26 can be blocked by transfecting the cells with an antisense heat-shock gene for the hsp (McGarry and Lindquist, 1986). During heat shock, the antisense gene is transcribed, producing RNA molecules complementary to the RNA transcribed from the normal hsp26 gene. The antisense RNA can then hybridize to the sense RNA and presumably prevent it from being translated. Although hsp26 synthesis is subsequently reduced, there is no apparent effect on the synthesis of the other hsps or on the recovery of normal cellular protein synthesis.

In other examples, no hsp synthesis appears to be required for the development of thermotolerance. Rat mammary adenocarcinoma cells, which normally require a conditioning heat shock to permit the accumulation of hsps, and hence develop thermotolerance, can bypass this requirement if gradually heated to the potentially lethal temperature of 42°C. Normally, under experimental conditions, these cells are heated from 37°C to 42°C within 3–5 min. If they are brought to the same temperature over 3 hr, however, they greatly reduce the synthesis of some of the hsps for several hours after reaching 42°C (Tomasovic *et al.*, 1983). Unlike cells that are rapidly heat-shocked, the reduction of hsp synthesis in these cells correlates with a doubling in their resistance to heat. These results suggest that the synthesis of hsps and the development of thermotolerance are only indirectly related, so that either (1) hsps are not required to protect the cell from heat-induced structural damage, or (2) the cell has other mechanisms by which it can achieve the same effect as hsps. Recently, some evidence has been accumulated to suggest that the function of the major hsps (e.g., hsp70) is to repair heat-induced cellular damage rather than serve a protective role.

2.3.2. Functions of hsps

The substantial body of correlative evidence indicates that there is some association between the development of thermotolerance and the homeostatic role of hsps. In this light, the functions of hsps are only just beginning to be elucidated. In this section, these recent findings are briefly reviewed and placed within the context of how hsps might function during heat shock. In order to discuss how hsps assist a cell during heat shock, however, we must first gain an understanding of what happens to a cell at elevated temperatures.

Heat shock affects almost all components of the cell. Perhaps the most susceptible cellular organelle is the plasma membrane, which often undergoes increases in fluidity (Lepock *et al.*, 1983) or develops lesions, resulting in the leakage of small molecules (reviewed by Nover et al., 1984). These changes in membrane structure can result in losses of critical membrane functions as well. For example, membrane destabilization is presumably responsible for heat-induced loss of plasma membrane Na^+, K^+-ATPase activity of HeLa cells (Burdon and Cutmore, 1982) and decreased uptake of amino acids by tomato cell cultures (Nover et al., 1984). Also, heat shock alters the ultrastructure of membranous organelles in some cell types. In rat embryo fibroblasts, the Golgi complex completely degenerates, and mitochondria swell and become rounded (Welch and Suhan, 1985; the heat-shock effects on mitochondria are discussed below), whereas in barley aleurone cells, heat shock causes the breakdown of the endoplasmic reticulum (Belanger et al., 1986).

Heat shock also alters the ultrastructure of the nucleus. In the fungus *Achlya ambisexualis*, heat shock causes the apparent condensation of the chromatin and the formation of bundles of 4-nm-thick filaments within the nucleus (Pekkala et al., 1984). Similar filaments, identified as being composed of actin, have also been observed in the nuclei of heat-shocked rat embryo fibroblasts (Welch and Suhan, 1985). In golden hamster embryo cells, DNA is also subject to breakage (Watanabe et al., 1984). Nucleoli appear to be even more sensitive to heat shock than the remainder of the nucleus. They disperse at elevated temperatures, followed by the formation of large ribonucleoprotein particles. During recovery from heat shock, nucleolar morphology returns to normal (Nover et al., 1984). Nucleoli also appear to be targeted by newly synthesized hsp70 during heat shock (Lewis and Pelham, 1985), which has led to some speculation on the function of hsp70.

In the cytoplasm of heat-shocked cells, there is evidence that the intermediate filament component of the cytoskeleton collapses around the nucleus (Falkner et al., 1981; Biessmann et al., 1982; Welch and Suhan, 1985). Interestingly, the distribution of microtubules is unaffected (Welch and Suhan, 1985), but there is an increase in the number of actin-containing stress fibers (Thomas et al., 1982). Polysomes are also rapidly dissociated (Lindquist McKenzie et al., 1975) and, in some cells, there is a transient acceleration in the rate of ubiquitin-dependent protein degradation followed by a slower phase of proteolysis (Parag et al., 1987). The mechanism of protein degradation may also play an integral part in the cell's heat-shock response (reviewed by Munro and Pelham, 1985).

Lastly, heat shock alters the pathways of energy metabolism in many cells. The major energy source for most cells is via oxidative phosphorylation, a series of mitochondrial reactions that generate ATP. During heat shock, however, mitochondria apparently become disabled. Investigators have observed heat-induced changes in mitochondrial morphology (Welch and Suhan, 1985) and an irreversible conformational change in intramembranous proteins of mitochondria (Lepock et al., 1983). Mitochondrial dysfunction is also inferred from reports that respiration and oxidative phosphorylation are rapidly inhibited during heat shock (Dickson and Oswald, 1976; Mondovi et al., 1969; Christiansen and Kvamme, 1969). Coupled with mitochondrial dysfunction is a rapid loss of cellular ATP. In *Tetrahymena*, this loss occurs within the first 2 min of heat shock and can be as great as 50% of the pre-existing levels of ATP

(Findly *et al.*, 1983). A similar ATP loss is also seen in *Drosophila* salivary gland cells (Leenders *et al.*, 1974) and in *Xenopus* embryos (see Section 3.6).

In response to ATP losses and the disabling of their mitochondria, heat-shocked cells must rely on alternative pathways of energy metabolism. One such alternative is anaerobic glycolysis, which can synthesize small amounts of ATP if the cell lacks oxygen or if the respiratory pathways in the mitochondria are disrupted. Predictably, several cell types shift their metabolism to anaerobic glycolysis (as determined by the accumulation of lactic acid) during heat shock. These include chicken embryo fibroblasts (Kelley and Schlesinger, 1982), rat heart muscle cells (Hammond *et al.*, 1982), and *Xenopus* embryonic cells (see Section 3.6).

Heat-induced changes in the energy metabolism of cells were believed to be the signals that stimulate hsp synthesis. This hypothesis was strengthened by evidence that uncouplers of oxidative phosphorylation and respiratory poisons could stimulate hsp synthesis (reviewed by Ashburner and Bonner, 1979; Landry, 1986). Several more recent lines of evidence suggest that changes in energy metabolism do not stimulate hsp synthesis. For example, the heat-shock response is not potentiated in mutant cell lines deficient in glycolysis or respiration (Landry *et al.*, 1986). More probably, the induction of hsp synthesis by respiratory inhibitors is caused by some mechanism that is also common to heat shock. One possibility is that crippling of the cellular energy metabolism may result in the accumulation of denatured proteins, the degradation of which is ATP dependent (Herschko, 1983); however, such a hypothesis remains to be tested.

It is evident that heat shock can result in extensive changes in the morphology and physiology of a cell. These changes occur in somatic cells during heat shock (e.g., in *Achlya* cells) (Pekkala *et al.*, 1984), even though they are synthesizing hsps. This suggests that the function of hsps is not to protect the cell from heat-induced damage. Rather, it may be more reasonable to predict that hsps function to repair heat-induced cellular damage. The ability to repair itself would give a cell an obvious advantage in being able to survive a heat shock, which would explain the many correlations between thermotolerance and hsp synthesis.

The reversal of heat-shock-induced damage appears to be the major function of the most conserved hsp, hsp70. The first indication of the function of hsp70 came from the isolation of a constitutive member of the mammalian family of hsp70-related proteins by ATP-agarose affinity chromatography. This protein, termed heat-shock cognate (hsc) protein 70, was identified as the previously described clathrin uncoating ATPase (Ungewickell, 1985; Chappell *et al.*, 1986). The function of hsc70 is to dissociate clathrin triskelia from the clathrin baskets surrounding endocytotic vesicles (clathrin uncoating). The actual mechanism of clathrin uncoating is unknown, but it is absolutely dependent on the hydrolysis of ATP (Patzer *et al.*, 1982; reviewed by Rothman and Schmid, 1986). The identification of this protein demonstrates that a member of the hsp70 family of proteins has a function in the disassembly of protein aggregates.

Other members of the hsp70 family have similar roles. Munro and Pelham

(1986) were able to isolate a cDNA clone encoding another hsp70-related protein by its homology to the *Drosophila* hsp70 gene. Upon characterization of the clone and its protein product, these investigators found that the gene encodes a protein identical to the 78-kDa mammalian glucose regulated protein (grp78), another stress protein that is induced by glucose starvation (Shiu *et al.*, 1977), and also to an immunoglobin (Ig) heavy-chain binding protein (BiP; Munro and Pelham, 1986; reviewed by Pelham, 1986). Grp78 is localized in the endoplasmic reticulum of cells (Munro and Pelham, 1986) and may have a function in the assembly or folding of secretory proteins (Pelham, 1986). Such a function is demonstrated by BiP, which appears to bind to the hydrophobic regions of Ig heavy chains before the attachment of light chains to this same region. This transient binding may prevent the aggregation of the heavy chains (Pelham, 1986). *In vitro*, BiP or grp78 can also be released from heavy chains by the addition of ATP, which suggests that this protein has a similar mechanism for attachment and release from proteins as hsc70.

It is an obvious suggestion that hsp70 has a similar role in heat-shocked cells, especially in light of the evidence that denatured or abnormal proteins stimulate hsp70 expression. The localization of hsp70 to the nucleolus suggests that its function is required for the reassembly of denatured ribosomes. This would explain the reversal of nucleolar morphology during heat shock and recovery and the observation of Welch and Suhan (1986) that hsp70 binds to partially assembled ribosomes. Also, transfection of cells with a plasmid that overproduces hsp70 accelerates the recovery of the nucleoli in these cells (Pelham, 1984). The function of hsp70 is also apparently ATP dependent, since it can be eluted from nucleoli by as little as 1 mM ATP *in vitro*, but not with ADP or ATP analogues (Lewis and Pelham, 1985).

Other than hsp70 and related proteins, little is known about the function of other hsps. The low-molecular-weight *Drosophila* hsps have homology to lens α-crystallin proteins (Ingolia and Craig, 1982), suggesting a role in cellular ultrastructure, especially with respect to intermediate filaments. Also, some hsps of yeast have been identified as glycolytic enzymes; examples include enolase (Iida and Yahara, 1985), glyceraldehyde 3-phosphate dehydrogenase (Lindquist, 1986), and phosphoglycerate kinase (Piper *et al.*, 1986). The synthesis of these enzymes during heat shock correlates with the observations that cells shift their metabolism to anaerobic glycolysis during heat shock. We have also identified some *Xenopus* hsps as glycolytic enzymes; their role in thermotolerance is discussed in more detail in Section 3.6.

3. Developmental Regulation

3.1. Developmental Acquisition of Thermotolerance

Thermotolerance is developmentally acquired. In general, very early embryos are highly susceptible to heat shock. As the embryos develop, however, they become more resistant to thermal stress, until they become thermotolerant. This effect is observed during the development of sea urchins (Roccheri et al., 1981; Giudice, 1985; Heikkila et al., 1985b); Drosophila (Graziosi et al., 1980; Bergh and Arking, 1984); locust (Schistocerca gregaria; Mee and French, 1986); mollusks (Spisula solidissima; Roosenburg et al., 1984), mammals, including pigs (Wildt et al., 1975), rats (Webster et al., 1985; Mirkes, 1985), and humans (Smith et al., 1978; Pleet et al., 1981); and amphibians, such as Xenopus laevis (Heikkila et al., 1985a,b; Nickells and Browder, 1985).

During development, heat shock can result in two distinct effects. The first is cell death, and the second is abnormal development. Usually, cell death is the result of heat shocking very early embryos. Later-stage embryos, however, may exhibit periods of thermolability in which heat shock does not kill the cells, but instead disrupts their normal program of differentiation. An example of this phenomenon is phenocopy induction during Drosophila development. In this instance, heat shock can induce nonheritable, abnormal phenotypes that often resemble mutant phenotypes (reviewed by Mitchell and Petersen, 1982a; Lindquist, 1986). It has been suggested that phenocopy induction is the result of altering the normal program of gene expression required for the differentiation of a cell into a specific phenotype. This is demonstrated by the phenocopy multihair (MtH), which alters bristle hair morphology so that a hair cell produces multiple branched hairs rather than a single hair (Mitchell and Petersen, 1982b; Petersen and Mitchell, 1982). Hair morphogenesis in Drosophila progresses in an anteroposterior direction. Consequently, hair morphogenesis of the wing occurs before the development of hairs on the dorsal thorax. This temporal pattern of morphogenesis reflects the susceptibility of the hair cells to the induction of the multihair phenocopy. For example, a 35-min heat shock at 40.5°C induces multihair formation on the wing if pupae are stressed at approximately 38 hr of development. The same heat-shock conditions can stimulate multihair formation in the hair cells of the dorsal thorax, but only when applied at approximately 42 hr of development (Mitchell and Petersen, 1982a). During this latter period, wing hair morphogenesis is unaffected. These periods of susceptibility to phenocopy induction correspond to dramatic changes in the pattern of RNA and protein synthesis by the hair cells in each respective region of the fly (Mitchell and Petersen, 1981). It is possible that heat shock may induce phenocopy formation by altering or delaying the production of new proteins, which in turn affects the normal morphogenesis of the cell (Petersen and Mitchell, 1982).

Phenocopy induction in Drosophila is not the only example of heat-induced developmental abnormalities. Similar examples exist in Xenopus embryos, in which heat shock can cause abnormal gastrulation of very early embryos (Nickells and Browder, 1985) (see Section 3.6) or partially interrupt the morphogenesis of paraxial mesoderm into somites of later stage embryos (Elsdale et al., 1976). Mammalian embryos are also susceptible to heat shock, which can cause congenital defects of the CNS in a variety of species (Smith et al., 1978; Pleet et al., 1981; Edwards, 1981; Webster and Edwards, 1984; Mirkes, 1985; Webster et al., 1985; Finnell et al., 1986). The effects of heat or other environmental stresses on mammalian development have significance to livestock management as well as to human health in situations involving maternal fever during pregnancy or exposure to other stresses that induce the heat-shock response (German, 1984).

3.2. Developmental Acquisition of the Heat-Shock Response

Since thermotolerance is developmentally acquired, it is reasonable to expect that the heat-shock response is subject to similar regulation. A wide spectrum of organisms are unable to synthesize hsps in response to stress during initial postfertilization development. These include *Drosophila* (Graziosi *et al.*, 1980; Dura, 1981), sea urchins (Roccheri *et al.*, 1981; Howlett *et al.*, 1983; Giudice, 1985), rabbit (Heikkila and Schultz, 1984), and mouse (Wittig *et al.*, 1983; Morange *et al.*, 1984; Muller *et al.*, 1985; Heikkila *et al.*, 1985b; and Hahnel *et al.*, 1986).

In addition to the delayed onset of hsp synthesis during development, complex chronological and spatial patterns of hsp induction occur throughout development. Although not exclusive to *Xenopus*, the complexity of developmental regulation is well illustrated by studies conducted with this species. We shall discuss the *Xenopus* work in detail, beginning with the oocyte and the conversion of oocyte to zygote through meiotic maturation, spawning, and fertilization. We then discuss the patterns of hsp synthesis through early development. Finally, we shall discuss the regulation of hsp induction in cells of the *Xenopus* erythroid cell lineage during terminal cell differentiation.

3.3. Heat-Shock-Induced Depression of Messengers in *Xenopus* Eggs

3.3.1. Stage Specificity

It has been established that hsps are synthesized in *Xenopus* oocytes after exposure to a brief sublethal temperature increase (Bienz and Gurdon, 1982; Bienz, 1984), but that fertilized eggs do not show a heat shock response (Heikkila *et al.*, 1985a; Browder *et al.*, 1987). Intermediate stages of gamete development show stage-to-stage variation in the spectrum of hsps that are synthesized (Browder *et al.*, 1987). We routinely maintain oocytes and coelomic eggs in modified Barth's solution (MBS), buffered with Hepes (Gurdon, 1976) and supplemented with approximately 1 mM each of pyruvate and oxaloacetate (Eppig and Steckman, 1976). Oocytes maintained *in vitro* in supplemented MBS synthesize four major heat-shock proteins in response to elevated temperature; named for their relative molecular masses, these are hsps83, 76, 70, and 62 (Fig. 3a). After maturation and ovulation, coelomic eggs recovered from the body cavity of hormone-stimulated females show heat shock-induced synthesis of hsps22 and 16 in addition to the four oocyte hsps (Fig. 3b). Since the possible presence of contaminating follicle cells on oocytes is a particular source of concern in research involving isolated *Xenopus* oocytes (King and

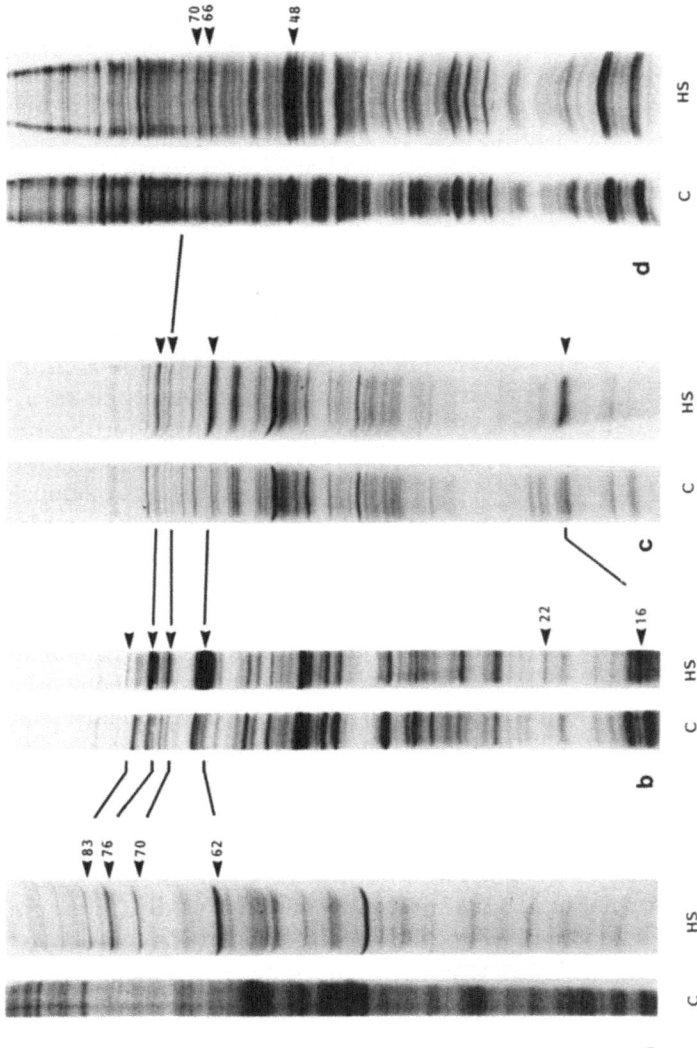

Figure 3. Fluorographs of proteins synthesized by control (C) and heat-shocked (HS) *Xenopus* gametes. Oocytes or eggs were labeled with [35S]methionine for 90 min at ambient temperature (C) or at 35–36°C (HS). Proteins were separated on 10% sodium dodecyl sulfate (SDS)–polyacrylamide gels (a,b,d) or on 7–17% gradient gels (c), with equal acid-precipitable radioactivity loaded onto corresponding control and heat-shock lanes. hsps are indicated by arrowheads with relative molecular masses shown in kDa; corresponding hsps are connected between lanes. (a) Oocytes incubated in supplemented modified Barth's solution (MBS) with pyruvate and oxaloacetate. (b) Coelomic eggs incubated in supplemented MBS. (c) Dejellied spawned eggs incubated in De Boer's solution (DB). (d) Activated eggs injected with [35S]methionine and incubated in DB. (Modified from Browder et al., 1987.)

Davis, 1987), it is significant that ovulated coelomic eggs show a heat shock response but are free from follicle cell contamination, as judged by scanning electron microscopy (SEM). Although this does not strictly prove that the hsps synthesized in enzymatically defolliculated oocytes are of oocyte origin, it does indicate that these proteins can be synthesized in the absence of follicle cells and that hsp synthesis is not dependent on the enzymatic defolliculation used to remove oocytes from the ovarian tissue. It is possible that the presence of hsps22 and 16 in heat-shocked coelomic eggs (but not in heat-shocked isolated oocytes) is related to the hormonal stimulus that induces maturation and ovulation. (We have ruled out the possibility that the enzymatic treatment of oocytes in some way prevents the synthesis of these two hsps by demonstrating that exposure to collagenase does not affect the heat-shock response of coelomic eggs.)

After maturation and release into the coelomic cavity, eggs are transported down the oviduct, where they are invested with their jelly coat, and are eventually spawned. (In the laboratory, spawned eggs are routinely dejellied to make experimental manipulation possible.) Since the primary functional property of spawned eggs that is of interest to researchers has been their fertilizability, and since spawned eggs in De Boer's (DB) solution remain fertilizable for some time (Katagiri, 1961), we have routinely used DB as an incubation medium for dejellied spawned eggs. In this medium, the heat-shock response in spawned eggs varies slightly from that seen in coelomic eggs in supplemented MBS; hsps76, 70, 62 and 22 are synthesized in response to heat shock, but hsps83 and 16 are not consistently seen (Fig. 3c).

Dejellied spawned eggs do not take up radiolabeled amino acids as effectively as do coelomic eggs; the extent to which this difference is attributable to the dejelling procedure is not clear. Microinjection of [^{35}S]methionine results in more efficient labeling of newly synthesized proteins, but also causes artificial activation of the egg. Such treatment produces striking changes in the heat shock response. Activated eggs exposed to elevated temperatures synthesize two unique protein species, hsps 66 and 48. Hsps 76, 62, and 16 are no longer inducible, whereas hsp70 is markedly reduced or absent (Fig. 3d).

In contrast to artificially activated eggs, fertilized eggs apparently lack a heat-shock response (Browder et al., 1987). The reasons for this difference are unclear. The ionic environments have a profound effect on the heat-shock response (see Section 3.3.2). Different ionic environments in fertilized and activated eggs could produce apparent differences in the capacity for heat-shock protein synthesis.

The changing patterns of hsp synthesis we have described are particularly interesting, in light of the mode of regulation of hsp synthesis. Bienz and Gurdon (1982) and Bienz (1984) have demonstrated that the induced synthesis of hsp70 in Xenopus oocytes uses pre-existing hsp70 messenger RNA (mRNA); in contrast to the situation in somatic cells, transcription of the hsp70 gene is not induced in response to heat shock. The pre-existing hsp70 messenger is one of the so-called oogenic messengers, which are synthesized in large quantities during oogenesis and stored for use during oocyte maturation, in unfertilized

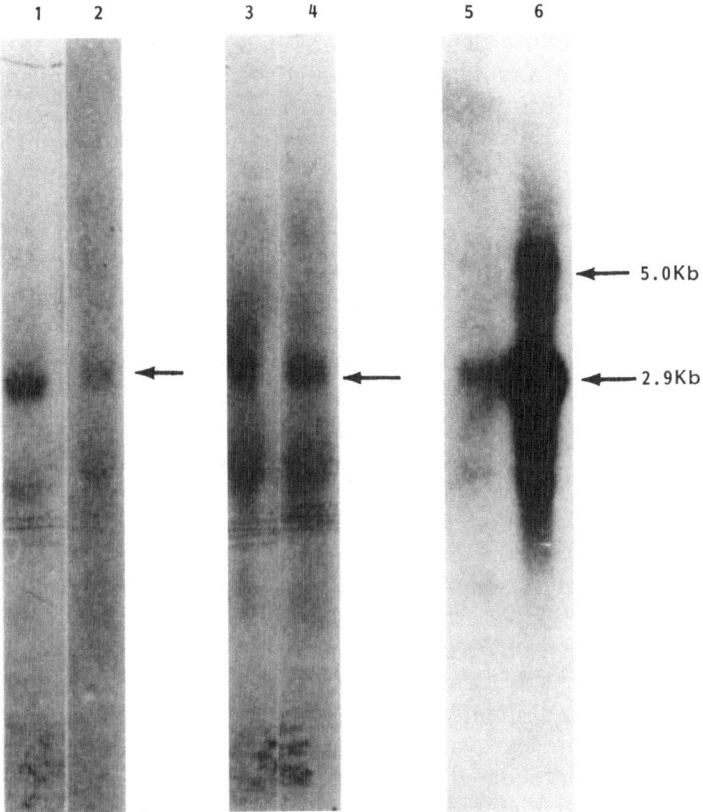

Figure 4. Detection of hsp70 mRNA in control and heat-shocked *Xenopus* eggs and embryos. Polyadenylated RNA was isolated from 500 control and heat-shocked eggs and 2-cell embryos and 200 control and heat-shocked neurulae, resolved by electrophoresis, transferred to nitrocellulose, and hybridized to the labeled subclone of hsp70. (For details, see Browder *et al.*, 1987.) Lane 1: unfertilized eggs, 22°C for 70 min; lane 2: unfertilized eggs, 35°C for 70 min; lane 3: 2-cell embryos, 22°C for 70 min; lane 4: 2-cell embryos, 35°C for 70 min; lane 5: neurulae, 22°C for 70 min; lane 6: neurulae, 35°C for 70 min. (From Browder *et al.*, 1987.)

eggs, and during initial embryonic development. Oogenic messengers are the sole templates for proteins synthesized during this time, since by the time of maturation the oocyte has ceased nuclear transcription (Davidson, 1986) and will remain essentially inactive in RNA synthesis until the mid-blastula transition (Kimelman *et al.*, 1987). As shown in the Northern blot of Fig. 4, hsp70 mRNA is present in unfertilized eggs and 2-cell embryos and shows no increase in steady-state level after heat shock. Since transcription does not occur during this interval, qualitative changes in protein synthesis result either from recruitment of specific messengers onto polysomes, from release of a block to translation of messengers already on polysomes, or from changes in the translational machinery leading to selective translation of specific messengers. The composition of the external medium, including variation in ion concentration and supplementation with the organic acids pyruvate and oxaloacetate, has dramatic effects on the pattern of induced protein synthesis. Since the patterns of protein synthesis in *Xenopus* change qualitatively during gametogenesis and early development (during which interval dramatic ionic fluxes are under way), these observations may lead to a fuller understanding of the modes of regulation over selective use of oogenic messengers in protein synthesis.

3.3.2. Modulation of Heat-Shock Protein Synthesis by Components of the External Medium

When spawned eggs are heat shocked in unsupplemented MBS, hsp synthesis is sharply limited or absent. If the medium is supplemented with pyruvate and oxaloacetate at approximately 1 mM each, however, labeling is much more robust, and a broad spectrum of heat-shock proteins is synthesized as described above (Fig. 5). Because of evidence identifying some of the heat-shock proteins in *Xenopus* embryos as glycolytic enzymes (see Section 4), this effect of glycolytic intermediates is particularly intriguing, although it must be stressed that we have not identified the mechanism for this striking change in protein synthetic patterns, nor have we established that these compounds enter the egg. An interesting observation was made in comparing the results of eggs incubated in MBS and DB. In unsupplemented DB, spawned eggs are capable of synthesizing a broad spectrum of hsps in response to heat treatment; this spectrum is similar, but not identical, to that seen in supplemented MBS. When DB is supplemented with pyruvate and oxaloacetate, however, the heat shock response is sharply attenuated or abolished altogether (Fig. 5). These results indicate that neither heat shock alone, nor supplementation alone, nor heat shock in the presence of supplementation is necessarily sufficient to permit the full expression of the heat-shock response in *Xenopus* gametes. Clearly, the ionic milieu has a pronounced effect on the translation of specific messengers in response to environmental influence.

Although the difference in ionic strength between DB and MBS is relatively minor, the concentrations of specific ions vary considerably (Table I). Most noteworthy is the higher NaCl concentration in DB (110 mM versus 88 mM for MBS). To determine whether the difference in NaCl concentration is responsible for the effects of the medium on hsp synthesis, spawned eggs were incubated in MBS with NaCl added to produce a concentration similar to that of DB. Under such conditions, a strong hsp response was seen in unsupplemented high-salt MBS; this response was similar, but not identical, to that seen under heat stress in unsupplemented DB (Fig. 6a). Specifically, the distribution of members of the hsp62 complex differ between these two media. In DB, the lowermost member of this complex is strongly labeled, but in high-salt MBS, two heat-shock proteins in this molecular-weight range are synthesized, with apparent molecular weights of approximately 62 and 64 kDa. We have not as yet determined the relationship between these two proteins. These results indicate that the salt difference is largely responsible for the difference in heat shock response between MBS and DB. Experiments involving MBS with addi-

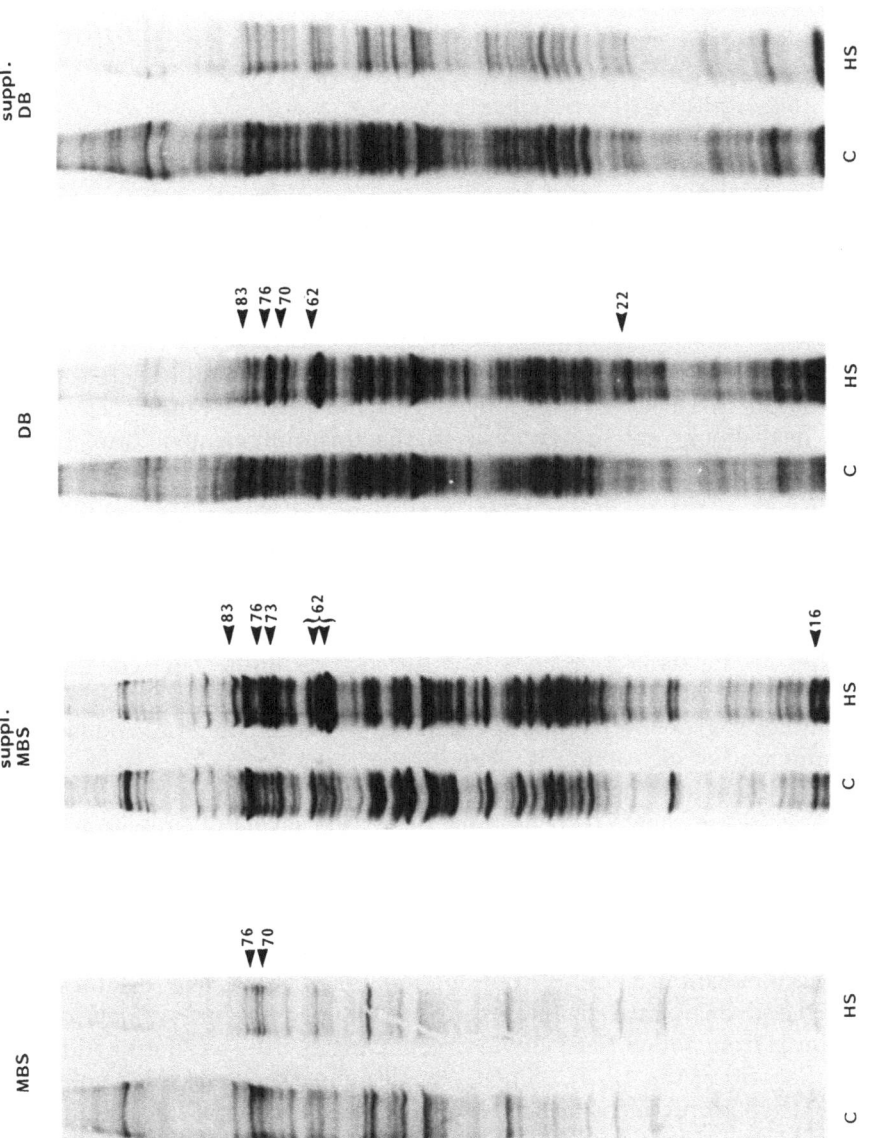

Figure 5. Fluorographs of proteins synthesized by control (C) and heat-shocked (HS) dejellied, spawned *Xenopus* eggs. Cells were maintained in the presence of [³⁵S]methionine in the media indicated for 90 min at ambient temperature (C) or at 35°C (HS). Proteins were separated on 10% sodium dodecyl sulfate (SDS)–polyacrylamide gels. Equal acid-precipitable radioactivity was loaded onto corresponding control and heat-shock lanes. hsps are indicated by arrowheads with relative molecular masses shown in kDa. (Modified from Browder et al., 1987.)

Table I. Composition of Modified Barth's
Solution and De Boer's Solution

	MBS[a] (mM)	De Boer's[b] (mM)
NaCl	88	110
KCl	1	1.3
$CaCl_2$	0.41	0.44
$Ca(NO_3)_2$	0.33	—
$MgSO_4$	0.82	—
$NaHCO_3$	2.4	< 1
HEPES	10	—
pH	7.6	7.2

[a]Modified Barth's solution. From Gurdon (1976).
[b]From Katagiri (1961).

tional choline chloride (Fig. 6b) have indicated that it is likely the increased Cl^- concentration in DB that is responsible for the permissiveness of this medium for heat-shock protein synthesis in the unsupplemented form, although elevated levels of extracellular Na^+ may have a modulating effect on this response.

Ionic effects appear to be involved in modulating the heat-shock response in other systems as well. Drummond et al. (1986) reported that, although cytoplasmic pH decreases and intracellular Ca^{2+} increases during heat shock, neither increased Ca^{2+} nor acidification of the cytoplasm will in itself stimulate the synthesis of heat-shock proteins in cultured *Drosophila* cells. The data reported by these investigators, however, indicate that, although pH does not *cause* hsp synthesis, it does affect the *spectrum* of hsps that are produced (see Fig. 6 in Drummond et al., 1986).

Increased extracellular concentrations of various Cl^- compounds (NaCl, $MgCl$, KCl, NH_4Cl) are reported to produce decreased levels of protein synthesis (Robbins et al., 1970; Saborio et al., 1974); such treatment apparently acts through the inhibition of translational initiation (Saborio et al., 1974). [Similar results with high levels of extracellular sucrose (Wengler and Wengler, 1972) have suggested to some investigators that these effects are due to hypertonicity of the incubation medium. It should be noted, however, that these findings are based on increased tonicity using 300 mM sucrose; effects on protein synthesis from added NaCl are observed using as little as 50 mM additional NaCl (Saborio et al., 1974). Since Wengler and Wengler (1972) noted that extremely hypertonic medium leads to shrinkage of cells, it may be that water loss from shrunken cells results in changes in intracellular ionic concentrations.] Previous studies of altered protein synthesis induced by ionic conditions have involved examination of cells incubated at normal temperatures. In heat-shocked *Xenopus* gametes, hsp synthesis can be altered by 12 mM added NaCl, although this degree of increased ionic concentration does not appear to affect protein synthetic patterns at ambient temperature. Sub-

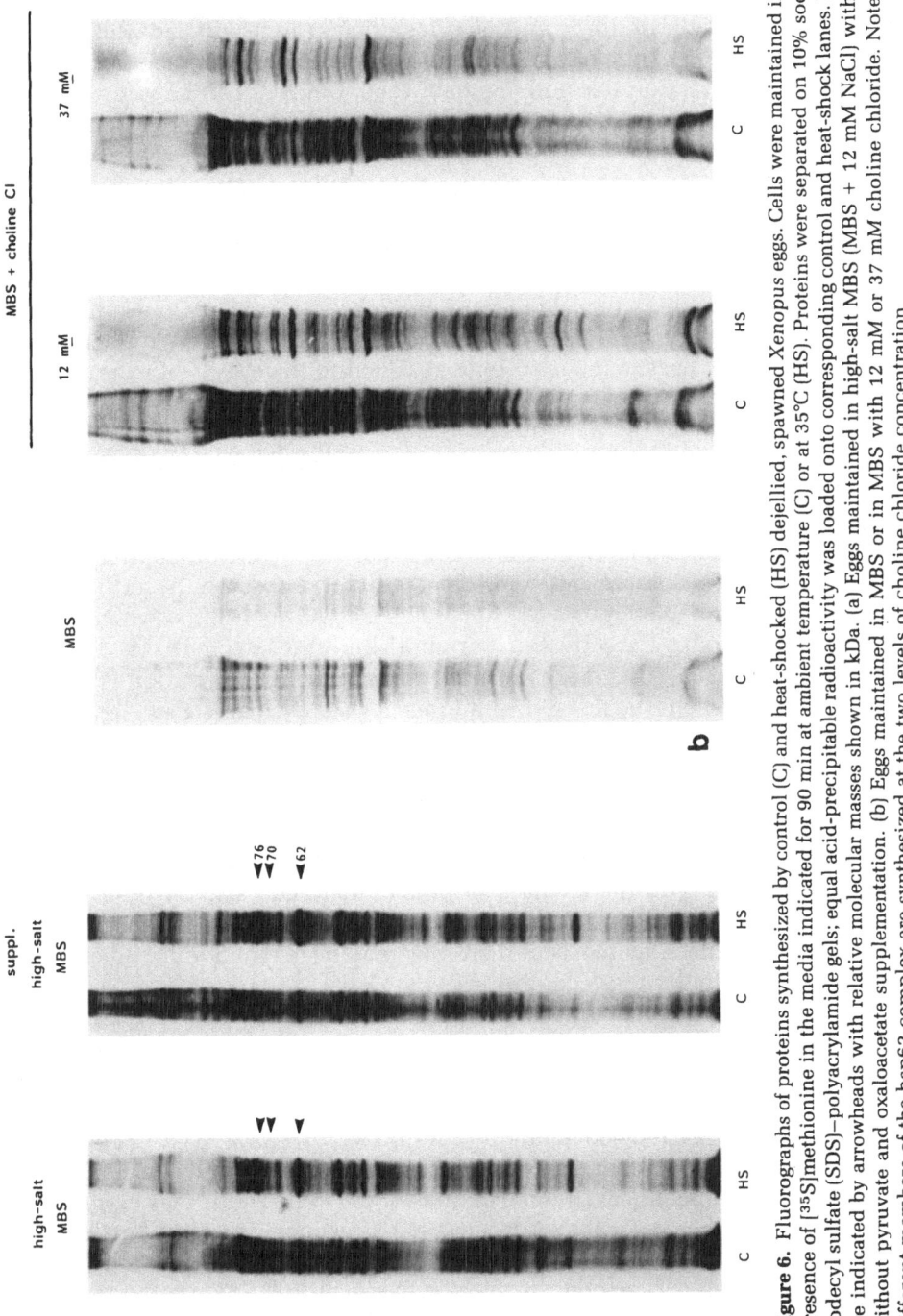

Figure 6. Fluorographs of proteins synthesized by control (C) and heat-shocked (HS) dejellied, spawned Xenopus eggs. Cells were maintained in the presence of [35S]methionine in the media indicated for 90 min at ambient temperature (C) or at 35°C (HS). Proteins were separated on 10% sodium dodecyl sulfate (SDS)—polyacrylamide gels; equal acid-precipitable radioactivity was loaded onto corresponding control and heat-shock lanes. Hsps are indicated by arrowheads with relative molecular masses shown in kDa. (a) Eggs maintained in high-salt MBS (MBS + 12 mM NaCl) with and without pyruvate and oxaloacetate supplementation. (b) Eggs maintained in MBS or in MBS with 12 mM or 37 mM choline chloride. Note that different members of the hsp62 complex are synthesized at the two levels of choline chloride concentration.

threshold stresses can act additively or synergistically to induce hsp synthesis (Rodenhiser et al., 1986); thus, it may be that decreased translational initiation resulting from low-level ionic effects is insufficient to produce detectable alterations in protein synthetic patterns but that additional inhibition resulting from heat stress (McCormick and Penman, 1969; Goldstein and Penman, 1973; Duncan and Hershey, 1984) allows the heat-induced alterations in translation that we observe.

We are also investigating the effects of extracellular pH on the heat-shock response in Xenopus eggs, because of our finding that the nominal pH of DB is maintained only briefly after preparation of the medium. The buffering capacity of this medium is provided solely by bicarbonate (Table I), and we have determined that the pH value of the solution drops by nearly a full pH unit over the time course of a typical heat-shock experiment. Although the intracellular pH in unfertilized eggs is reported to be resistant to the effects of external pH over a wide range (Nuccitelli et al., 1981), we are examining the effects of high- and low-pH media on the Xenopus heat-shock response.

The apparent effect of Cl^- on the synthesis of hsps in Xenopus gametes is of particular interest because of the importance of Cl^- fluxes in hormone-induced maturation and fertilization of amphibian species. During maturation in oocytes of Ambystoma mexicanum and Pleurodeles waltlii, an increase in intracellular Cl^- is observed after germinal vesicle breakdown. At fertilization and/or activation of Xenopus eggs, a massive Cl^- efflux occurs, which is responsible for carrying the activation current (Kline and Nuccitelli, 1985; Webb and Nuccitelli, 1985); this mechanism for membrane depolarization is apparently common among freshwater amphibians (e.g., Maéno, 1959; Ito, 1972; Cross and Elinson, 1980). The extent to which these ion fluxes influence translation is unknown, although it should be noted that increased protein synthesis is associated both with maturation and fertilization in Xenopus (Woodland, 1974). Changes in intracellular pH also occur at maturation (Lee and Steinhardt, 1981) and activation (Webb and Nuccitelli, 1981) and are temporally correlated with the changes in protein synthetic activity observed at these times. It may be that alterations in intracellular pH are not in themselves a modulator for protein synthesis; rather, they may result from ionic fluxes that alter translation under these conditions.

An additional effect of the ionic milieu on hsp synthesis is observed in the heat-shock response of artificially activated eggs. Eggs undergo activation when injected with radiolabeled amino acids; if such eggs are subsequently heat-shocked in De Boer's solution, they demonstrate a loss of the egg-type heat-shock response, with the concomitant introduction of a new spectrum of hsps: hsp66 and hsp48. hsp48 is apparently specific for activation and is synthesized in activated eggs under heat stress regardless of the ionic conditions employed; hsp66, however, has been shown to change its responsiveness to heat stress, depending on the extracellular concentration of Ca^{2+} ions (Fig. 7). hsp66 is synthesized specifically upon heat stress in media with $[Ca^{2+}]$ greater than 0.4 mM, whereas this protein is synthesized constitutively in media with $[Ca^{2+}]$ less than 0.3 mM.

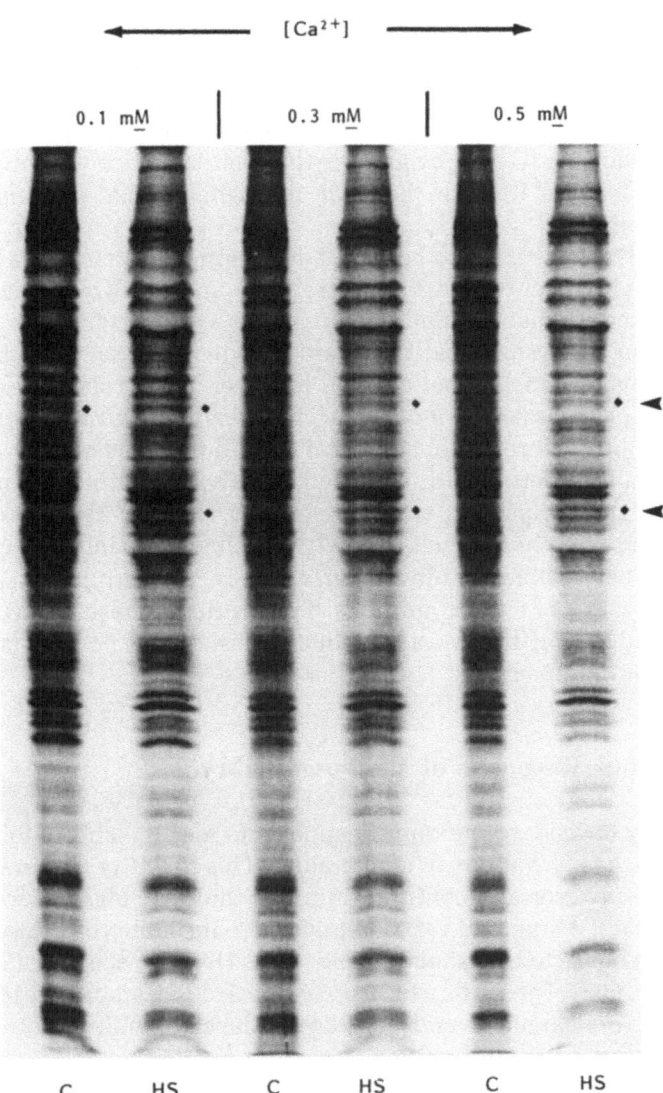

Figure 7. Fluorographs of proteins synthesized by control (C) and heat-shocked (HS) activated *Xenopus* eggs. Eggs were simultaneously activated and microinjected with [^{35}S]methionine, and incubated for 90 min at ambient temperature or at 35°C in F_1 medium (Hollinger and Corton, 1980) containing the indicated concentration of Ca^{2+}. Proteins were separated on 10% sodium dodecyl sulfate (SDS)–polyacrylamide gels; equal acid-precipitable radioactivity was loaded onto corresponding control and heat-shock lanes. hsp66 and hsp48 are indicated by diamonds; note that at [Ca^{2+}] <0.3 mM hsp66 is synthesized constitutively.

Modification of intracellular $[Ca^{2+}]$ (i.e., $[Ca^{2+}]_i$) will also result in changes in intracellular pH, since these two ionic factors are inextricably linked (Busa and Nuccitelli, 1984); the possible significance of pH_i changes during development has been noted above. Changes in $[Ca^{2+}]_i$ figure prominently in activation at fertilization in *Xenopus* (Busa and Nuccitelli, 1985) and are correlated with a variety of other events occurring at this time, including the Cl^- ion flux responsible for the activation potential (Kline and Nuccitelli, 1985). Thus, changing $[Ca^{2+}]_i$ by altering the external concentration of that ion may affect translation through the same mechanisms that modulate protein synthesis at activation.

Although there is reason to believe that the ionic effects we see are related to events normally occurring during maturation and fertilization in *Xenopus*, we cannot be certain that the modification of hsp synthesis by manipulation of the ionic *milieu* is physiologically relevant. For the purposes of understanding the mechanisms of translational control, however, it is probably more important that these changes can be *made* to occur than that they are observed under normal conditions. (Normal physiological conditions in laboratory-reared animals are difficult to define with precision.) Analysis of proteins translated *in vitro* from isolated polysomes would determine whether hsp messengers are newly recruited upon heat shock. If newly recruited messengers can be identified, it should be possible to follow the use of these transcripts in heat-shocked eggs under a variety of ionic conditions. The results of these experiments will help us to understand the ways in which translational control is exercised during early development.

3.4. Heat-Shock Response of *Xenopus* Embryos

The newly formed zygote rapidly appears to lose the ability to synthesize hsps in response to elevations in temperature. This inability is characteristic of cleavage-stage embryos and continues through the early blastula stages (Heikkila *et al.*, 1985a). At the mid-blastula transition, the embryos begin to synthesize hsps in response to heat shock (Bienz, 1984; Heikkila *et al.*, 1985a; Nickells and Browder, 1985). The pattern of hsps synthesized by embryos can vary with the conditions of heat shock and the stage of development.

Heat-shocked mid-cell blastulae commonly synthesize two hsps of 87 and 70 kDa (Fig. 8). hsp70 separates into two or more species on some gels as shown here. There is no apparent regional difference in the synthesis of these hsps, since the pattern of synthesis is identical in both the animal and vegetal hemispheres. The pattern of hsp synthesis becomes more complex at the fine-cell blastula stage or early gastrula stage of development (Fig. 9). These embryos also synthesize hsps 87 and 70, as well as hsps of 62, 48, and 35 kDa. [hsps 62 and 48 were formerly measured as 57 and 43 kDa, respectively; (Nickells and Browder, 1985).] In addition, these embryos often strongly synthesize a 76 kDa hsp. These additional hsps are synthesized predominantly in the vegetal pole cells, with the synthesis of hsp35 apparently restricted to these cells. This

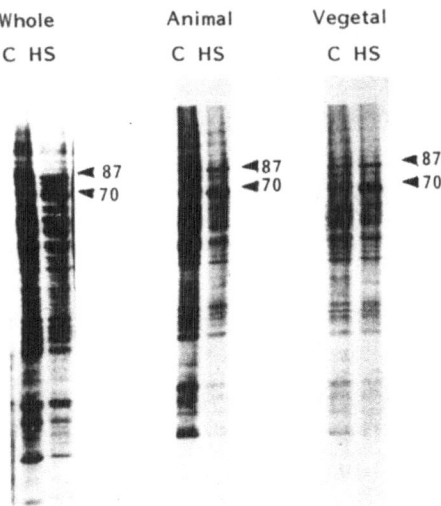

Figure 8. Fluorographs of the proteins synthesized by control (C) and heat-shocked (HS) mid-cell blastulae. For these experiments, mid-cell blastula embryos were cut into animal and vegetal halves using a pair of iridectomy scissors. The embryos and embryo halves were heat-shocked for 20 min at 35.5°C before the addition of [^{35}S]methionine, followed by a further 70-min heat shock with label. Equal acid-precipitable counts were loaded for each lane. Control and heat-shocked samples are indicated. The molecular weights of the hsps are given in kDa. (From Nickells and Browder, 1985.)

regional specification of hsp synthesis has relevance to the ability of all the cell types of the early gastrula to survive heat shock. This phenomenon is discussed at length in Section 3.6. In all embryonic stages after early gastrulation, heat shock induces the synthesis of hsps 87, 76, 70, and 62. This is shown in Fig. 10, which shows a developmental series of control and heat-shocked *Xenopus* embryos.

The synthesis of hsp35 is quite variable. It synthesis is dependent on the temperature and duration of heat shock as well as the developmental stage of the embryo. This stage specificity is shown in Fig. 10. Fine-cell blastulae

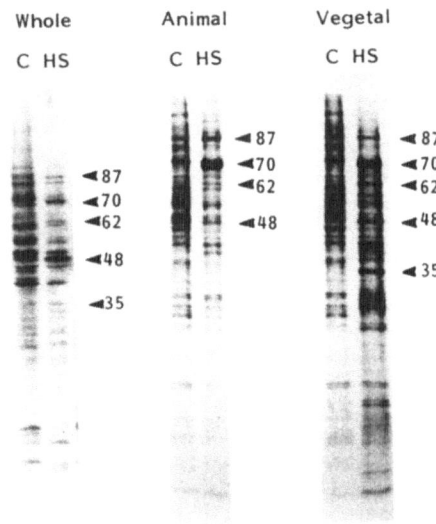

Figure 9. Fluorographs of the proteins synthesized by control (C) and heat-shocked (HS) early gastrulae. The acquisition of thermotolerance by the vegetal hemisphere cells of the early gastrulae correlates with the synthesis of additional hsps. hsp35 synthesis appears to be restricted to the cells of the vegetal hemisphere. (From Nickells and Browder, 1985.)

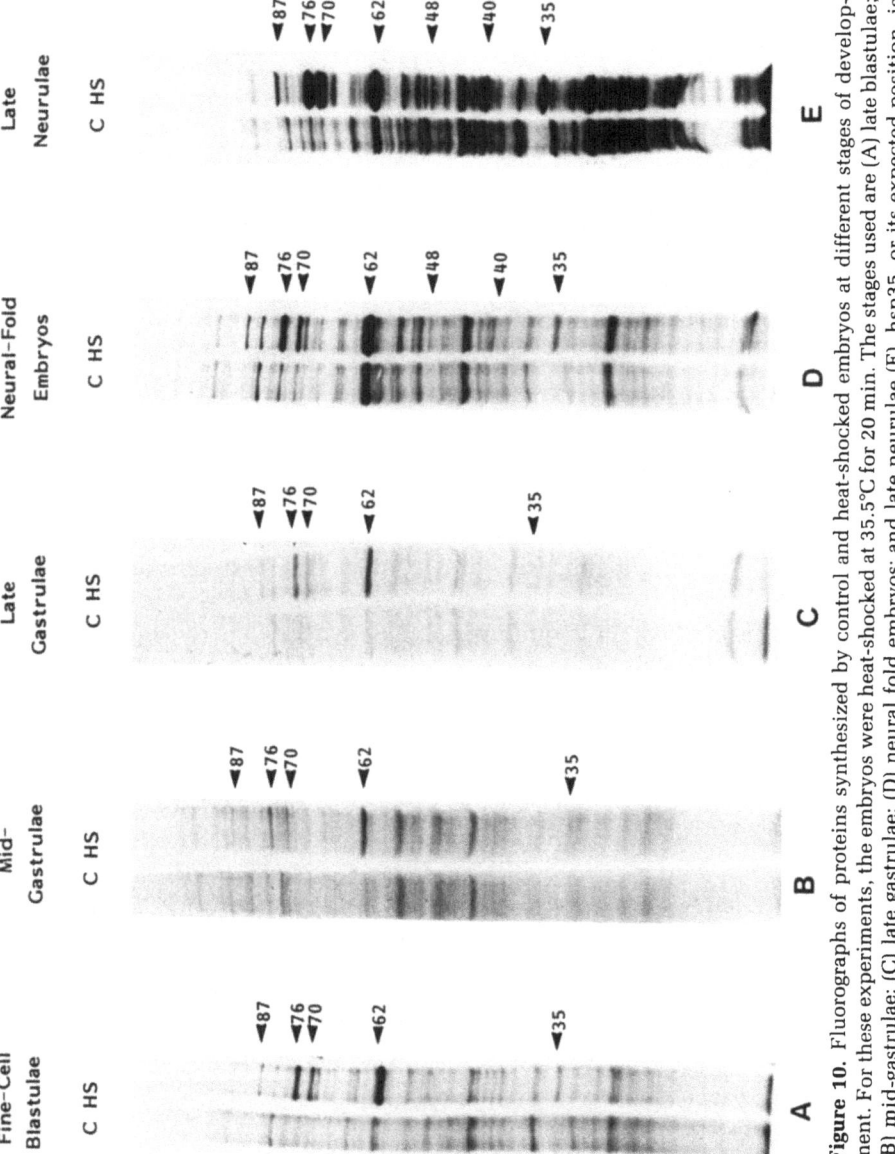

Figure 10. Fluorographs of proteins synthesized by control and heat-shocked embryos at different stages of development. For these experiments, the embryos were heat-shocked at 35.5°C for 20 min. The stages used are (A) late blastulae; (B) mid-gastrulae; (C) late gastrulae; (D) neural fold embryos; and late neurulae (E). hsp35, or its expected position, is marked with an arrow for each heat-shock lane. hsp35 is easily detected in blastulae and neurulae, but not in gastrulae. (From Nickells, 1987.)

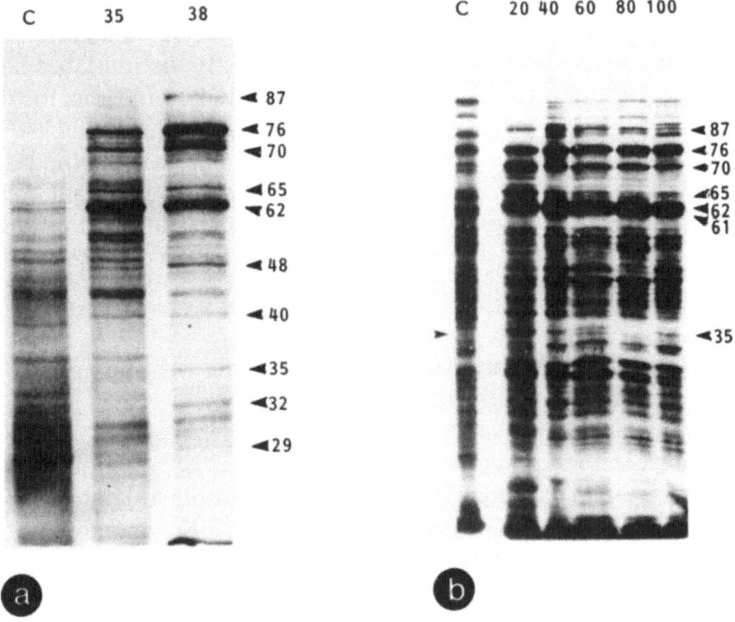

Figure 11. Fluorographs showing the characteristics of hsp35 synthesis in neurulae. (a) Temperature series of neurulae heat-shocked for 20 min at 35°C or 38°C. The embryos synthesize most of the hsps at the two temperatures, except for hsp35 and hsp32, which are only detected at the higher temperature. (b) Time-course experiment of neurulae heat-shocked at 37°C. Neurulae were heat-shocked for the period shown above each lane (in minutes) and exposed to label for the last 20 min of each period. hsp35 synthesis is strong during the first 40 min of heat shock, after which it declines. In the later stages of the experiment, its synthesis is restimulated. (From Nickells, 1987.)

strongly synthesize hsp35 during heat shock, but this synthesis is greatly reduced or shut off in heat-shocked mid- to late gastrulae. Once the embryos neurulate, hsp35 synthesis is again strongly inducible; however, hsp35 is not always synthesized by neurulae either (see Section 3.6). However, in all groups of embryos examined, hsp35 synthesis was greatly reduced in gastrulae, even if sibling embryos were able to synthesize it at the late blastula stage or during neurulation.

Figure 11 shows the temperature and temporal characteristics of hsp35 synthesis, respectively. Figure 11a shows the [^{35}S]methionine labeled proteins synthesized by neurulae heat-shocked at different temperatures. At the lower temperature (35°C), virtually all the hsps are synthesized except hsp35, which is only detected at the higher temperature (38°C). At these higher temperatures, hsp35 is rapidly synthesized during heat shock. This is demonstrated by time-course labeling experiments such as the one shown in Fig. 11b. Neurulae were heat-shocked for up to 100 min at 37°C. Some neurulae were labeled with [^{35}S]methionine for 20-min intervals during this heat shock and then processed immediately for sodium dodecyl sulfate (SDS) gel electrophoresis after the labeling period. hsp35 is detected within the first 20 min of heat shock, along with all the other major hsps. During the course of the experiment, however, hsp35 synthesis appears to decline so that it is not detected during the interval between 60 and 80 min. Its synthesis is subsequently restimulated between 80 and 100 min. Although we have defined the optimal conditions for the induction of hsp35 synthesis, it is not always synthesized by heat-shocked embryos. These variable characteristics of hsp35 synthesis can be explained by the fact that hsp35 is the glycolytic enzyme glyceraldehyde 3-phosphate dehydrogenase (GAPDH). As a consequence, the physiological state of the embryo just before heat shock dictates the requirement for hsp35 synthesis. This phenomenon is discussed in more detail in Section 3.6.

3.5. Developmental Acquisition of Competence of hsp Gene Transcription in *Xenopus*

Heat-shock-induced synthesis of most hsps is not detectable during *Xenopus laevis* embryogenesis until after the mid-blastula stage of development. The following evidence suggests that this phenomenon is controlled primarily at the transcriptional level:

1. Elevated synthesis of hsp70 was evident in the *in vitro* translation products derived from mRNA of heat-shocked neurulae but not of mRNA from heat-shocked cleavage-stage embryos (Heikkila *et al.*, 1985a).
2. Using S_1 nuclease (Bienz, 1984) and RNase protection analysis (Krone and Heikkila, 1989), it has been shown that hsp 70 mRNA was not heat-inducible in *Xenopus* embryos until the late blastula/early gastrula stage.
3. RNA blot assays using heterologous and homologous recombinant DNA probes have shown that neither heat shock, sodium arsenite, nor ethanol exposure induces the accumulation of a 2.7-kb hsp70 mRNA in *Xenopus* embryos until after the mid-blastula transition (Fig. 12) (Heikkila *et al.*, 1985a, 1987a).
4. Microinjection experiments have demonstrated that a *Xenopus* hsp70 promoter/ chloramphenicol acetyl transferase fusion gene is not heat inducible until after the mid-blastula transition (Krone and Heikkila, 1989).
5. Although ubiquitin and a putative hsp87 mRNA are detectable throughout early development, heat shock-induced accumulation of these mRNAs does not occur until the blastula/gastrula stage (Fig. 13) (Heikkila *et al.*, 1987a; Ovsenek and Heikkila, 1988). Similarly, the heat-shock-induced stage-dependent synthesis of hsps during the early development of sea urchin (Howlett *et al.*, 1983; Heikkila *et al.*, 1985b), *Drosophila* (Zimmerman *et al.*, 1983), mouse (Heikkila *et al.*, 1985b; Hahnel *et al.*, 1986), and rabbit (Heikkila *et al.*, 1985b) is controlled at the level of transcription. Thus, it appears that transcriptional regulation of the developmental acquisition of the heat-shock response may be widespread.

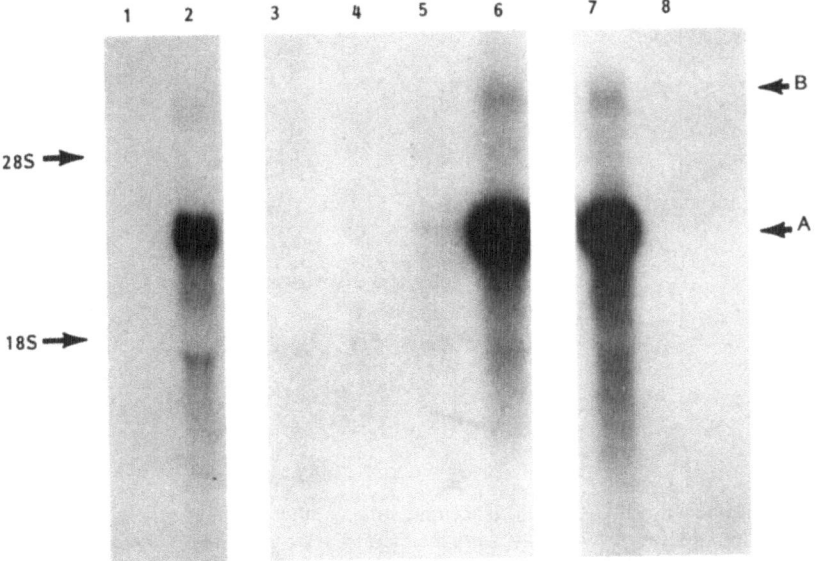

Figure 12. Heat-induced hsp70 mRNA accumulation in *Xenopus* A6 cells and embryos. *Xenopus laevis* A6 cells and embryos were maintained at either 22°C or 33°C for 1 hr. Polyadenylated RNA (2 µg) was resolved by electrophoresis on formaldehyde agarose gels, transferred to nitrocellulose and then hybridized to the ^{32}P nick-translated subclone of the *Xenopus* hsp70 gene. Transcript sizes: A = 2.7 kb; B = 5.0 kb. Lane 1: A6 cells at 22°C; lane 2: A6 cells at 33°C; lane 3: cleavage embryos at 22°C; lane 4: cleavage embryos at 33°C; lane 5: neurula embryos at 22°C; lane 6: neurula embryos at 33°C; lane 7: neurula embryos at 33°C; lane 8: RNA from neurula embryos at 33°C shown in lane 7 treated with 10 µg/ml RNase A. (From Heikkila *et al.*, 1987a.)

Heat-shock gene activation in *Drosophila* is mediated by *trans*-acting factors that bind to a heat-shock element (HSE) in the 5′ upstream region of the gene (see Section 1.1). The fact that *Xenopus* oocytes can transcribe an injected *Drosophila* hsp70 gene (Bienz and Pelham, 1982) argues for the presence of these factors in *Xenopus*. Furthermore, the *Xenopus* hsp70 gene, which has been cloned and sequenced, has also been studied with respect to its expression in *Xenopus* oocytes and monkey COS cells (Bienz, 1984; Bienz and Pelham, 1986). These studies have shown the presence of three HSEs in the promoter region of hsp70; at least two of the HSEs are required for maximal heat-induced expression in transfected COS cells. One possible explanation for the stage-dependent expression of hsp genes is that the levels of *trans*-acting factors may be limiting in cleavage embryos and do not begin to increase to functional levels until the mid-blastula transition.

hsp70 and 87 gene expression is heat inducible at the mid-blastula stage (see Section 3.4). This phenomenon coincides with the generalized activation of embryonic gene expression, which is part of a number of changes collectively called the mid-blastula transition (MBT) (Gerhart, 1980; Shiokawa *et al.*,

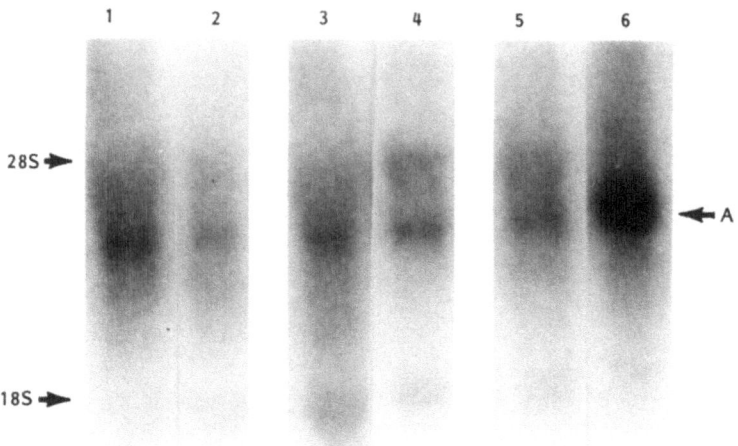

Figure 13. Stage-dependent heat-induced accumulation of putative hsp87 mRNA in *Xenopus* eggs and embryos. *Xenopus* eggs or embryos were maintained at 22°C for 1 hr. Polyadenylated RNA (2 μg) was electrophoresed, transferred to nitrocellulose, and hybridized to the ^{32}P nick-translated subclone of the *Drosophila* hsp83 gene. Transcript size: A = 3.2 kb. Lane 1: unfertilized eggs at 22°C; lane 2: unfertilized eggs at 35°C; lane 3, cleavage embryos at 22°C; lane 4, cleavage embryos at 35°C; lane 5: neurula embryos at 22°C; lane 6, neurula embryos at 35°C. (From Heikkila *et al.*, 1987 *a*.)

1981*a,b*; Newport and Kirschner, 1982*a,b*; Etkin, 1988). Kimelman *et al.* (1987) showed that transcription can be activated before the MBT by inhibiting the mitotic cell cycle with cycloheximide or by inhibiting DNA replication with aphidicolin. The implication of these results is that the rapid mitotic divisions of cleavage do not allow sufficient time for transcription. Treatment with either of these inhibitors would permit transcription, which is otherwise prevented. It would be instructive to test whether the hsp70 or 87 genes are inducible after activation of premature transcription. This would indicate whether transcriptional competence of the genome is both necessary and sufficient for induction of these genes during development.

The effects of temperature on hsp70 mRNA accumulation have been characterized by Heikkila *et al.* (1987*a*). Neurulae were exposed to temperatures ranging from 27° to 37°C for 1 hr (Fig. 14). Heat-induced hsp70 mRNA accumulation was first detectable at 27°C, with relatively greater levels at 30–35°C and lower levels at 37°C. A more complex effect of temperature was observed in time-course experiments; whereas continuous exposure of neurulae to heat shock (27–35°C) induced a transient accumulation of hsp70 mRNA, the temporal pattern of hsp70 mRNA accumulation was temperature dependent. Exposure of embryos to 33–35°C induced maximum levels of hsp70 mRNA within 1–1.5 hr (Fig. 15), whereas at 30° and 27°C, peak hsp70 mRNA accumulation occurred at 3 and 12 hr, respectively (Fig. 16). Thus, the rapidity of hsp70 mRNA induction increases with the severity of the temperature stress. Assuming that this phenomenon is regulated primarily at the transcriptional

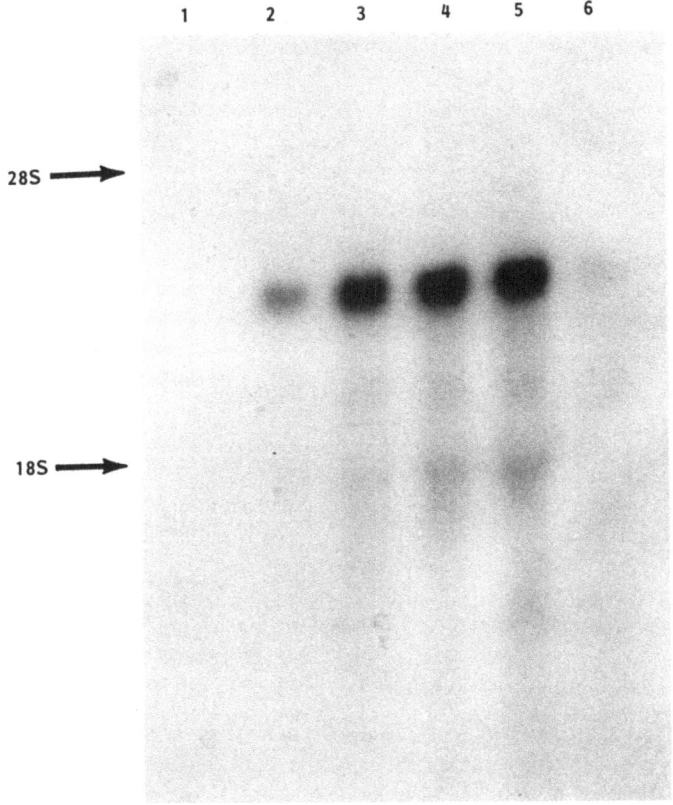

Figure 14. Effect of temperature on hsp70 mRNA accumulation in *Xenopus* neurulae. Embryos were maintained at temperatures ranging from 22°C to 37°C for 1 hr. Total RNA (10 μg) was electrophoresed, transferred to nitrocellulose, and hybridized. Lane 1: 22°C; lane 2: 27°C; lane 3: 30°C; lane 4: 33°C; lane 5: 35°C; lane 6: 37°C. (From Heikkila *et al.*, 1987a.)

level, it is possible that *Xenopus* HSF activation is more pronounced at higher heat-shock temperatures. This study also examined the relative levels of hsp70 mRNA during recovery at 22°C after a 1-hr heat shock at 33°C. An initial decrease in hsp70 mRNA levels within 15–30 min was followed by a transient increase in hsp70 mRNA at 1–2 hr before decaying to background levels by 7 hr. This secondary transient increase in relative hsp70 mRNA levels during recovery in embryos may be attributable to an increase in hsp70 gene transcription or a transient increase in the stability of hsp70 mRNA. Recently, it has been found that ubiquitin mRNA levels mimic the accumulation pattern of hsp70 mRNA during continuous heat shock and during recovery in neurulae (Ovsenek and Heikkila, 1988).

Heat-shocked *Xenopus* somatic cells transcribe a number of heat-shock genes, including a family of hsp30 genes (Heikkila *et al.*, 1988; Darasch *et al.*, 1988). However, during embryogenesis, the heat-induced transcript is not de-

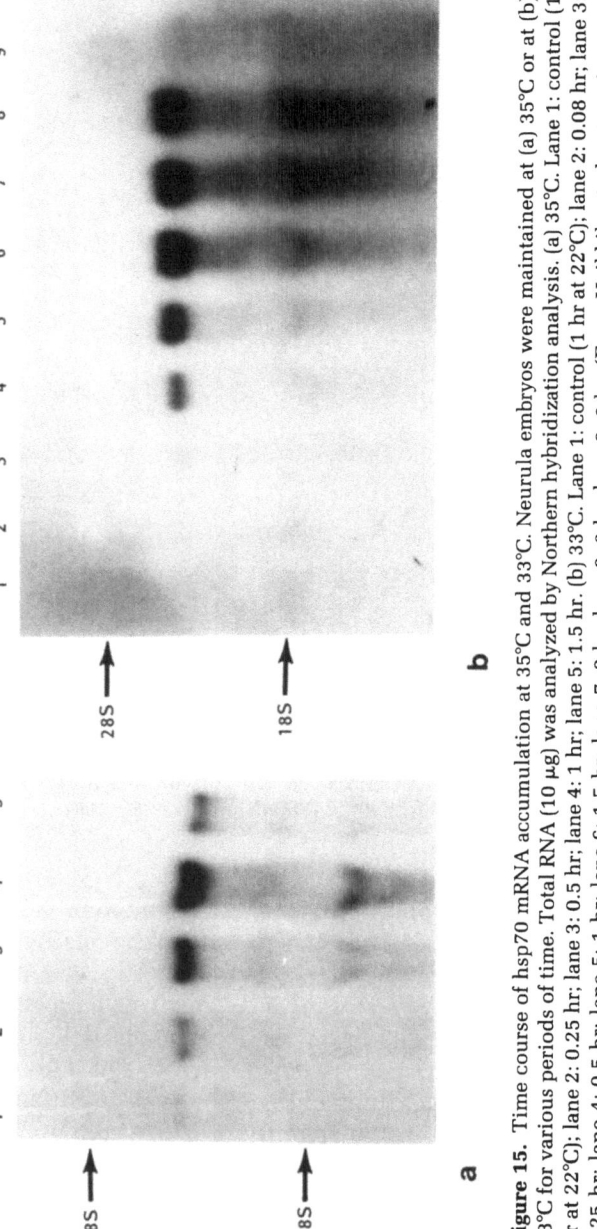

Figure 15. Time course of hsp70 mRNA accumulation at 35°C and 33°C. Neurula embryos were maintained at (a) 35°C or at (b) 33°C for various periods of time. Total RNA (10 μg) was analyzed by Northern hybridization analysis. (a) 35°C. Lane 1: control (1 hr at 22°C); lane 2: 0.25 hr; lane 3: 0.5 hr; lane 4: 1 hr; lane 5: 1.5 hr. (b) 33°C. Lane 1: control (1 hr at 22°C); lane 2: 0.08 hr; lane 3: 0.25 hr; lane 4: 0.5 hr; lane 5: 1 hr; lane 6: 1.5 hr; lane 7: 2 hr; lane 8: 3 hr; lane 9: 6 hr. (From Heikkila et al., 1987a.)

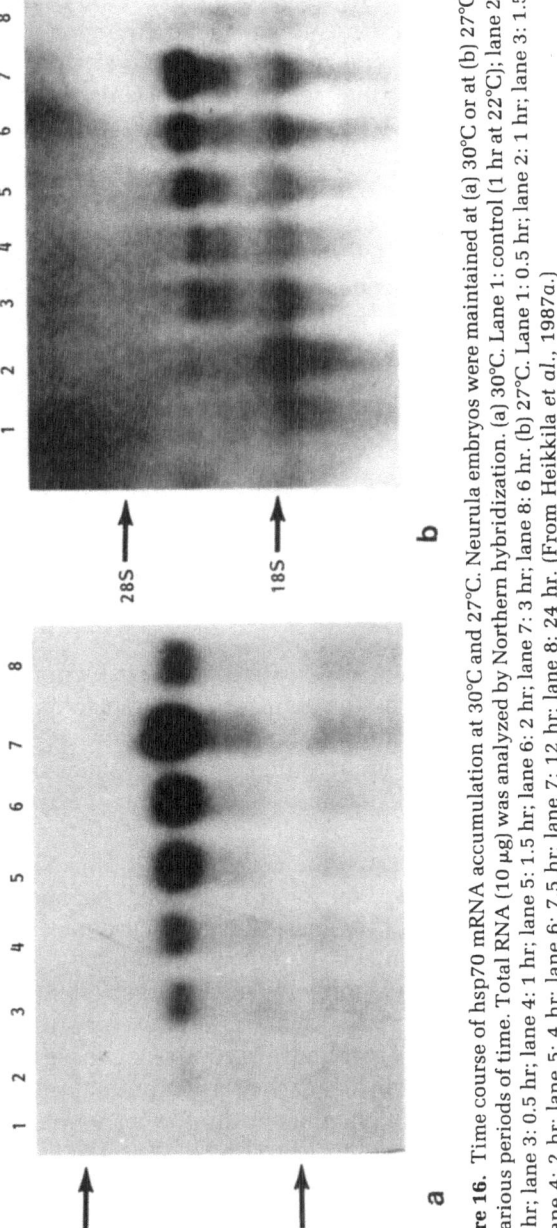

Figure 16. Time course of hsp70 mRNA accumulation at 30°C and 27°C. Neurula embryos were maintained at (a) 30°C or at (b) 27°C for various periods of time. Total RNA (10 μg) was analyzed by Northern hybridization. (a) 30°C. Lane 1: control (1 hr at 22°C); lane 2: 0.25 hr; lane 3: 0.5 hr; lane 4: 1 hr; lane 5: 1.5 hr; lane 6: 2 hr; lane 7: 3 hr; lane 8: 6 hr. (b) 27°C. Lane 1: 0.5 hr; lane 2: 1 hr; lane 3: 1.5 hr; lane 4: 2 hr; lane 5: 4 hr; lane 6: 7.5 hr; lane 7: 12 hr; lane 8: 24 hr. (From Heikkila et al., 1987a.)

tectable until the tadpole stage (Bienz, 1984; Krone and Heikkila, 1988). Is there an inhibitory substance that suppresses the transcription of hsp30 in post-MBT embryos, or does heat-induced hsp30 transcription require an additional factor(s) that is also developmentally regulated? Bienz (1984) suggested that hsp70 may have a higher affinity for *trans*-acting factors than for the hsp30 promoter. Thus, if pretadpole *Xenopus* embryos have a relatively low or limited level of heat-shock-activated *trans*-acting factors, it is possible that the factors may be sequestered by sequences (e.g., hsp70 HSEs) having a high affinity for them. However, this seems unlikely given the recent estimate of at least 2000 molecules of heat-shock factor per eukaryotic cell (Wu *et al.*, 1987).

Since hsp70, hsp30, and ubiquitin are regulated quite differently during development, it is of interest to examine their pattern of expression at the tadpole stage when all these genes are heat inducible. Continuous exposure of tadpoles to heat shock (33°C) results in a coordinate transient accumulation of hsp30 and 70 mRNA (Fig. 17). A similar finding was made for ubiquitin mRNA accumulation. A coordinate temporal pattern of the three types of transcripts is also observed in embryos recovering from brief heat shock. Thus, although hsp70, hsp30, and ubiquitin genes are inducible at different developmental stages, the genes in the tadpole are expressed coordinately during heat shock (Krone and Heikkila, 1988).

3.6. Thermotolerance in *Xenopus* Embryos

Xenopus zygotes lack the ability to synthesize hsps in response to thermal stress, possibly as a result of the ionic fluxes associated with fertilization. *Xenopus* embryos do not become competent to synthesize hsps until the mid-blastula transition, when the embryonic genome becomes transcriptionally active (see Section 3.5). It is also at this stage that the embryos begin to acquire thermotolerance.

Before the mid-blastula transition, embryos are subject to a variety of lethal effects induced by heat shock. For example, 20-min heat shock at 35°C induces almost immediate cytolysis of the blastomeres of early cleavage-stage embryos (Fig. 18). At later stages, the effects are not as immediately severe. Instead, heat shock of early blastula-stage embryos results in their abnormal development. These embryos fail to gastrulate, as their cells become organized into three discrete clusters (Fig. 19a). Histological sections of these abnormal embryos exhibit a cluster of pigmented cells at the former animal pole, a second cluster of cells containing large yolk platelets at the former vegetal pole, and a third cluster of cells at the former equatorial region of the embryo that do not contain pigment and have smaller yolk platelets (Fig. 19b). These abnormal embryos survive approximately 1 day postfertilization and then die.

Xenopus embryos begin to acquire thermotolerance at the mid-blastula transition. This thermotolerance is acquired in two distinct phases. That is, when heat-shocked as mid-cell blastulae, only the ectodermal and mesodermal derivatives of these embryos develop normally, whereas the endoderm does

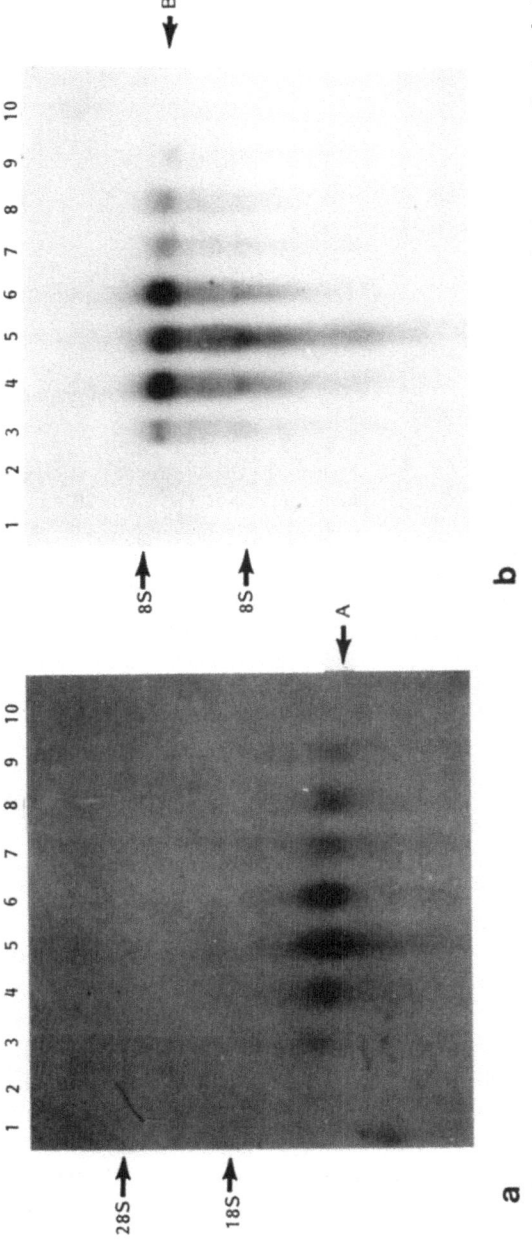

Figure 17. Time course of tadpole hsp30 and hsp70 mRNA accumulation at 33°C. Eight-day-old tadpoles were maintained at 33°C for various periods. Total lithium chloride precipitable RNA (10 μg) was analyzed by Northern hybridization. (a) RNA blot was hybridized against the labeled Xenopus hsp30 genomic subclone to yield this autoradiogram. (b) The radioactive signal on the nitrocellulose blot was allowed to decay, after which the blot was rehybridized against the Xenopus hsp70 genomic probe. Size of transcripts: A = 1.4 kb; B = 2.7 kb. Lane 1: control (22°C for 1 hr); lane 2: 33°C for 0.25 hr; lane 3: 33°C for 0.5 hr; lane 4: 33°C for 1 hr; lane 5: 33°C for 1.5 hr; lane 6: 33°C for 2 hr; lane 7: 33°C for 3 hr; lane 8: 33°C for 4 hr; lane 9: 33°C for 5 hr; lane 10: 33°C for 21 hr. (From Krone and Heikkila, 1988.)

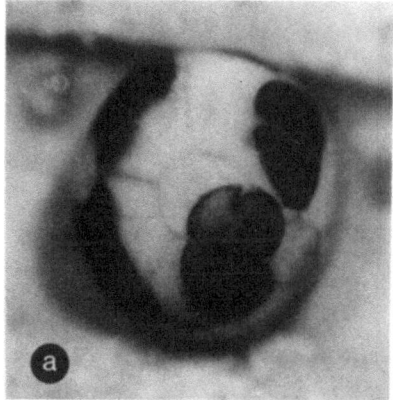

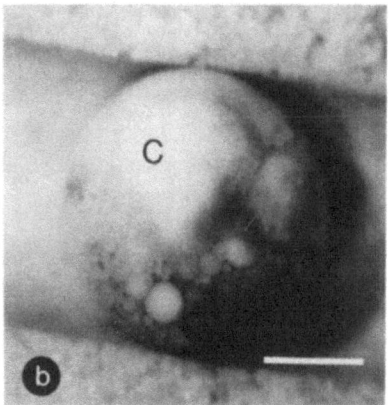

Figure 18. Cleavage stage embryos that had been heat-shocked at 35.5°C for 20 min. (a) Embryo 1 hr after heat shock. The embryo has been oriented to expose the animal pole to demonstrate the contraction of the apical pigment away from the cell boundaries. (b) Embryo left to develop overnight after heat shock. Small cell-like objects are visible in the lower regions of the embryo. Areas of complete cell lysis (C) are also apparent. Scale bar = 0.5 mm for both embryos. (From Nickells, 1987.)

not. This is shown in Fig. 19c,d, which shows the external and internal morphology of mid-cell blastulae that have been heat-shocked for 20 min at 35°C and allowed to develop overnight. Both the ectodermal (i.e., epidermis and neural tube) and mesodermal (i.e., somites, notochord, and lateral plate mesoderm) derivatives appear to have undergone normal morphogenesis, but some endoderm cells have moved out through the unclosed blastopore toward the dorsal surface of the embryo (exogastrulation).

Shortly after the mid-cell blastula stage of development, at the late blastula stage or very early gastrula stage (dorsal lip blastopore formation), the embryos are able to withstand the heat shock treatment and develop normally (Nickells and Browder, 1985). We have investigated the molecular basis for this biphasic acquisition of thermotolerance in *Xenopus* embryos. Because previous evidence in other cell types had indicated that hsps confer thermotolerance (see Section 2.3.1), we were particularly interested in the pattern of hsps synthesized by cells giving rise to ectoderm and mesoderm derivatives and by the cells giving rise to the endoderm derivatives. These two cell populations were crudely separated from each other by cutting the vegetal pole cells (presumptive endoderm: vegetal half) from the rest of the embryo (animal half). The

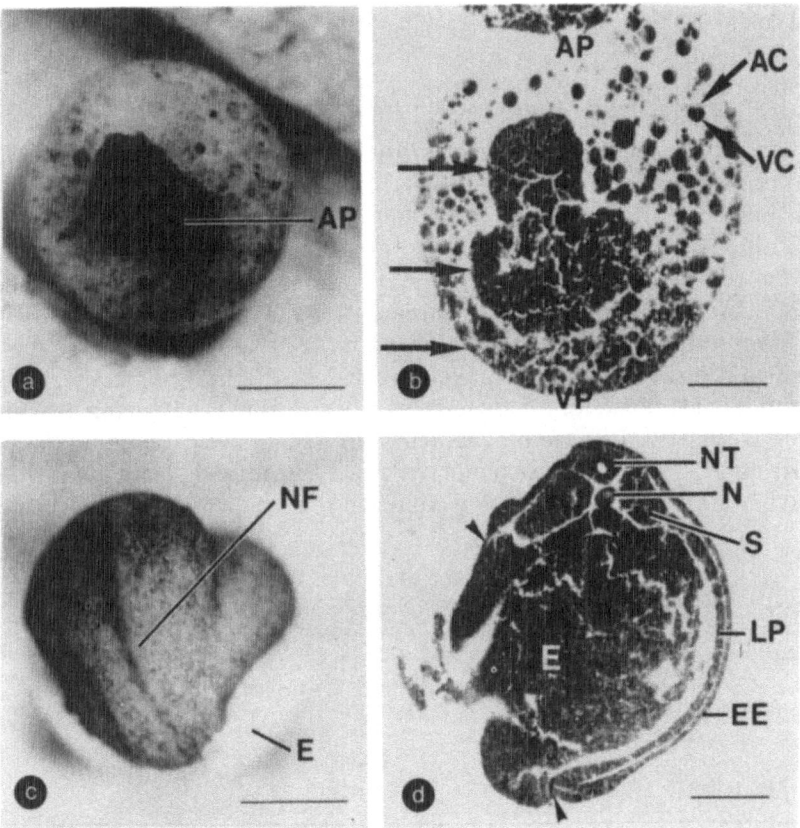

Figure 19. External and internal morphology of embryos heat-shocked and then allowed to develop overnight at ambient temperature. (a,b) Embryos developed from heat-shocked large cell blastulae showing external and internal morphology, respectively. (b) Arrows indicate three distinct cell clusters (see text). These embryos also exhibit large cells containing cytoplasm characteristic of animal hemisphere (AC, animal cytoplasm) and vegetal hemisphere (VC, vegetal cytoplasm). The animal pole (AP) and vegetal pole (VP) are indicated for each embryo. (c,d) Embryos developed from heat-shocked mid-cell blastulae showing external and internal morphology, respectively. These embryos exhibit endoderm (E) flowing outward toward the neural fold (NF) region of the embryo. (d) Arrowheads indicate the limits of the unclosed blastopore. The ectoderm derivatives (NT, neural tube; EE, epidermal ectoderm) and the mesoderm derivatives (N, notochord; S, somites, LP, lateral plate mesoderm) all appear to have developed normally. Scale bar for external morphology = 0.5 mm; for internal morphology = 250 µm. (Modified from Nickells and Browder, 1985.)

two embryo halves were then heat-shocked, and the patterns of hsp synthesis in the two halves were determined. Both halves of the mid-cell blastula synthesize the 87- and 70-kDa hsps (see Fig. 12). This result is surprising, considering that the cells of the vegetal hemisphere develop abnormally, and suggests that the synthesis of these two hsps is not sufficient to confer thermotolerance, at least on this cell type. The complete thermotolerance of the vegetal cells is instead correlated with the synthesis of a second group of hsps of 62, 48, and 35 kDa (see Fig. 10). hsps 62 and 48 are synthesized primarily in the vegetal hemisphere cells, whereas the synthesis of hsp35 appears to be restricted to these cells (see Section 3.4).

These results suggest that these additional hsps may be vital for the development of thermotolerance by the vegetal hemisphere cells. Further investigation of these hsps has shown that hsp35 and the glycolytic enzyme GAPDH share a number of properties in addition to identical molecular weights. GAPDH shows increased specific activity in homogenates from heat-shocked embryos, which correlates with its accumulation during heat shock as analyzed by immunoblotting (Fig. 20a,b). In addition, the peptide map of hsp35 matches the predicted peptide map of chicken muscle GAPDH (Table II). Perhaps the most dramatic evidence for the *de novo* synthesis of GAPDH during heat shock, however, is seen in cleavage-blocked embryos (coenocytic embryos) (Newport and Kirschner, 1982b). These embryos often synthesize large quantities of hsp35 when heat-shocked. We have analyzed the isozymes of GAPDH in these

Table II. Comparison of the Molecular Weights of the NCS Digestion Products of hsp 35 and GAPDH

Relative molecular weight (kDa)		Predicted molecular weight (kDa)	
hsp35	GAPDH[a] (rabbit)	GAPDH (chicken)	Amino acid length
32	33	35	333
31	29	32.3	308
27	27	25.7	245
24	23	23.7	226
20	17	20.4	194
15	14	14.6	139
—[b]	12[c]	12.3[b]	117
13	—	11.5[c]	109
		8.9[c]	85
		2.3[c]	22

[a]Fragment sizes of rabbit muscle GAPDH were determined from a 12% polyacrylamide gel, rather than a 15% gel used for hsp35.

[b]No methionine residues are present in this fragment of chicken muscle GAPDH.

[c]The N′-chlorosuccinimide digestion of rabbit muscle GAPDH and hsp35 yields seven fragments, whereas the predicted number of fragments for GAPDH is 10. It is unlikely that the gel system used for these experiments resolves the 2.3-kDa fragment. It is also possible that the three fragments of 12.3, 11.5, and 8.9 kDa migrate as one band. The only discrepancy between the fragment sizes of GAPDH and hsp35 is in the first fragment. This fragment is predicted to be undigested GAPDH monomer, and it comigrates with undigested GAPDH for both rabbit muscle GAPDH and hsp35. Presumably the discrepancy in the determination of the first fragment of GAPDH is due to error produced by extrapolating this molecular weight from standards. The molecular weights of the predicted fragments of chicken muscle GAPDH were estimated by multiplying the length of each fragment (in amino acids) by a constant of 0.105 kDa per amino acid.

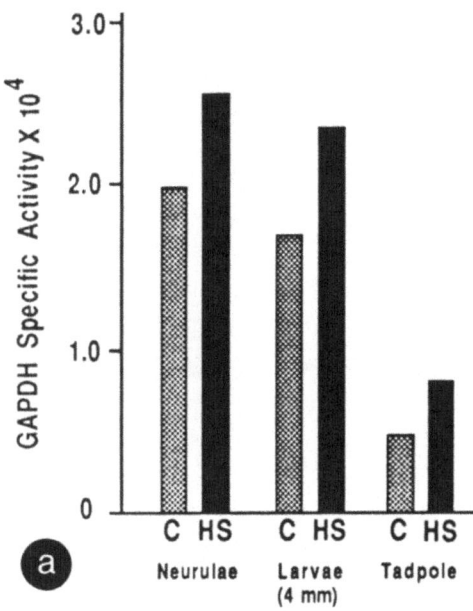

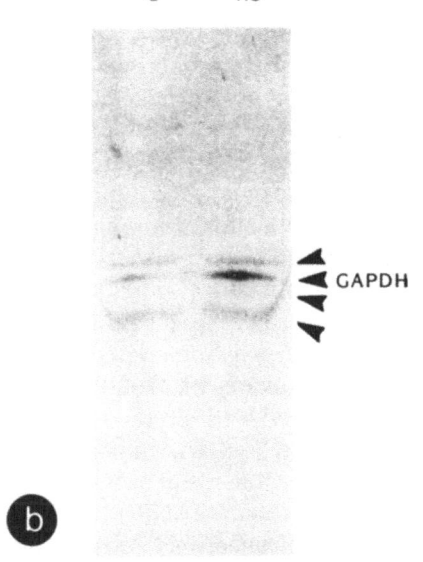

Figure 20. The level of GAPDH increases during heat shock. (a) Histogram showing the specific activity of GAPDH in control (C) and heat-shocked (HS) neurulae and postneurula stages of development. Specific activity is shown as the mean of four determinations for each sample and is expressed as the change in absorbance at 340 nm/sec per μg protein. All heat-shocked samples show a significant increase in specific activity over the respective control samples (Mann-Whitney U test, $p < 0.025$ for each experiment shown). (b) Immunoblot of equal amounts of protein from control (C) and heat-shocked (HS) neurulae probed with GAPDH antiserum. Closeup of the immunoblot in the 35-kDa region of the gel. The antiserum recognizes four peptides (small arrowheads) in each sample. The protein showing the greatest increase during heat shock comigrates with rabbit muscle GAPDH, whose position relative to the *Xenopus* proteins is indicated. (From Nickells and Browder, 1988.)

embryos by separating them on nondenaturing polyacrylamide gels and staining the gels for GAPDH enzymatic activity. Heat-shocked coenocytic embryos exhibit two isozymes of GAPDH, whereas control embryos only exhibit one. In addition, an hsp comigrates with the heat-shock-specific isozyme of GAPDH (Fig. 21). These data suggest that the synthesis of GAPDH is enhanced in these embryos, possibly in an attempt to stimulate anaerobic glycolysis by increasing

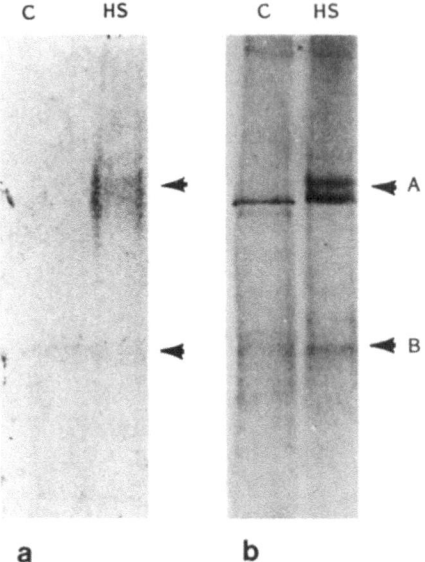

C   HS   C   HS

◄   ◄ A

◄   ◄ B

a   b

Figure 21. Heat-shocked coenocytic embryos show increases of an isozyme of GAPDH. Comparison of the GAPDH enzymatic activity staining pattern (a) and fluorograph (b) of control (C) and heat-shocked (HS) coenocytic embryo proteins separated on a nondenaturing 5% polyacrylamide gel. After electrophoresis, the gel was stained for GAPDH enzymatic activity. The heat-shocked extracts have a unique band of enzymatic activity, which corresponds to the position of a heat-shock protein (arrowheads marked by A). Both samples show the enzymatic activity staining of a second isozyme of GAPDH (arrowheads marked by B). The coenocytic embryos were heat-shocked when sibling normal embryos had reached the dorsal lip stage of development. Equal amounts of acid precipitable cpm were loaded for each lane. (From Nickells and Browder, 1988.)

the level of some glycolytic enzymes. In support of this hypothesis, we have demonstrated that another glycolytic enzyme, pyruvate kinase (PK), is immunologically related to hsp62. As with GAPDH, an isozymic form of PK accumulates during heat shock, and an increase in PK specific activity correlates with its accumulation (Marsden *et al.*, 1989).

The synthesis of glycolytic enzymes suggests that heat shock stresses the energy metabolism of *Xenopus* embryonic cells as it does other cell types (see Section 2.3.2). We have observed several heat-induced changes in the normal energy metabolism of *Xenopus* embryos, including the rapid decrease in ATP and the increase of lactic acid (Nickells and Browder, 1988). This response by *Xenopus* embryos is remarkably like that of maize seedlings exposed to anaerobic stress (Sachs *et al.*, 1980). Upon exposure to an anaerobic environment (buffer saturated with argon), the seedlings rapidly synthesize a class of proteins of 35.5- to 31.5-kDa molecular weights (the transition proteins). After the synthesis of the transition proteins, the seedlings synthesize an additional class of anaerobic proteins (ANPs) of 87, 77, 65, 64, 55, 45, 40, and 35.5–31.5 kDa. This pattern of anaerobic protein synthesis is similar to the pattern of hsps synthesized by heat-shocked *Xenopus* embryos (see Fig. 11a). The anaerobic proteins are also enzymes of the glycolytic pathway such as two isoforms of alcohol dehydrogenase (ANP40A and B) (Sachs and Freeling, 1978; Ferl *et al.*, 1979), fructose 1,6-diphosphate aldolase (Kelly and Freeling, 1984a), pyruvate decarboxylase (Wignarajah and Greenway, 1976), and glucose phosphate isomerase (Kelley and Freeling, 1984b). Kelley and Freeling (1982) suggest that the anaerobic proteins may assist the seedling to survive anaerobiosis by stimulating anaerobic glycolysis via ethanol synthesis. We propose a similar function of the hsps in *Xenopus* embryos, except that in *Xenopus*, glycolysis apparently results in the production of lactic acid.

An increase in glycolysis in response to heat shock may not be required by all organisms. Maize seedlings, although they appear to mount a stress response to anaerobiosis identical to the *Xenopus* heat-shock response, only synthesize the 35.5- to 31.5 kDa-transition anaerobic proteins when heat shocked (Kelley and Freeling, 1982). This suggests that heat shock does not induce complete anaerobiosis in the seedlings. The lack of glycolytic enzyme synthesis in other heat-shocked cell types suggests that an increase in these enzymes is not essential for the development of thermotolerance.

Alternatively, some glycolytic enzymes may have been synthesized by other heat-shocked cells, but their detection and characterization have been difficult. A good example of this is seen in *Drosophila* embryos, larvae, and tissue culture cells. Occasionally, these cell types synthesize hsps of 60- and 37-kDa (Lindquist, 1980). A 34-kDa hsp has also been detected in *Chironomous* salivary glands (Vincent and Tanguay, 1979). In our laboratory, we have detected a minor hsp in *Drosophila* larvae that co-migrates with purified GAPDH standard on one-dimensional SDS-polyacrylamide gels (See Fig. 2). The synthesis of this hsp appears to be stronger in larval brains than in salivary glands or wing discs, suggesting regional specification of its synthesis. Also, we have been unable to detect the synthesis of this hsp in larvae that are grown under crowded conditions. These data agree with earlier observations that the expression of minor hsps in *Drosophila* is highly variable (Lindquist, 1980), which may explain why they have not been characterized.

A potential explanation for the lack of heat-shock-induced synthesis of glycolytic enzymes in other organisms is obtained from *Xenopus* embryos. The heat-shock response of *Xenopus* embryos can also be highly variable, especially with respect to hsp35 (GAPDH). The apparent inability of *Xenopus* mid- to late gastrulae to synthesize hsp35 is an example of this variability (see Fig. 10). Even *Xenopus* neurulae, which can synthesize large amounts of hsp35, do not always reliably synthesize detectable quantities of it during heat shock. Knowing that hsp35 is GAPDH, we have measured the specific activities of GAPDH in control and heat-shocked embryos from different developmental stages. Surprisingly, we have observed that heat-shocked embryos only show an increase in GAPDH specific activity if they have low pre-existing levels of enzymatic activity. This inverse relationship is shown graphically in Fig. 22, which represents data from six experiments using different groups of neurula-stage embryos and one experiment using free-swimming tadpoles. The heat-induced increases in GAPDH specific activity also correlate with the synthesis of hsp35. As one might predict, only heat-shocked neurulae that show relatively large increases in GAPDH specific activity synthesize detectable amounts of hsp35 (Fig. 23).

This inverse relationship may explain why mid- to late gastrulae do not exhibit the synthesis of GAPDH during heat shock. We have measured the constitutive levels of GAPDH enzymatic activity throughout this period of development. Figure 24 shows that the activity of GAPDH increases dramatically during gastrulation and then drops when the embryos begin to neurulate.

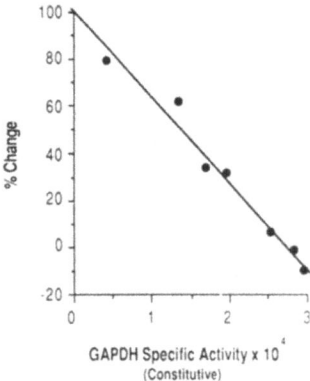

Figure 22. Increases in GAPDH specific activity are inversely proportional to the level of constitutive GAPDH specific activity. Plot of seven experiments compares the constitutive level of GAPDH specific activity and the percentage change in activity of the corresponding heat-shocked sample. Included with the plotted points is the calculated best-fit line. There is a significant correlation between control GAPDH specific activity and the subsequent change induced by heat shock (Spearman test, $p < 0.001$). It is evident that only embryos with relatively lower constitutive levels of GAPDH specific activity exhibit heat-induced increases in enzymatic activity. These data are from experiments using neurulae and early larval stages. (From Nickells and Browder, 1988.)

This peak of GAPDH activity correlates precisely with the loss of GAPDH expression by heat-shocked embryos (compare Figs. 24 and 10). From these results, we suggest that some *Xenopus* embryos have sufficient constitutive levels of GAPDH to cope with the heat-shock-induced demands for increases in anaerobic glycolysis. This example serves as a possible explanation as to why glycolytic enzyme synthesis is not detected more often in other heat-shocked cells: Perhaps they also have sufficient pre-existing levels of glycolytic enzymes. This would explain why other investigators have observed increases in lactate concentration in heat-shocked cells without a corresponding increase in glycolytic enzyme activity (Kelley and Schlesinger, 1982). With more investigation, it might be possible to establish a universal thermoprotective role for

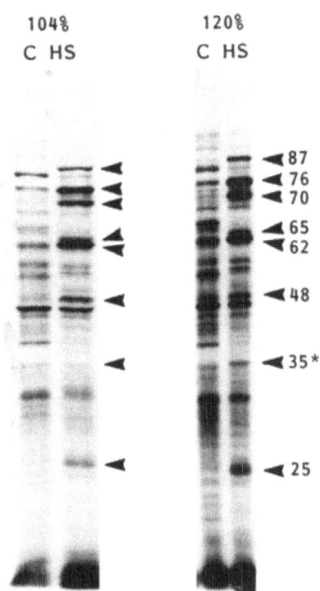

Figure 23. hsp35 synthesis correlates with increases in GAPDH specific activity. Fluorograph of [35S]methionine-labeled proteins of control (C) and heat-shocked (HS) neurulae from two separate spawnings of eggs separated on a SDS–10% polyacrylamide gel. Samples of the same neurulae were also tested for heat shock-induced changes in GAPDH specific activity. The change in specific activity is shown at the top of each set of samples as a percentage of control GAPDH activity (i.e., control embryos exhibit 100% enzymatic activity). The molecular weights of the hsps are shown in kDa. hsp35 (asterisk) is detected in the sample with the relatively high increase in GAPDH specific activity. Equal acid-precipitable cpm were loaded for each lane. (From Nickells and Browder, 1988.)

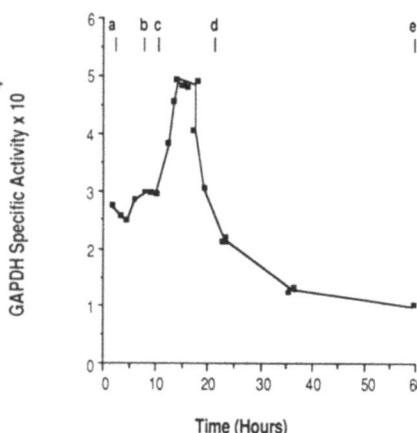

Figure 24. Graphic representation of the constitutive GAPDH specific activity of embryos during development. Letters at top of graph indicate important events in early *Xenopus* development. a, fertilization; b, mid-blastula transition; c, dorsal lip formation (the beginning of gastrulation); d, neural plate embryos (and the end of gastrulation); e, 2½-day larvae. It is apparent that the level of constitutive GAPDH specific activity increases dramatically and transiently during gastrulation. This period correlates with reduced or no hsp35 synthesis by heat-shocked embryos. (From Nickells, 1987.)

glycolytic enzymes, which may complement the role that the other hsps (e.g., hsp70) play in thermotolerance.

3.7. Latent Transcriptional Activity in *Xenopus* Erythrocytes Evoked by Heat Shock

In Section 3.4 we documented the changing patterns of induced hsp synthesis during embryonic development of *Xenopus*. This phenomenon is not restricted to the embryonic phase of development, but also occurs during development of a defined cell lineage as in erythropoiesis, an ongoing process of cellular differentiation during the life of the organism. As erythropoiesis proceeds, there is a progressive decrease in the number of hsps induced (Winning and Browder, 1988). One exception to this pattern is the appearance of a novel erythroid hsp, hsp66, at the penultimate stage of differentiation. It is unknown whether this protein is related to the hsp66 observed in artificially activated eggs. In spite of the general decrease in hsp synthesis, mature erythrocytes respond to heat shock by synthesizing one hsp: hsp70 (Fig. 25). Interestingly, erythrocytes have a very narrow range of hsp70 inducibility, with peak induction at 33°C. The induced synthesis of hsp70 by erythrocytes is surprising, since these cells have been described as being transcriptionally inert with very low levels of protein synthesis (Maclean *et al.*, 1973). It has been proposed that the limited protein synthesis that does occur uses messenger, ribosomal, and transfer RNA (mRNA, rRNA, and tRNA) synthesized during earlier stages of erythroid differentiation.

Since the heat-shock response in *Xenopus* oocytes and eggs relies on preexisting transcripts, it is possible that stored hsp70 transcripts could be used in the erythrocyte heat shock response. Alternatively, these cells could retain latent transcriptional potential that can be induced by heat shock. The results of experiments with actinomycin D suggest that the latter possibility actually pertains: The drug inhibits hsp70 synthesis by heat-shocked erythrocytes, al-

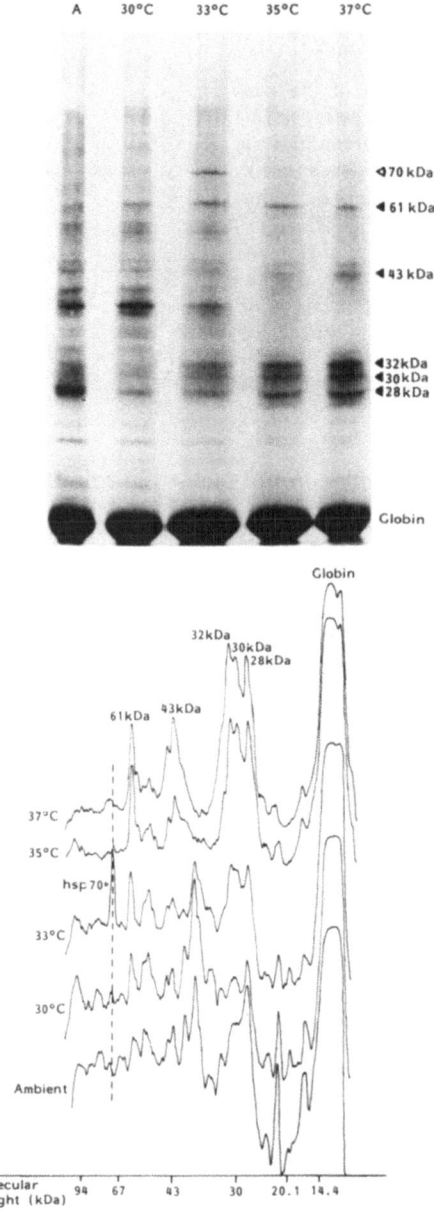

Figure 25. Heat-shock protein synthesis in *Xenopus* erythrocytes. On the top is a fluorograph of proteins labeled with [^{35}S]methionine in erythrocytes incubated at ambient temperature (A) and at the heat-shock temperatures indicated. Open arrow indicates the position of hsp70. Closed arrowheads indicate major proteins that incorporate [^{35}S]methionine at 37°C. Equal amounts of acid-insoluble radioactivity were loaded onto each lane. Densitometer scans of the fluorograph lanes are shown on the bottom. The scan of each lane is indicated by its incubation temperature. Positions of protein markers are shown at the bottom. The 70-kDa position is indicated by a dotted line. Peaks of [^{35}S]methionine incorporation are labeled by molecular mass. (From Winning and Browder, 1988.)

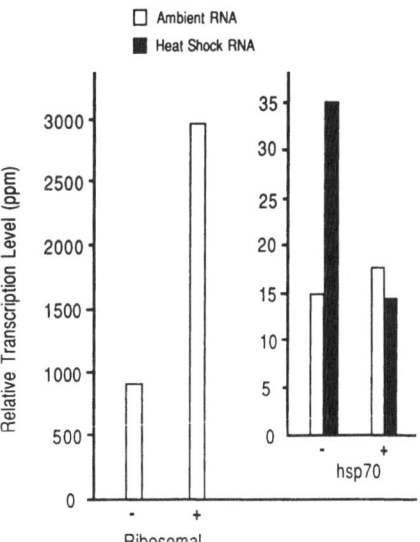

Figure 26. Erythrocyte hsp70 gene transcription during heat shock is inhibited by α-amanitin. Nuclei were isolated from erythrocytes that had been incubated at ambient temperature or at a heat-shock temperature of 33°C. [^{32}P]-UTP was incorporated into runoff transcripts in the absence (−) or presence (+) of 5 μg/ml α-amanitin. Labeled RNA from ambient nuclei (open bars) was hybridized to dotted ribosomal and hsp70 gene probes. Labeled RNA from heat-shock nuclei (closed bars) was hybridized to the dotted hsp70 probe. (From Winning and Browder, 1988.)

though constitutive protein synthesis (including globin synthesis) is unaffected. Thus, although protein synthesis at ambient temperature does not rely to any significant extent on newly synthesized transcripts, the induction of hsp70 synthesis is dependent upon transcription.

In order to determine whether hsp70 gene transcription is induced by heat shock, we conducted nuclear runoff transcription assays. Nuclear runoff experiments measure the level of transcription existing in the nuclei at the time of nuclear isolation. [^{32}P]-UTP is incorporated into nascent transcripts; no initiation of transcription occurs in the isolated nuclei (Hadjiolov and Milchev, 1974). The level of transcription for specific genes is determined by hybridizing the in vitro-synthesized RNA to cloned Xenopus genes immobilized on nitrocellulose filters. The data are then expressed as the transcription level relative to total transcription. Our results (Winning and Browder, 1988) demonstrate extreme temperature sensitivity of hsp70 gene transcription. Transcription rose dramatically at 33°C and fell to insignificant levels at 37°C. These results mirror the temperature dependence for the synthesis of hsp70. The increase in relative transcription of the hsp70 gene at 33°C for all assays ranged from 2.5-fold to 33-fold. This increase was inhibited by addition of α-amanitin at a concentration that inhibits transcription by RNA polymerase II, confirming that we were observing actual RNA polymerase II-mediated transcription (Fig. 26). Clearly, erythrocytes possess latent transcriptional capacity; the hsp70 gene is not irreversibly inactivated.

The induction of hsp70 gene transcription by heat shock in Xenopus erythrocytes contrasts with the situation in heat-shocked chicken erythroid cells. Banerji et al. (1984, 1987) showed that induced hsp70 synthesis in definitive red cells of embryos and in adult reticulocytes results from activation of translation of constitutively synthesized hsp70 mRNA.

4. Conclusions

The capacity for heat-shock protein synthesis is subject to extensive developmental regulation. As cells differentiate, the ongoing process of physiological change modifies both the cellular requirements to survive stress and the mode of response. The linkage between overt differentiation and stress-inducible protein synthesis must therefore be direct. The regulation of hsp synthesis can occur at either the transcriptional or post-transcriptional levels. Additional study of this developmental system will further our understanding of how the utilization of reserve capacity for gene expression is regulated.

ACKNOWLEDGMENTS. This work was supported by research grants to L. W. B. and J. J. H. from NSERC (Canada), a fellowship to M. P. from The Alberta Heritage Foundation for Medical Research (AHFMR), and a graduate studentship to R. S. W. and a research allowance to R. W. N. from AHFMR. The authors are grateful to Dr. D. I. Rodenhiser for his comments on the manuscript.

References

Ashburner, M., 1970, Patterns of puffing activity in the salivary gland chromosomes of *Drosophila*. V. Responses to environmental treatments, *Chromosoma* **31**:356–376.

Ashburner, M., and Bonner, J. J., 1979, The induction of gene activity in *Drosophila* by heat shock, *Cell* **17**:241–254.

Atkinson, B. G., and Walden, D. B. (eds.), 1985, *Changes in Eukaryotic Gene Expression in Response to Environmental Stress*, Academic, Orlando, Florida.

Banerji, S. S., Theodorakis, N. G., and Morimoto, R. I., 1984, Heat shock-induced translational control of hsp70 and globin synthesis in chicken reticulocytes, *Mol. Cell. Biol.* **4**:2437–2438.

Banerji, S. S., Laing, K., and Morimoto, R. I., 1987, Erythroid lineage-specific expression and inducibility of the major heat shock protein HSP70 during avian embryogenesis, *Genes Dev.* **1**: 946–953.

Belanger, F. C., Brodl, M. R., and Ho, T.-H., 1986, Heat shock causes destabilization of specific mRNAs and destruction of endoplasmic reticulum in barley aleurone cells, *Proc. Natl. Acad. Sci. USA* **83**:1354–1358.

Berger, E. M., and Woodward, M. P., 1983, Small heat shock proteins in *Drosophila* may confer thermal tolerance, *Exp. Cell Res.* **147**:437–442.

Bergh, S., and Arking, R., 1984, Developmental profile of the heat shock response in early embryos of *Drosophila*, *J. Exp. Zool.* **231**:379–391.

Bienz, M., 1984, Developmental control of the heat shock response in *Xenopus*, *Proc. Natl. Acad. Sci. USA* **81**:3138–3142.

Bienz, M., and Gurdon, J. B., 1982, The heat shock response in *Xenopus* oocytes is controlled at the translational level, *Cell* **29**:811–819.

Bienz, M., and Pelham, H. R. B., 1982, Expression of a *Drosophila* heat shock protein in *Xenopus* oocytes: Conserved and divergent regulatory signals, *EMBO J.* **1**:1583–1588.

Bienz, M., and Pelham, H. R. B., 1986, Heat shock regulatory elements function as an inducible enhancer in the *Xenopus* hsp70 gene and when linked to a heterologous promoter, *Cell* **45**: 753–760.

Biessmann, H., Falkner, F. G., Saumweber, H., and Walter, M. F., 1982, Disruption of the vimentin cytoskeleton may play a role in heat-shock response, in: *Heat Shock from Bacteria to Man* (M. J. Schlesinger, M. Ashburner, and A. Tissières, eds.), pp. 227–234, Cold Spring Harbor Laboratory, Cold Spring Harbor, New York.

Bond, U., and Schlesinger, M. J., 1985, Ubiquitin is a heat shock protein in chick embryo fibroblasts, *Mol. Cell. Biol.* **5**:949–956.

Bond, U., and Schlesinger, M. J., 1987, Heat-shock proteins and development, *Adv. Genet.* **24**:1–29.

Bonner, J. J., and Kerby, R. L., 1982, RNA polymerase II transcribes all of the heat shock induced genes of *Drosophila melanogaster, Chromosoma* **85**:93–108.

Browder, L. W., Pollock, M., Heikkila, J. J., Wilkes, J., Wang, T., Krone, P., Ovsenek, N., and Kloc, M., 1987, Decay of the oocyte-type heat shock response of *Xenopus laevis, Dev. Biol.* **124**:191–199.

Burdon, R. H., and Cutmore, C. M. M., 1982, Human heat shock gene expression and the modulation of plasma membrane sodium-potassium ATPase activity, *FEBS Lett.* **140**:45–48.

Busa, W. B., and Nuccitelli, R., 1984, Metabolic regulation via intracellular pH, *Am. J. Physiol.* **246**: R409–R438.

Busa, W. B., and Nuccitelli, R., 1985, An elevated free cytosolic Ca^{2+} wave follows fertilization in the eggs of the frog *Xenopus laevis, J. Cell. Biol.* **100**:1325–1329.

Chappell, T. G., Welch, W. J., Schlossman, D. M., Palter, K. B., Schlesinger, M. J., and Rothman, J. E., 1986, Uncoating ATPase is a member of the 70 kilodalton family of stress proteins, *Cell* **45**: 3–13.

Christiansen, E. N., and Kvamme, E., 1969, Effects of thermal treatment on mitochondria of brain, liver, and ascites cells, *Acta Physiol. Scand.* **76**:472–484.

Cross, N. L., and Elinson, R. P., 1980, A fast block to polyspermy in frogs mediated by changes in the membrane potential, *Dev. Biol.* **75**:187–198.

Darasch, S., Mosser, D. D., Bols, N. C., and Heikkila, J. J., 1988, Heat shock gene expression in *Xenopus laevis* A6 cells in response to heat shock and sodium arsenite treatments, *Biochem. Cell Biol.* **66**:862–870.

Davidson, E. H., 1986, *Gene Activity in Early Development*, 3rd ed., Academic, Orlando, Florida.

Dickson, J. A., and Oswald, B. E., 1976, The sensitivity of a malignant cell line to hyperthermia (42°C) at low intracellular pH, *Br. J. Cancer* **34**:262–271.

DiDomenico, B. J., Bugaisky, G. E., and Lindquist, S., 1982, The heat shock response is self-regulated at both the transcriptional and posttranscriptional levels, *Cell* **31**:593–603.

Drummond, I. A. S., McClure, S. A., Poenie, M., Tsien, R. Y., and Steinhardt, R. A., 1986, Large changes in intracellular pH and calcium observed during heat shock are not responsible for the induction of heat shock proteins in *Drosophila melanogaster, Mol. Cell. Biol.* **6**:1767–1775.

Duncan, R., and Hershey, J. W. B., 1984, Heat shock-induced translational alterations in HeLa cells, Initiation factor modifications and the inhibition of translation, *J. Biol. Chem.* **259**:11882–11889.

Dura, J.-M., 1981, Stage dependent synthesis of heat shock induced proteins in early embryos of *Drosophila melanogaster, Mol. Gen. Genet.* **184**:381–385.

Edwards, M. J., 1981, Clinical disorders of fetal brain development: Defects due to hyperthermia, in: *Fetal Brain Disorders—Recent Approaches to the Problem of Mental Deficiency* (B. S. Hetzl and R. M. Smith, eds.), pp. 335–364, Elsevier, Amsterdam.

Elsdale, T., Pearson, M., and Whitehead, M., 1976, Abnormalities in somite segmentation following heat shock to *Xenopus* embryos, *J. Embryol. Exp. Morphol.* **35**:625–635.

Eppig, J. J., and Steckman, M. L., 1976, Comparison of exogenous energy sources for *in vitro* maintenance of follicle cell-free *Xenopus laevis* oocytes, *In Vitro* **12**:173–179.

Etkin, L. D., 1988, Regulation of the mid-blastula transition in amphibians, in: *Developmental Biology: A Comprehensive Synthesis*, Vol. 5: *The Molecular Biology of Cell Determination and Cell Differentiation* (L. W. Browder, ed.), pp. 209–225, Plenum, New York.

Falkner, F. G., Saumweber, H., and Biessmann, H., 1981, Two *Drosophila melanogaster* proteins related to intermediate filament proteins of vertebrate cells, *J. Cell Biol.* **91**:175–183.

Ferl, R. J., Dlouhy, S. R., and Schwartz, D., 1979, Analysis of maize alcohol dehydrogenase by native-SDS two dimensional electrophoresis and autoradiography, *Mol. Gen. Genet.* **169**:7–12.

Findly, R. C., Gillies, R. J., and Shulman, R. G., 1983, *In vivo* phosphorus-31 nuclear magnetic resonance reveals lowered ATP during heat shock of *Tetrahymena, Science* **219**:1223–1225.

Finnell, R. H., Moon, S. P., Abbott, L. C., Golden, J. A., and Chernoff, G. F., 1986, Strain differences in heat-induced neural tube defects in mice, *Teratology* **33**:247–252.

Gerhart, J. G., 1980, Mechanisms regulating pattern formation in the amphibian egg and early embryo, in: *Biological Regulation and Development* (R. F. Goldberger, ed.), pp. 133–315, Plenum, New York.

German, J., 1984, Embryonic stress hypothesis of teratogenesis, *Am. J. Med.* **76**:293–301.

Gerner, E. W., and Schneider, M. J., 1975, Induced thermal resistance in HeLa cells, *Nature (Lond.)* **256**:500–502.

Giudice, G., 1985, Heat shock proteins in sea urchin development, in: *Changes in Eukaryotic Gene Expression in Response to Environmental Stress* (B. G. Atkinson and D. B. Walden, eds.), pp. 115–133, Academic, Orlando, Florida.

Goldstein, E. S., and Penman, S., 1973, Regulation of protein synthesis in mammalian cells. V. Further studies on the effect on actinomycin D on translation control in HeLa cells, *J. Mol. Biol.* **80**:243–254.

Graziosi, G., Micali, F., Marzari, R., DeCristini, F., and Savioni, A., 1980, Variability of response of early *Drosophila* embryos to heat shock, *J. Exp. Zool.* **214**:141–145.

Gurdon, J. B., 1976, Injected nuclei in frog oocytes: Fate, enlargement, and chromatin dispersal, *J. Embryol. Exp. Morphol.* **36**:523–540.

Hadjiolov, A. A., and Milchev, G. I., 1974, Synthesis and maturation of ribosomal ribonucleic acids in isolated HeLa cell nuclei, *Biochem. J.* **142**:263–272.

Hahnel, A. C., Gifford, D. J., Heikkila, J. J., and Schultz, G. A., 1986, Expression of the major heat shock protein (hsp 70) family during early mouse embryo development, *Teratogenesis, Carcinog. Mutagen.* **6**:493–510.

Hammond, G. L., Lai, Y.-K., and Markert, C. L., 1982, Diverse forms of stress lead to new patterns of gene expression through a common and essential pathway, *Proc. Natl. Acad. Sci. USA* **79:** 3485–3488.

Heikkila, J. J., Kloc, M., Bury, J., Schultz, G. A., and Browder, L. W., 1985a, Acquisition of the heat shock response and thermotolerance during early development of *Xenopus laevis*, *Dev. Biol.* **107**:483–489.

Heikkila, J. J., Miller, J. G. O., Schultz, G. A., Kloc, M., and Browder, L. W., 1985b, Heat shock gene expression during early animal development, in: *Changes in Eukaryotic Gene Expression in Response to Environmental Stress* (B. G. Atkinson and D. B. Walden, eds.), pp. 135–158, Academic, Orlando, Florida.

Heikkila, J. J., Ovsenek, N., and Krone, P., 1987a, Examination of heat shock protein mRNA accumulation in early *Xenopus laevis* embryos, *Biochem. Cell Biol.* **65**:87–94.

Heikkila, J. J., Darasch, S. P., Mosser, D. D., and Bols, N. C., 1987b, Heat and sodium arsenite act synergistically on the induction of heat shock gene expression in *Xenopus laevis* A6 cells, *Biochem. Cell Biol.* **65**:310–316.

Heikkila, J. J., and Schultz, G. A., 1984, Different environmental stresses can activate the expression of a heat shock gene in rabbit blastocysts, *Gamete Res.* **10**:45–56.

Henle, K. J., and Leeper, D. B., 1976, Interaction of hyperthermia and radiation in CHO cells: Recovery kinetics, *Radiat. Res.* **66**:505–518.

Hershko, A., 1983, Ubiquitin: Roles in protein modification and breakdown, *Cell* **34**:11–12.

Hollinger, T. G., and Corton, G. L., 1980, Artificial fertilization of gametes from the South African Clawed Frog, *Xenopus laevis*, *Gamete Res.* **3**:45–57.

Howlett, S., Miller, J., and Schultz, G. A., 1983, Induction of heat shock proteins in early embryos of *Arbacia punctulata*, *Biol. Bull.* **165**:500.

Iida, H., and Yahara, I., 1985, Yeast heat-shock protein of M_r 48,000 is an isoprotein of enolase, *Nature (Lond.)* **315**:688–690.

Ingolia, T. D., and Craig, E. A., 1982, Four small *Drosophila* heat shock proteins are related to each other and mammalian α-crystallin, *Proc. Natl. Acad. Sci. USA* **79**:2360–2364.

Ito, S., 1972, Effects of media of different ionic composition on the activation potential of anuran egg cells, *Dev. Growth Diff.* **14**:217–220.

Johnston, R. N., and Kucey, B. L., 1988, Competitive inhibition of hsp70 gene expression causes thermosensitivity, *Science* **242**:1551–1554.

Katagiri, C., 1961, On the fertilizability of the frog egg. I, *J. Fac. Sci. Hokkaido Univ. Ser. VI, Zool.* **14**:607–613.

Kelley, P. M., and Freeling, M., 1982, A preliminary comparison of maize anaerobic and heat-shock proteins, in: *Heat Shock. From Bacteria to Man* (M. J. Schlesinger, M. Ashburner, and A. Tissières, eds.), pp. 315–319, Cold Spring Harbor Laboratory, Cold Spring Harbor, New York.

Kelley, P. M., and Freeling, M., 1984a, Anaerobic expression of maize fructose-1,6-diphosphate aldolase, *J. Biol. Chem.* **259:**14180–14183.

Kelley, P. M., and Freeling, M., 1984b, Anaerobic expression of maize glucose phosphate isomerase I, *J. Biol. Chem.* **259:**673–677.

Kelley, P. M., and Schlesinger, M. J., 1982, Antibodies to two major chicken heat shock proteins cross-react with similar proteins in widely divergent species, *Mol. Cell. Biol.* **2:**267–274.

Kimelman, D., Kirschner, M., and Scherson, T., 1987, The events of the midblastula transition in *Xenopus* are regulated by changes in the cell cycle, *Cell* **48:**399–407.

King, M.-L., and Davis, R., 1987, Do *Xenopus* oocytes have a heat shock response?, *Dev. Biol.* **119:** 532–539.

Klemenz, R., Hultmark, D., and Gehring, W. J., 1985, Selective translation of heat shock mRNA in *Drosophila melanogaster* depends on sequence information in the leader, *EMBO J.* **4:**2053–2060.

Kline, D., and Nuccitelli, R., 1985, The wave of activation current in the *Xenopus* egg, *Dev. Biol.* **111:**471–487.

Krone, P. H., and Heikkila, J. J., 1988, Analysis of hsp 30, hsp 70 and ubiquitin gene expression in *Xenopus laevis* tadpoles, *Development* **103:**59–67.

Krone, P. H., and Heikkila, J. J., 1989, Expression of microinjected HSP30-CAT and HSP70-CAT chimeric genes in development of *Xenopus laevis* embryos, *Development* (in press.)

Krüger, C., and Benecke, B. J., 1981, In vitro translation of *Drosophila* heat-shock and non-heat-shock mRNAs in heterologous and homologous cell-free systems, *Cell* **23:**595–603.

Landry, J., 1986, Heat shock proteins and cell thermotolerance, in: *Hyperthermia in Cancer Treatment,* Vol. I (L. J. Anghileri and J. Robert, eds.), pp. 37–58, CRC Press, Boca Raton, Florida.

Landry, J., Bernier, D., Chretién, P., Nicole, L. M., Tanguay, R. M., and Marceau, N., 1982, Synthesis and degradation of heat shock proteins during development and decay of thermotolerance, *Cancer Res.* **42:**2457–2461.

Landry, J., Samson, S., and Chretién, P., 1986, Hyperthermia-induced cell death, thermotolerance and heat shock proteins in normal, respiration-deficient and glycolysis-deficient Chinese hamster cells, *Cancer Res.* **46:**324–327.

Lee, H. C., and Steinhardt, R. A., 1981, Observations on intracellular pH during cleavage of eggs of *Xenopus laevis, J. Cell. Biol.* **91:**414–419.

Leenders, H. J., Kemp. A., Koninkx, J. F. J. G., and Rosing, J., 1974, Changes in cellular ATP, ADP and AMP levels following treatments affecting cellular respiration and the activity of certain nuclear genes in *Drosophila* salivary glands, *Exp. Cell Res.* **86:**25–30.

Lepock, J. R., Cheng, K.-H., Al-Qysi, H., and Kruuv, J., 1983, Thermotropic lipid and protein transitions in Chinese hamster lung cell membranes: Relationship to hyperthermic cell killing, *Can. J. Biochem. Cell Biol.* **61:**428–437.

Lewis, M. J., and Pelham, H. R. B., 1985, Involvement of ATP in the nuclear and nucleolar functions of the 70 kd heat shock protein, *EMBO J.* **4:**3137–3143.

Li, G. C., and Hahn, G. M., 1980, A proposed operational model of thermotolerance based on effects of nutrients and initial temperature, *Cancer Res.* **40:**4501–4508.

Li, G. C., and Laszlo, A., 1985, Thermotolerance in mammalian cells: A possible role for heat shock proteins, in: *Changes in Eukaryotic Gene Expression in Response to Environmental Stress* (B. G. Atkinson, and D. B. Walden, eds.), pp. 227–256, Academic, Orlando, Florida.

Li, G. C., and Mivechi, N. F., 1986, Thermotolerance in mammalian systems: A review, in: *Hyperthermia in Cancer Treatment,* Vol. I. (L. J. Anghileri and J. Robert, eds.), pp. 59–77, CRC Press, Boca Raton, Florida.

Li, G. C., and Werb, Z., 1982, Correlation between synthesis of heat shock proteins and development of thermotolerance in Chinese hamster fibroblasts, *Proc. Natl. Acad. Sci. USA* **79:**3268–3272.

Lindquist, S., 1980, Varying patterns of protein synthesis in *Drosophila* during heat shock: Implications for regulation, *Dev. Biol.* **77:**463–479.

Lindquist, S., 1986, The heat-shock response, *Annu. Rev. Biochem.* **55:**1151–1191.

Lindquist, S., and DiDomenico, B., 1985, Coordinate and noncoordinate gene expression during heat shock: A model for regulation, in: *Changes in Eukaryotic Gene Expression in Response to Environmental Stress* (B. G. Atkinson, and D. B. Walden, eds.), pp. 71–90, Academic, Orlando, Florida.

Lindquist McKenzie, S., Henikoff, S., and Meselson, M., 1975, Localization of RNA from heat-induced polysomes at puff sites in *Drosophila melanogaster, Proc. Natl. Acad. Sci. USA* **72:** 1117–1121.

Maclean, N., Hilder, V. A., and Baynes, Y. A., 1973, RNA synthesis in *Xenopus* erythrocytes, *Cell Diff.* **2:**261–269.

Maéno, T., 1959, Electrical characteristics and activation potential of *Bufo* eggs, *J. Gen. Physiol.* **43:** 139–157.

Marsden, M., Nickells, R. W., Wang, T. I., Kapoor, M., and Browder, L. W., 1989, The induction of pyruvate kinase synthesis by heat shock in *Xenopus laevis* embryos. Manuscript submitted.

McCormick, W., and Penman, S., 1969, Regulation of protein synthesis in HeLa cells: Translation at elevated temperatures, *J. Mol. Biol.* **39:**315–333.

McGarry, T. J., and Lindquist, S., 1985, The preferential translation of *Drosophila* hsp70 mRNA requires sequences in the untranslated leader, *Cell* **42:**903–911.

McGarry, T. J., and Lindquist, S., 1986, Inhibition of heat shock protein synthesis by heat-inducible antisense RNA, *Proc. Natl. Acad. Sci. USA* **83:**399–403.

Mee, J. E., and French, V., 1986, Disruption of segmentation in a short germ insect embryo. I. The location of abnormalities induced by heat shock, *J. Embryol. Exp. Morphol.* **92:**245–266.

Mirault, M.-E., Goldschmidt-Clermont, M., Moran, L., Arrigo, A. P., and Tissières, A., 1978, The effect of heat shock on gene expression in *Drosophila melanogaster, Cold Spring Harbor Symp. Quant. Biol.* **42:**819–827.

Mirkes, P. E., 1985, Effects of acute exposure to elevated temperatures on rat embryo growth and development *in vitro, Teratology* **32:**259–266.

Mitchell, H. K., and Petersen, N. S., 1981, Rapid changes in gene expression in differentiating tissues of *Drosophila, Dev. Biol.* **85:**233–242.

Mitchell, H. K., and Petersen, N. S., 1982a, Heat shock induction of abnormal morphogenesis in *Drosophila*, in: *Heat Shock. From Bacteria to Man* (M. J. Schlesinger, M. Ashburner, and A. Tissières, eds.), pp. 337–344, Cold Spring Harbor Laboratory, Cold Spring Harbor, New York.

Mitchell, H. K., and Petersen, N. S., 1982b, Development abnormalities in *Drosophila* induced by heat shock, *Dev. Genet.* **3:**91–102.

Mondovi, B., Stron, R., Rotilio, G., Argo, A. F., Cavaliere, R., and Rossi Fanelli, A., 1969, The biochemical mechanism of selective heat sensitivity of cancer cells. I. Studies on cellular respiration, *Eur. J. Cancer* **5:**129–136.

Morange, M., Dui, A., Bensaude, O., and Babinet, C., 1984, Altered expression of heat shock proteins in embryonal carcinoma and mouse early embryonic cells, *Mol. Cell. Biol.* **4:**730–735.

Muller, W. U., Li, G. C., and Goldstein, L. S., 1985, Heat does not induce synthesis of heat shock proteins or thermotolerance in the earliest stages of embryo development, *Int. J. Hypertherm.* **1:** 97–102.

Munro, S., and Pelham, H. R. B., 1985, What turns on heat shock genes?, *Nature (Lond.)* **317:**477–478.

Munro, S., and Pelham, H. R. B., 1986, An hsp70-like protein in the ER: Identity with the 78 kd glucose-regulated protein and immunoglobulin heavy chain binding protein, *Cell* **46:**291–300.

Newport, J., and Kirschner, M., 1982a, A major developmental transition in early *Xenopus* embryos. I. Characterization and timing of cellular changes at the midblastula stage, *Cell* **30:** 675–686.

Newport, J., and Kirschner, M., 1982b, A major developmental transition in early *Xenopus* embryos. II. Control of the onset of transcription, *Cell* **30:**687–696.

Nickells, R. W., 1987, The developmental acquisition and mechanism of thermotolerance in *Xenopus laevis* embryos, Ph.D. thesis, University of Calgary, Calgary, Alberta, Canada.

Nickells, R. W., and Browder, L. W., 1985, Region-specific heat shock protein synthesis correlates with a biphasic acquisition of thermotolerance in *Xenopus laevis* embryos, *Dev. Biol.* **112:**391–395.

Nickells, R. W., and Browder, L. W., 1988, A role for glyceraldehyde-3-phosphate dehydrogenase in the development of thermotolerance in *Xenopus laevis* embryos, *J. Cell Biol.* **107:**1901–1909.

Nover, L., Hellmund, D., Neumann, D., Scharf, K.-D., and Serfling, E., 1984, The heat shock response of eukaryotic cells, *Biol. Zentralbl.* **103:**357–435.

Nuccitelli, R., Webb, J. D., Lagier, S. T., and Matson, G. B., 1981, ^{31}P NMR reveals an increase in intracellular pH after fertilization in *Xenopus* eggs, *Proc. Natl. Acad. Sci. USA* **78:**4421–4425.

Ovsenek, N., and Heikkila, J. J., 1988, Heat shock-induced accumulation of ubiquitin mRNA in *Xenopus laevis* embryos is developmentally regulated, *Dev. Biol.* **129:**582–585.

Parag, H. A., Raboy, B., and Kulka, R. G., 1987, Effect of heat shock on protein degradation in mammalian cell: Involvement in the ubiquitin system, *EMBO J.* **6:**55–61.

Parker, C. S., and Topol, J., 1984, A *Drosophila* RNA polymerase II transcription factor binds to the regulatory site of an hsp70 gene, *Cell* **37:**273–283.

Patzer, E. J., Schlossman, D. M., and Rothman, J. E., 1982, Release of clathrin from coated vesicles dependent upon a nucleoside triphosphate and cytosol fraction. *J. Cell Biol.* **93:**230–236.

Pekkala, D., Heath, I. B., and Silver, J. C., 1984, Changes in chromatin and the phosphorylation of nuclear proteins during heat shock in *Achlya ambisexualis*, *Mol. Cell. Biol.* **4:**1198–1205.

Pelham, H. R. B., 1982, A regulatory upstream promoter element in the *Drosophila* hsp70 heat-shock gene, *Cell* **30:**517–528.

Pelham, H. R. B., 1984, hsp70 accelerates the recovery of nuclear morphology after heat shock, *EMBO J.* **3:**3095–3100.

Pelham, H. R. B., 1986, Speculations on the functions of the major heat shock and glucose-regulated proteins, *Cell* **46:**959–961.

Pelham, H. R. B., and Bienz, M., 1982, A synthetic heat-shock promoter element confers heat-inducibility on the herpes simplex virus thymidine kinase gene, *EMBO J.* **1:**1473–1477.

Petersen, N. S., and Mitchell, H. K., 1981, Recovery of protein synthesis after heat shock: Prior heat treatment affects the ability of cells to translate mRNA, *Proc. Natl. Acad. Sci. USA* **78:**1708–1711.

Petersen, N. S., and Mitchell, H. K., 1982, Effects of heat shock on gene expression during development: Induction and prevention of the multihair phenocopy in *Drosophila*, in: *Heat Shock. From Bacteria to Man* (M. J. Schlesinger, M. Ashburner, and A. Tissières, eds.), pp. 345–352, Cold Spring Harbor Laboratory, Cold Spring Harbor, New York.

Piper, P. W., Curran, B., Davies, M. W., Lockheart, A., and Reid, G., 1986, Transcription of the phosphoglycerate kinase gene of *Saccharomyces cerevisiae* increases when fermentive cultures are stressed by heat shock, *Eur. J. Biochem.* **161:**525–531.

Pleet, H., Graham, J. M., Jr., and Smith, D. W., 1981, Central nervous system and facial defects associated with maternal hyperthermia at four to 14 weeks' gestation, *Pediatrics* **67:**785–789.

Riabowol, K. T., Mizzen, L. A., and Welch, W. J., 1988, Heat shock is lethal to fibroblasts microinjected with antibodies against hsp70, *Science* **242:**433–436.

Ritossa, F. M., 1962, A new puffing pattern induced by heat shock and DNP in *Drosophila*, *Experientia* **18:**571–573.

Robbins, E., Pederson, T., and Klein, P., 1970, Comparison of mitotic phenomena and effects induced by hypertonic solutions in HeLa cells, *J. Cell Biol.* **44:**400–416.

Roccheri, M. C., Di Bernardo, M. G., and Giudice, G., 1981, Synthesis of heat shock proteins in developing sea urchin, *Dev. Biol.* **83:**173–177.

Rodenhiser, D. L., Jung, J. H., and Atkinson, B. G., 1986, The synergistic effect of hyperthermia and ethanol on changing gene expression of mouse lymphocytes, *Can. J. Genet. Cytol.* **28:**1115–1124.

Roosenburg, W. H., Wright, D. A., and Castagna, M., 1984, Thermal tolerance by embryos and larvae of the surf clam *Spisula solidissima*, *Environ. Res.* **34:**162–169.

Rothman, J. E., and Schmid, S. L., 1986, Enzymatic recycling of clathrin from coated vesicles, *Cell* **46:**5–9.

Saborio, J. L., Pong, S.-S., and Koch, G., 1974, Selective and reversible inhibition of protein synthesis in mammalian cells, *J. Mol. Biol.* **85:**195–211.

Sachs, M. M., and Freeling, M., 1978, Selective synthesis of alcohol dehydrogenase during anaerobic treatment of maize, *Mol. Gen. Genet.* **161:**111–115.

Sachs, M. M., Freeling, M., and Okimoto, R., 1980, The anaerobic proteins of maize, *Cell* **20**:761–767.

Schlesinger, M. J., 1986, Heat shock proteins: The search for functions, *J. Cell Biol.* **103**:321–325.

Schlesinger, M. J., Ashburner, M., and Tissières, A. (eds.), 1982, *Heat Shock. From Bacteria to Man,* Cold Spring Harbor Laboratory, Cold Spring Harbor, New York.

Scott, M. P., and Pardue, M. L., 1981, Translational control in lysates of *Drosophila melanogaster* cells, *Proc. Natl. Acad. Sci. USA* **78**:3353–3357.

Shiokawa, K., Misumi, Y., and Yamana, K., 1981a, Demonstration of rRNA synthesis in pre-gastrular embryos of *Xenopus laevis, Dev. Growth Diff.* **23**:579–587.

Shiokawa, K., Tashiro, K., Misumi, Y., and Yamana, K., 1981b, Non-coordinated synthesis of RNA's in pre-gastrular embryos of *Xenopus laevis, Dev. Growth Diff.* **23**:589–597.

Shiu, R. P. C., Pouyssegur, J., and Paston, I., 1977, Glucose depletion accounts for the induction of two transformation-sensitive membrane proteins in Rous sarcoma virus-transformed chick embryo fibroblasts, *Proc. Natl. Acad. Sci. USA* **74**:3840–3844.

Shuey, D. J., and Parker, C. S., 1986, Binding of *Drosophila* heat-shock gene transcription factor to the hsp70 promoter. Evidence for symmetric and dynamic interactions, *J. Biol. Chem.* **261**:7934–7940.

Smith, D. W., Clarren, S. K., and Sedgwick, Harvey, M. A., 1978, Hyperthermia as a possible teratogenic agent, *J. Pediatr.* **92**:878–883.

Spradling, A., Penman, S., and Pardue, M. L., 1975, Analysis of Drosophila mRNA by in situ hybridization: Sequences transcribed in normal and heat shocked cultured cells, *Cell* **4**:395–404.

Storti, R. V., Scott, M. P., Rich, A., and Pardue, M.-L., 1980, Translational control of protein synthesis in response to heat shock in D. melanogaster cells, *Cell* **22**:825–834.

Thomas, G. P., Welch, W. J., Mathews, M. B., and Feramisco, J. R., 1982, Molecular and cellular effects of heat shock and related treatments of mammalian tissue culture cells, *Cold Spring Harbor Symp. Quant. Biol.* **46**:985–996.

Tissières, A., Mitchell, H. K., and Tracy, U. M., 1974, Protein synthesis in salivary glands of *Drosophila melanogaster:* Relation to chromosome puffs, *J. Mol. Biol.* **84**:389–398.

Tomasovic, S. P., Rosenblatt, P. L., Johnston, D. A., Tang, K., and Lee, P. S. Y., 1984, Heterogeneity in induced heat resistance and its relation to synthesis of stress proteins in rat tumor cell clones, *Can. Res.* **44**:5850–5856.

Tomasovic, S. P., Steck, P. A., and Heitzman, D., 1983, Heat stress proteins and thermal resistance in rat mammary tumor cells, *Radiat. Res.* **95**:399–413.

Ungewickell, E., 1985, the 70-kd mammalian heat shock proteins are structurally and functionally related to the uncoating protein that releases clathrin triskelia from coated vesicles, *EMBO J.* **4**:3385–3391.

Vincent, M., and Tanguay, R. M., 1979, Heat-shock induced protins present in the cell nucleus of *Chironomus tentans* salivary gland, *Nature (Lond.)* **281**:501–503.

Watanabe, M., Nikaido, O., and Sugahara, T., 1984, Simultaneous hyperthermia at 43°C reduces radiation-induced malignant transformation frequencies in golden hamster embryo cells, *Int. J. Cancer* **33**:483–489.

Webb, D. J., and Nuccitelli, R., 1981, Direct measurement of intracellular pH changes in *Xenopus* eggs at fertilization and cleavage, *J. Cell Biol.* **91**:562–567.

Webb, D. J., and Nuccitelli, R., 1985, Fertilization potential and electrical properties of the *Xenopus laevis* egg, *Dev. Biol.* **107**:395–406.

Webster, W. S., and Edwards, M. J., 1984, Hyperthermia and the induction of neural tube defects in mice, *Teratology* **29**:417–425.

Webster, W. S., Germain, M. A., and Edwards, M. J., 1985, The induction of microphthalmia, encephalocele and other head defects following hyperthermia during gastrulation process in the rat, *Teratology* **31**:73–82.

Welch, W. J., and Suhan, J. P., 1985, Morphological study of the mammalian stress response: Characterization of changes in cytoplasmic organelles, cytoskeleton, and nucleoli, and appearance of intranuclear actin filaments in rat fibroblasts after heat-shock treatment, *J. Cell Biol.* **101**:1198–1211.

Welch, W. J., and Suhan, J. P., 1986, Cellular and biochemical events in mammalian cells during and recovery from physiological stress, *J. Cell Biol.* **103**:2035–2052.

Wengler, G., and Wengler, G., 1972, Medium hypertonicity and polyribosome structure in HeLa cells. The influence of hypertonicity of the growth medium on polyribosomes in HeLa cells, *Eur. J. Biochem.* **27**:162–173.

Wiederrecht, G., Shuey, D. J., Kibbe, W. A., and Parker, C. S., 1987, The Saccharomyces and Drosophila heat shock transcription factors are identical in size and DNA binding properties, *Cell* **48**:507–515.

Wignarajah, K., and Greenway, H., 1976, Effect of anaerobiosis on activities of alcohol dehydrogenase and pyruvate decarboxylase in roots of *Zea mays, New Phytol.* **77**:575–584.

Wildt, D. E., Riegle, G. D., and Dukelow, W. R., 1975, Physiological temperature response and embryonic mortality in stressed swine, *Am. J. Physiol.* **229**:1471–1475.

Winning, R. S., and Browder, L. W., 1988, Changes in heat shock protein synthesis and hsp70 gene transcription during erythropoiesis of *Xenopus laevis, Dev. Biol.* **128**:111–120.

Wittig, S., Hensse, S., Keitel, C., Elsner, C., and Wittig, B., 1983, Heat shock gene expression is regulated during teratocarcinoma cell differentiation and early embryonic development, *Dev. Biol.* **96**:507–514.

Woodland, H. R., 1974, Changes in polysome content of developing *Xenopus laevis* embryos, *Dev. Biol.* **40**:90–101.

Wu, C., 1985, An exonuclease protection assay reveals heat-shock element and TATA box DNA-binding proteins in crude nuclear extracts, *Nature (Lond.)* **317**:84–87.

Wu, C., Wilson, S., Walker, B., Dawid, II, Paisley, T., Zimarino, V., and Ueda, H., 1987, Purification and properties of *Drosophila* heat shock activator protein. *Science* **238**:1247–1253.

Zimarino, V., and Wu, C., 1987, Induction of sequence-specific binding of *Drosophila* heat shock activator protein without protein synthesis, *Nature (Lond.)* **327**:727–730.

Zimmerman, J. L., Petri, W., and Meselson, M., 1983, Accumulation of a specific subset of *Drosophila melanogaster* heat shock messenger RNA in normal development without heat shock, *Cell* **32**:1161–1170.

Chapter 7

Transdifferentiation in Animals

A Model for Differentiation Control

DAVID S. McDEVITT

1. Introduction

Transdifferentiation may be generally defined as the change of one recognizable cell type to another different cell type. The term was first used by Selman
and Kafatos (1974) to denote the change of the cuticular cells of the moth larval
silk gland to those producing HCO_3 during metamorphosis/development and
has since been used in many different contexts. So as not to produce a welter of
semantics to replace the term already in use, I shall instead categorize the
phenomenon of transdifferentiation by levels: primary, secondary, and tertiary
transdifferentiation. Primary (or true) transdifferentiation would include the
cell-type conversion or cell metaplasia that is so well documented to occur in
some amphibian eye tissues *in vitro* and in amphibian (newt) eye tissues *in situ*
(Fig. 1). This level is characterized by verifiably postmitotic cells, terminally
differentiated and producing a specific cell product, transforming into a completely different cell type with differing cell product(s). Secondary transdifferentiation is marked by the conversion of those cells or tissues not definite-
ly demonstrable as terminally differentiated, i.e., from an embryonic or
possible stem-cell source. Also included is the concept of transdetermination
(Hadorn, 1965), in which certain groups of cells in *Drosophila* occasionally
become determined or committed to a developmental fate different from that
expected. Tertiary transdifferentiation would encompass other purported/
reported changes of tissue types, e.g., that of muscle to cartilage (Namenwirth,
1974), and of striated to smooth muscle in Anthomedusa as reported by Schmid
and Alder (1984) and Weber *et al.* (1987). The well-known plasticity of plant
tissues, especially *in vitro*, is the rule, rather than the exception, and as a topic
of transdifferentiation is beyond the scope of this chapter.

These three levels of transdifferentiation are reviewed and the relevant

DAVID S. McDEVITT • Department of Animal Biology, School of Veterinary Medicine, University of Pennsylvania, Philadelphia, Pennsylvania 19104-6045.

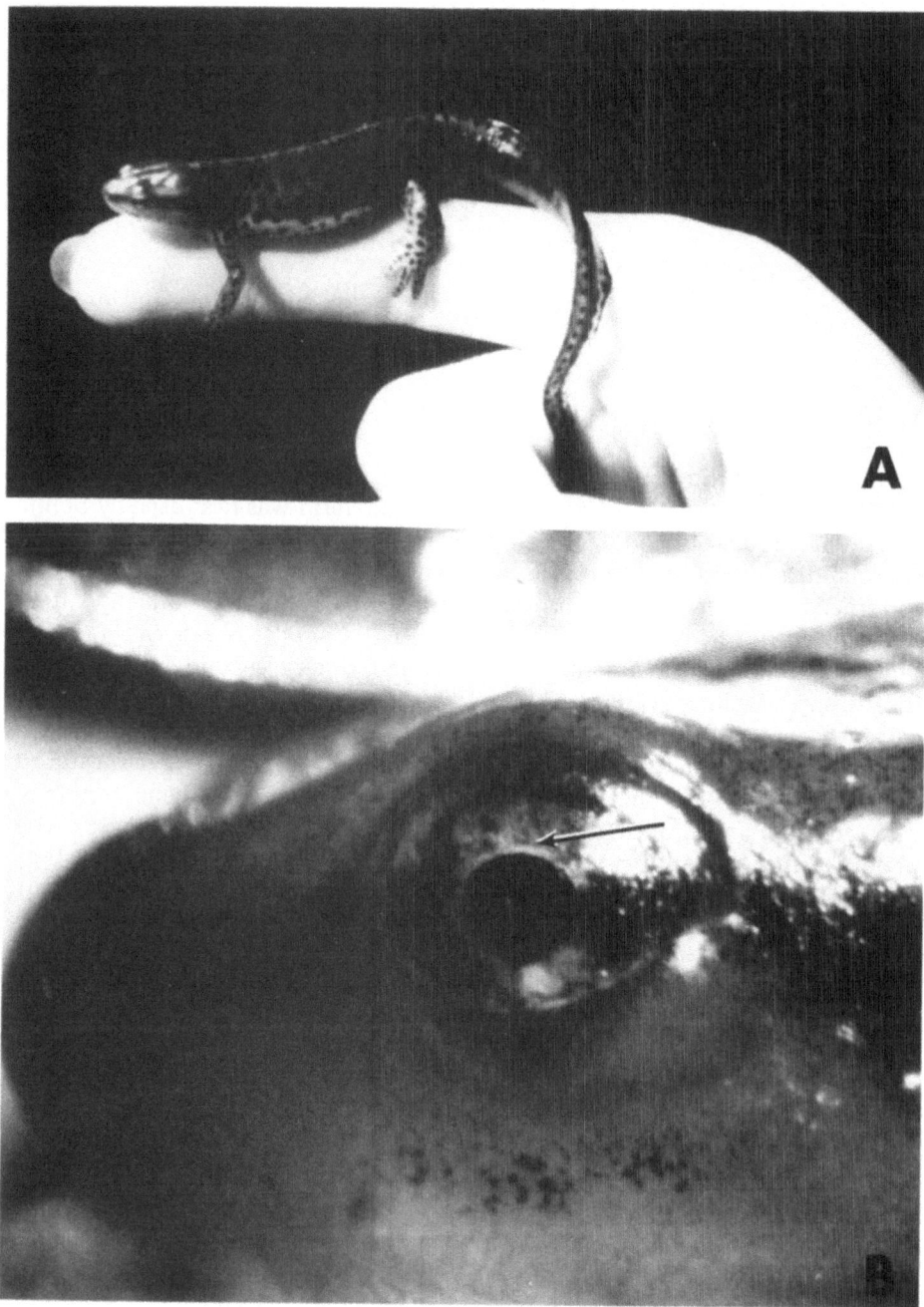

Figure 1. (A) Adult *Notophthalmus viridescens*, the Eastern (U.S.) spotted newt. (B) Eye of *N. viridescens*. Arrow denotes area (dorsal iris) from which a new eye lens will be regenerated upon removal of the original eye lens.

experiments critically examined. Because of greater relevance to problems of cell differentiation, with the least equivocal cell source, most of the discussion centers upon primary (true) transdifferentiation. Possible mechanisms permitting this reprogramming of differentiated cells are explored, with emphasis on the changes in gene expression that accompany, or are causal to, this phenomenon (for a general review of the activation of dormant genes, see DiBerardino et al., 1984).

2. Primary Transdifferentiation

Ocular tissue, for unknown reasons, is the site of most documented cases of transdifferentiation. In vitro, this capability is shared by neural retina cells, pigmented retina cells, and pigmented epithelia of other eye tissues, all derived from adult newt, and embryonic/fetal tissue from other vertebrates. Primary transdifferentiation, involving pigmented iris epithelial cells in vitro, has been documented, as well as in vivo/in situ primary transdifferentiation. The latter is restricted in adult vertebrates to the transdifferentiation of (1) pigmented epithelium of the dorsal iris to lens, after removal of the eye lens (reviewed by Reyer, 1954; Yamada, 1967); and (2) pigmented epithelia of the retina and ciliary body epithelia to neural retina, after removal of neural retina (reviewed by Reyer, 1977). The histology of the tissues in question, as classically studied in the adult newt eye, can be seen in Fig. 2.

Recent research has centered chiefly on the regeneration of the eye lens after lentectomy. Because of its primacy in the field of transdifferentiation research, some background follows. This phenomenon, in which complete removal of the lens is followed by formation of a new lens at the dorsal pupillary margin of the iris epithelium, was first reported by Colucci (1891) and later by Wolff (1895), for whom this (Wolffian) regeneration is named (reviewed by Reyer, 1954, 1977). The capacity for lens regeneration from iris is restricted to a small number of genera, all urodeles, e.g., Triturus, Diemyctilus, Cynopus, and Notophthalmus (synonymous genera); also in Taricha granulosus, Pleurodeles waltlii, Typhlotriton spelaeus, Eurycea lucifuga, Eurycea bislineata (Berardi and McDevitt, 1982), Salamandra s. salamandra, and Salamandra perspicillata (Stone, 1967). Lens regeneration has never been confirmed to occur in birds or mammals, despite many attempts to demonstrate this phenomenon in higher organisms (Reyer, 1954, 1977; Stone, 1967; Yamada, 1967).

The classic idea that the source of the regenerated lens is the dorsal iris epithelial cells (reviewed extensively by Yamada, 1967, 1977, 1982) has been confirmed experimentally. Organ culture of explanted pure populations of dorsal iris epithelial cells has demonstrated that these cells are transformed into lens cells now possessing the lens-specific structural proteins, the crystallins*

---

*The protein content of the vertebrate eye lens is very high (approximately 35%) 80–90% of which is composed of the lens-specific structural proteins; the α, β, and γ—γ replaced by δ in sauropsidans-crystallins (for complete reviews, see Harding and Dilley, 1976; Bloemendal, 1981; Piatigorsky, 1984; Lindley et al., 1985). These water-soluble proteins are classified generally on

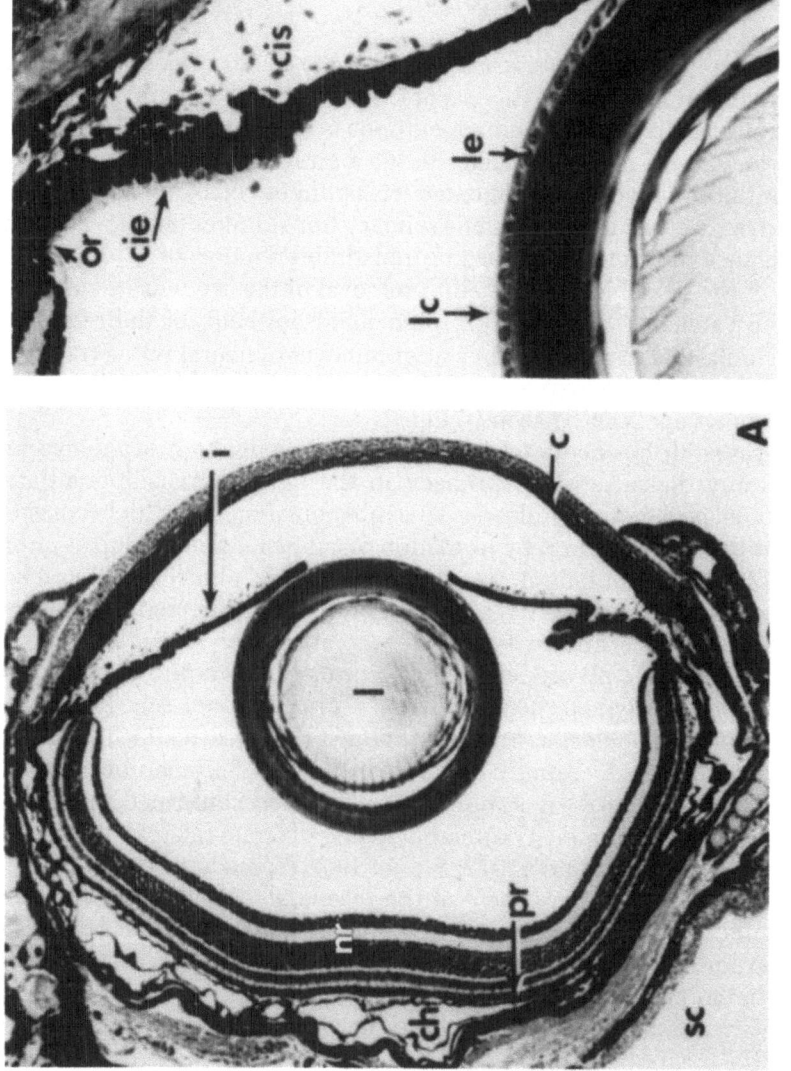

Figure 2. (A) Vertical meridional section of the eye of the adult newt, *Notophthalmus viridescens*. c, Cornea; l, lens; i, iris; nr, neural retina; pr, pigmented retina; ch, choroid; sc, sclera. (B) Detail of cornea, lens, dorsal iris, and ciliary body. le, Lens epithelium; lc, lens capsule; ie, iris epithelium; is, iris stroma; cie, ciliary epithelium; cis, stroma of ciliary body; or, ora serrata; ce, corneal epithelium; cs, corneal stroma (substantia propria). (From Reyer, 1977.)

(Yamada *et al.*, 1973; Yamada and McDevitt, 1974). Immunofluorescence of antibodies, specific for total crystallins and γ-crystallins, applied to sections through such cultures were positive, confirming longstanding microscopic evidence.

The morphogenesis of the regenerating and regenerated lens rudiment is superficially similar to that of the embryonic lens rudiment in normal lens development, despite the fact that it is produced by adult tissue cells (see Fig. 2). Some differences, however, should be noted between embryonic and regenerative formation of the lens. In normal development, a single layer of head ectoderm gives rise to the lens, whereas in regeneration, the double-layered iris, derived from neural ectoderm, is the source of the lens. Also, in normal development, the lens rudiment separates early from surrounding tissue and becomes an independent system, in the sense of no further cell contribution by other tissues, whereas in regeneration, the lens rudiment remains connected with the iris epithelium via a "stalk" and continues to receive cells from it until late in lens development. It has been well documented for several decades that lens regeneration from the dorsal iris *in vivo* is dependent on a stimulus coming from the neural retina (for reviews, see Yamada, 1967; Reyer 1977). Despite much research, such a "factor(s)" has not been identified or characterized directly. Iris epithelial cells in cell culture, however, can be found to transform into lens cells without the presence of retina (Eguchi *et al.*, 1974; McDevitt and Yamada, in Yamada, 1977); the effect of neural retina thus appears to be permissive, rather than instructive, and its presence is not an absolute condition for transformation of iris epithelial cells into lens cells. Yamada and Beauchamp (1978) suggest that the "factor" involved with lens regeneration in situ acts indirectly, to preferentially shorten the cell cycle time in the dorsal iris epithelial cells *in vivo* as compared with ventral iris, permitting the dorsal iris epithelial cells to go through the requisite number of cell cycles for conversion into lens cells before mitosis ceases once again in the iris epithelium.

Concerning the events of lens regeneration per se, the first cellular alterations after lentectomy, in condensation of chromatin and in ultrastructure of the nucleoli of iris epithelial cells (Karasaki, 1964; Dumont *et al.*, 1970), can first be recognized on day 2 after lentectomy [i.e., lens regeneration stages (Yamada, 1967, after Sato, 1940), cf. Fig. 3]. The process continues to days 4–5, with the condensed chromatin characteristic of these cells progressively decreasing and nuclear volume increasing and nucleoli becoming larger.

These structural alterations of nuclei and nucleoli are associated with enhanced incorporation of labeled precursors into ribosomal and heterogeneous nuclear RNA (Reese *et al.*, 1969; Reese, 1973). These events are followed immediately by replication of DNA (Eisenberg and Yamada, 1966), starting on days 3–4.

---

the basis of their net surface charge and molecular weight; most studies have been done on mammalian, especially bovine, lenses because of their size and availability. Thus, α-crystallins have the highest molecular weight and greatest electrophoretic mobility and the γ-crystallins the lowest molecular weight and electrophoretic mobility, with the β-crystallins occupying a rather indeterminate middle ground.

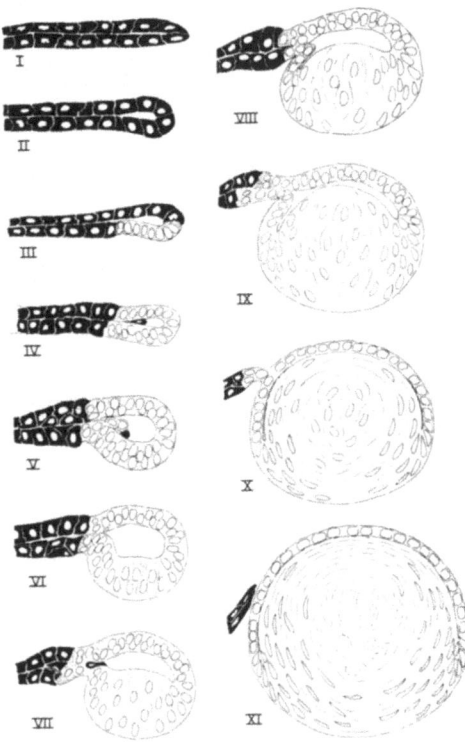

Figure 3. Diagrams showing morphological stages of wolffian lens regeneration (from dorsal iris) in adult *Notophthalmus viridescens*. Sections are through the mid-dorsal pupillary margin of the iris, oriented perpendicular to the main body axis. The cornea is above, the retina below, and dorsal is to the left. The pigmented iris cells are indicated by black cytoplasm, and depigmented regenerate cells by white cytoplasm. Incomplete depigmentation is shown as dotted areas. Iris stroma cells and ameboid cells are not pictured. (From Yamada, 1967.)

Although the iris epithelium of adult newts is normally a nondividing tissue, mitotic activity can be demonstrated in the tissue from day 4 postlentectomy, soon after the start of DNA replication (Yamada and Roesel, 1971). A high level of mitotic activity and DNA replication can be seen throughout the iris epithelium from day 5 to approximately day 20.

When iris epithelial cells are in the cell cycle, the number of melanosomes per cell decreases. During the phase of depigmentation, the volume of cytoplasm increases, extensive intercellular space appears, and cytoplasmic processes are produced in iris epithelial cells (Dumont and Yamada, 1972). Melanosomes are extruded via exocytosis as a membrane-bound complex or individually (Dumont and Yamada, 1972). Macrophages invade the iris epithelium and take up the discharged melanosomes. Depigmentation is most rapid at the dorsal margin of the epithelium. The subsequent increase in nonpigmented iris epithelial cells is due to replication of nonpigmented cells as well as completion of depigmentation in other cells.

Only those iris epithelial cells that become completely depigmented participate in formation of the lens rudiment (vesicle, i.e., with a central cavity), which appears around days 14–16 at the mid-dorsal iris margin. Those proliferating iris epithelial cells located outside the dorsal marginal zone only partially depigment, withdraw from the cell cycle, and resynthesize melanosomes. With iris epithelial cells that have not replicated, they reconstitute the iris epithelium.

In the internal wall of the newly formed lens vesicle, replication of cells soon stops, and they enter the terminal, nondividing phase (Reyer, 1971a). The internal wall thickens by elongation of individual cells, now called primary fibers (Euguchi, 1964). They begin to synthesize crystallins. The external wall of the lens vesicle expands and thins, forming the lens epithelium. In contrast to lens fibers, the lens epithelium retains its proliferative capacity. The ontogeny and localization of specific crystallin classes in lens regeneration have now been determined (McDevitt and Brahma, 1982). β-Crystallins are first detectable in the thickening internal layer of the stage V (cf. Fig. 3) lens (vesicle); γ-crystallins also first appear, albeit erratically, at stage V in the same location. α-Crystallins do not appear until stage VII, in a few cells of the developing lens fiber region, and later in the external layer/lens epithelium.

During subsequent lens regeneration, as the vesicle grows, the connection with the remaining iris epithelium (stalk) forms a pathway for depigmenting iris epithelial cells to join the vesicle epithelium. The secondary lens fibers, which form to the periphery of the primary lens fibers, also cease to divide and begin to elongate. Nuclei of the primary fibers begin to disappear, and secondary fibers derived from lens epithelium continue to be added to the core of primary fibers. The lens epithelium is thus solely responsible for the increased cell numbers in the fiber area. The stalk then ceases as a source of iris epithelial cells into the regenerating lens by losing its connection. The lens is now independent of other eye tissues, and subsequently grows as does the normal lens. The similarity of developmental events with the embryonic lens is thus striking.

Examination of the crystallins produced by the regenerating/regenerated lens indicates a similar, yet not identical, profile. The ontogeny and localization of the crystallins in regenerating lens differs (McDevitt and Brahma, 1982) from that seen in the normal development of the newt lens (McDevitt and Brahma, 1981). Chromatographic and isoelectric focusing analysis of fully regenerated lenses (6 months to 1 year postlentectomy) demonstrates a decrease in high-molecular-weight components (α- and $β_{HMW}$-crystallins) and an increase in low-molecular-weight γ-crystallins (McDevitt, 1982; Borst and McDevitt, 1987). This tendency becomes more marked upon successive or serial lentectomies (Fig. 4)—up to six times—as shown by thin-layer isoelectric focusing and sensitive high-performance liquid chromatography (HPLC) (Borst et al., 1981; Borst, 1984). These results suggest a reactivation of the dormant crystallin genes (DiBerardino et al., 1984) in the newt dorsal iris (pigmented) epithelium, with varying gene expression. Implicit in the relentectomies described is the interesting developmental question of why some dorsal iris epithelial cells will not participate in the formation of an initially regenerated lens, yet upon a successive lentectomy will contribute to another, new lens regenerate—certainly diametric differentiative decisions.

Research on the molecular events of in situ lens regeneration in the newt has been sparse, probably because of logistical problems due to the small amounts of available tissue present. In fact, no molecular biology/nucleic acids studies have been reported since before the advent of recombinant DNA technology. The last reported (Collins, 1974), in an analysis of DNA per se from

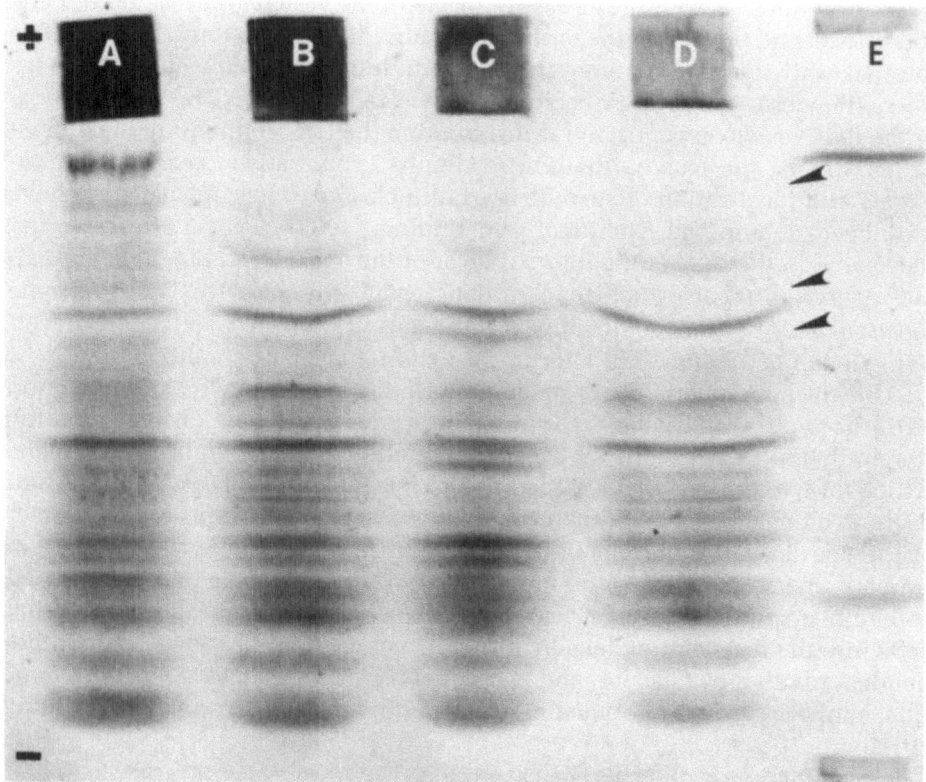

Figure 4. Thin-layer isoelectric focusing on precast 5% polyacrylamide gel (LKB), of normal adult and sequentially regenerated *Notophthalmus viridescens* total soluble lens proteins. The samples were electrophoresed in a pH-gradient 3.5–9.5, anode (+) to cathode (−), according to Brahma and van den Starre (1976). Bours (1971) demonstrated a general separation of the α-, β-, and γ-crystallins at lower (α), intermediate (β), and higher (γ) pH. (A) Normal adult (original) lens proteins. (B) Once-regenerated lens proteins. (C) Twice-regenerated lens proteins. (D) Three-times regenerated lens proteins. (E) Standards, from anode to cathode: lactoglobulin, pH 5.2, and myo-globin, pH 6.85 and 7.35. Arrowheads indicate areas of progressive loss of α- and β-crystallins.

regenerating newt irises (up to day 7), a change in DNA configuration from duplex to single-stranded or nicked, and an increase in template activity. Collins himself notes the difficulty in interpreting these experiments. Earlier reports (Reese *et al.*, 1969; Reese, 1973) on ribosomal RNA enhancement in regenerating newt iris constitute the meager background on newt lens regeneration in this area. The prospects for the use of recombinant techniques in this system are explored at the end of this chapter.

Assuredly, *in vivo* lens regeneration in the newt has the major relevance to problems of health, i.e., replacement of lenses in mammals caused by cataract, as well as to the control of differentiation. The system, however, is complex, and several laboratories have therefore examined transdifferentiation in vertebrate ocular tissues *in vitro*. This has enabled the use of modern research techniques, including recombinant DNA, and, albeit with some provisos discussed below, has provided us with a clearer picture of crystallin gene expression in the transdifferentiative process in cultured cells (reviewed by Okada, 1983; Clayton et al., 1986). This approach began during the 1970s with the tissue, primary, and clonal culture of newt dorsal iris epithelial cells for the first time (Yamada and McDevitt, 1974; Eguchi et al., 1974; Abe and Eguchi, 1977). The conversion of these cells into lens cells (lentoid bodies) involved depigmentation as in situ; however, several important differences became apparent. Unlike the in situ situation, in which only dorsal iris epithelial cells (from the dorsal pupillary margin) participate in lens regeneration, cells derived from both dorsal and ventral iris will transdifferentiate into lens cells with comparable ability. In addition, the time interval needed for appearance of lens phenotype is almost twice as long as that needed in situ (28 versus 15 days). Also, a factor elicited by the neural retina has long been known to be necessary for lens regeneration in situ (Reyer, 1954); iris epithelial cells in vitro have no such requirement.

Yamada (1977, 1982) used these and other observations concerning cell cycle times in newt lens regeneration to hypothesize that the number of cell cycles undergone is a primary determinant of lens regenerative ability in iris epithelial cells. In situ, the dorsal iris epithelium undergoes $\geqq 6$ cell cycles during the same period that the ventral iris cells traverse $\leqq 4$ cell cycles. At the end of this period, lens phenotype (detection of crystallins) appears in the dorsal iris, but the ventral iris cells stop dividing and repigment. In vitro, there is no such block to ventral iris cells, and both dorsal and ventral iris epithelial cells have similar cell cycle times, both can undergo at least 6 cell cycles, and thus ventral iris cells are capable of transdifferentiating into lens cells (Yamada and Beauchamp, 1978). The neural retina in vivo supplies a mitogenic factor that enables the population of dorsal iris to progress through at least 6 cell cycles; in vitro, both populations can traverse the critical number of cell cycles, obviating the need for the neural retina. Grigoryan et al. (1987) recently confirmed this putative relationship between cell cycle number (6–7) and ability for transdifferentiation, with newt ventral iris in organ culture with a mitogenic factor. Although support for this hypothesis is largely circumstantial, it provides a framework for discussion and future research heretofore unavailable. Such a possibility, in the context of this and other mechanisms, is discussed in Section 5.2.

The retina of the adult newt closely resembles that of birds and mammals, with a single outer layer of pigmented epithelium closely adherent to an inner multilayer neural or sensory retina (Fig. 2a). If these layers are merely separated (Stone, 1950), or the neural retina is surgically removed or caused to degenerate (Mitashov, 1968; Reyer, 1971b; Keefe, 1973), the pigmented epithelium of the retina (pigmented retina) will undero in situ transdifferentiation into neural retina.

Although not as extensively investigated as transdifferentiation of iris epithelial cells into lens in the adult newt, there is good autoradiographic and functional evidence for this phenomenon (reviewed by Reyer, 1977; Stroeva and Mitashov, 1983). Restricted in adults to newts (although negative in other adult urodeles and anurans, some evidence may exist for tadpoles), the source of the complete new neural retina has been found by autoradiography to be not only from centrally located pigmented retinal epithelial cells but also partially from the "anterior complex" (inner epithelial layer at the junction of the *pars ciliaris retinae* and *ora serrata*). The latter component apparently contributes to the peripheral part of the regenerated neural retina.

Analogous to regeneration of lens from iris epithelial cells, the pigmented retinal epithelial cells will, during the course of transdifferentiation, depigment, extrude melanosomes, and re-enter the cell cycle (from which they had withdrawn), with an outer layer then redifferentiating into pigmented epithelium and an inner layer becoming increasingly stratified and histologically differentiated as neural retina (Fig. 5). This is reminiscent of the divergent fate of the dorsal iris epithelial cells upon lentectomy. Also strikingly similar is the reported acceleration of the cell cycle in the regenerating neural retina compared with the pigmented epithelium (Mitashov, 1969), creating two populations of cells with differing proliferative capabilities. After several months, the resulting neural retina exhibits a normal electroretinogram and vision (Cronly-Dillon, 1968; Dabagyan and Oganesyan, 1971).

Thus, in many ways transdifferentiation *in situ* of the histologically simple pigmented retinal epithelial cells into the complex neural retina resembles that of iris epithelial cells into lens. That the resultant tissue exhibits even greater diversity of cell type than does lens recommends this rather neglected system for further study.

3. Secondary Transdifferentiation

Included in secondary transdifferentiation are reported cases of transdifferentiation that (1) are not unequivocal, as is *in situ* regeneration of neural retina from retinal pigmented epithelium or lens from dorsal iris epithelium, or *in vitro* conversion of iris epithelial cells to lens; (2) cannot document/demonstrate restricted cell lineage; or (3) arise from cells not withdrawn from the cell cycle. It is clear, however, that the following cell types would never normally express the resulting altered differentiated state. The conversion from one phenotype to another is striking and will provide us with insight into the molecular events of transdifferentiation.

The system most commonly used and most valuable (reviewed by Clayton *et al.*, 1986; Moscona and Linser, 1983) is that involving *in vitro* culture of embryonic chick neural retina and retinal pigmented epithelium (usually 3.5–9 days of incubation age). Coulombre *et al.* (1963) reported for neural retina, and Stroeva and Mitashov (1983) for pigmented retina, that mitosis is still present in some sectors of these tissues of the embryonic chick eye until 10–12 days; thus, the source of cells in most reports is not totally withdrawn from the cell cycle. In addition, when embryonic chick neural retina cells are clonally cultured, two sets of cells are eventually produced: those expressing the lens

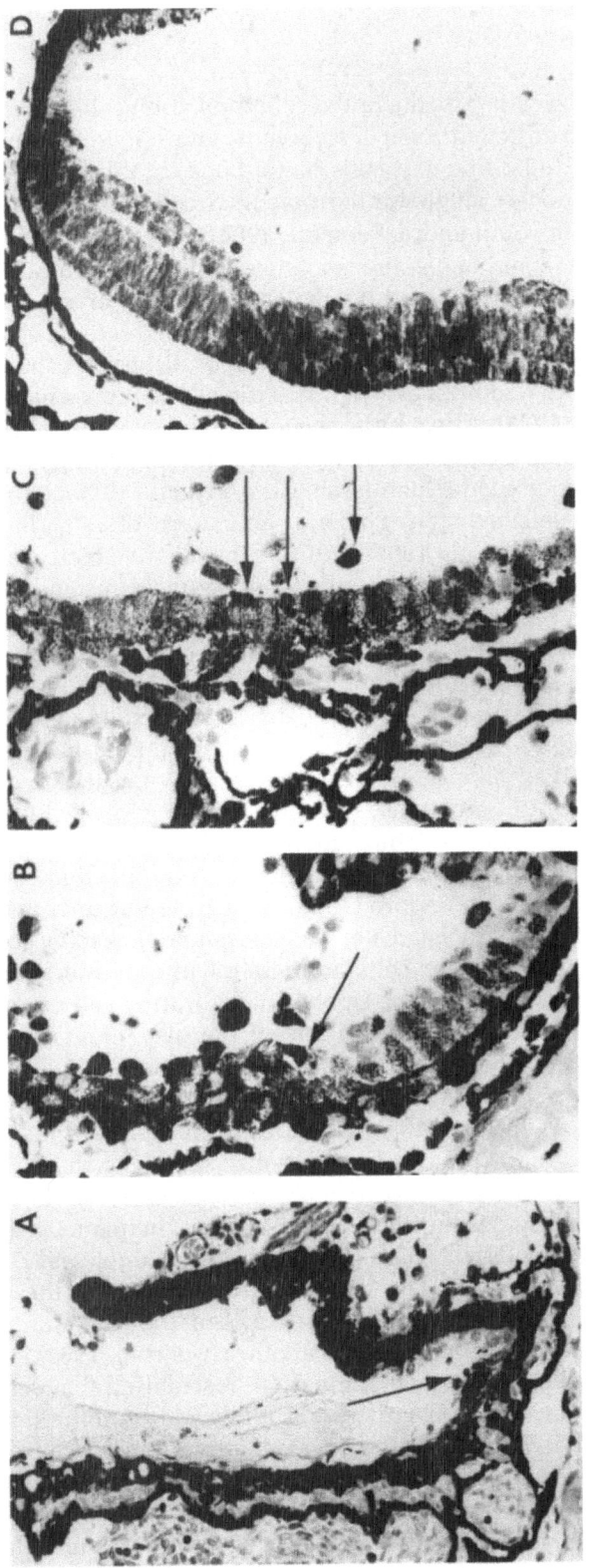

Figure 5. Autoradiographs of neural retina regeneration following retinectomy in adult *Notophthalmus viridescens*. (A) Six days after operation and 2 days after i.p. injection of [³H]thymidine at 4 days after retinectomy. Arrow indicates a group of heavily labeled cells among the unpigmented cells remaining adjacent to the ventral *ora serrata*. Pigmented epithelium is thickened with beginning depigmentation over nuclei but is unlabeled. (B) Fifteen days after operation and 3 hr after isotope injection. Label is present over many cells in both unpigmented ventral regenerate and depigmenting pigmented epithelium. Arrow indicates junction between these two components. (C) Twenty-two days after operation and 2 days after isotope injection. Pigmented epithelium has formed layer of low columnar depigmented cells, most of which are labeled. Scattered pigmented cells are still present (arrows). (D) Differentiating neural retina 35 days after operation and 3 hr after isotope injection. Labeled cells are still present in peripheral growth zone adjacent to *ora serrata* and in parts of retina where nuclear layers have not yet become separate. (From Reyer, 1977.)

phenotype, and pigmented cells (i.e., the initial cell population exhibits hetero-geneity). Thus, in the transdifferentiation of embryonic chick neural retina glial (Muller) cells (Moscona, 1983) into lens cells/lentoid bodies, it is not possible to demonstrate unambiguously single Muller precursor cells because of their lack of a definitive marker (Galli and dePomerai, 1984).

In a series of elegant experiments, the laboratories of Clayton in Scotland and of Okada in Japan have investigated the molecular biology of the *in vitro* transdifferentiation of embryonic neural retina and pigmented retina cells into lens. Most commonly, δ-crystallin mRNA accumulation (δ is the most abundant embryonic chick crystallin) had been assayed; specific cDNA probes have also been used (Yasuda *et al.*, 1983). Apparently, the crystallin genes, especially δ-crystallin, are leaky (Moscona and Linser, 1983), and extremely low levels of transcripts can be found in nonlenticular tissue (Clayton *et al.*, 1986), although translation cannot be detected (de Pomerai and Carr, 1987). *In situ* hybridization analysis (Jeanny *et al.*, 1985) confirms the massive increase in δ-transcripts reported in transdifferentiating cultures of neural retina and retinal pigmented epithelium in 3.5- and 8-day chick embryos (Thomson *et al.*, 1979, 1981). Localized almost exclusively in the nucleus, these δ-transcripts are high molecular weight, indicating that they are unprocessed. Not until 14 days of neural retina culture (7 days for αA) and even later in pigmented epithelium cultures, are these δ mRNAs detectable. Errington *et al.* (1985) have never detected any αA mRNA other than mature, however. It is obvious that there are varying transcriptional controls in transdifferentiating embryonic chick eye tissues, the possible mechanisms for which are discussed in Section 5.3.

Another example of secondary transdifferentiation is the phenomenon of transdetermination in *Drosophila* (Hadorn, 1965). When the imaginal discs of *Drosophila* are fragmented or dissociated and reaggregated, followed by culture in the abdomens of adult females, the cells proliferate without differentiating (hemolymph serves as the culture media). Upon serial culturing and reimplan-tation into a host larva, the developmental state of the disc, elicited by ec-dysone of the host, may be assayed. Such analyses demonstrate that discs are highly determined with disc-specific developmental states reproducible for as much as 55 generations, e.g., genital disc structures (Gehring, 1978). Despite this stability, changes in groups of cells from one disc-specific state of deter-mination to another may occasionally occur, hence transdetermination. These switches are generally to recognizable alternative disc types, in frequencies and directions characteristic for different discs (e.g., leg discs to wing), and *Droso-phila* species (Steiner *et al.*, 1981). Occurring in groups of cells, these are epigenetic rather than somatic mutational changes, and transdetermination involves a switch from one determined state to another (Gehring, 1967). There is multiple cell cooperativity in transdetermination, resembling that seen in homeotic mutants (in which one body structure is replaced by another); mini-mally, polyclones are involved in these switches. [Certainly most transdeter-minative events resemble homeotic mutations, and the relationship between homeotic gene loci (McGinnis *et al.*, 1984) and transdetermination should be investigated.] The transdetermining cells are thus committed, yet undifferentiated, in the sense that their original cell-type specificity has not been expressed at the time the cells are shifted from their original pathway.

4. Tertiary Transdifferentiation

Tertiary transdifferentiation includes those examples of a switch in phenotype from tissue composed of progenitor cells without overt terminal differentiation to one exhibiting properties of the differentiated state. These are frequently the least well documented of cases involving transdifferentiation, without the assurance of known cell lineages in vivo/in situ and without clonal culture of the progenitor cells in vitro. That "stem cells" are the origin of the transdifferentiation observed in these systems is usually clear. Further research, as seen in examples of primary and secondary transdifferentiation, to quantify and delineate is necessary in most cases, so as to attain possible resolution of the cellular pedigree and elucidation of the cell/molecular biology of the transformation. A discussion of several representative cases follows.

When *Xenopus laevis* larvae are lentectomized, a new lens regenerates from the inner cell layer of the outer cornea (Freeman, 1963; Waggoner, 1973). Regeneration of the lens, with respect to histology and crystallin content, is reminiscent of normal lens development in *Xenopus*, albeit itself somewhat aberrant of other anurans (McDevitt and Brahma, 1973; Brahma and McDevitt, 1974). The necessity of a factor emanating from the neural retina, as in Wolffian lens regeneration, for successful transplant or in vivo corneal regeneration has also been demonstrated (Freeman, 1963; Waggoner, 1973; Filoni et al., 1982). Unlike the dorsal iris epithelium in Wolffian lens regeneration, however, the transdifferentiating source is undifferentiated and still in the cell cycle (Waggoner and Reyer, 1975). It is not possible to elicit lens regeneration from the cornea in *adult X. laevis*.

It is known that skeletal muscle (embryonic rat) can form hyaline cartilage in vitro and bone in vivo, when demineralized bone in some form is present (Urist, 1965; Nogami and Urist, 1974; Nathanson et al., 1978). This can be considered as differentiated cells capable of altering their phenotype with the proper stimulus and is an inherently interesting phenomenon. Nathanson and Hay (1980) present evidence, however, that suggest both skeletal muscle and cartilage arise from a common cell precursor. Even though the cells would not have formed cartilage if the extract were not present, the transformation is viewed as a continued differentiation, elicited by a signal resulting in a phenotypic change.

Other examples of tertiary transdifferentiation are as follows:

1. Transformation of rat pancreatic cells (of questionable origin) into hepatocytes through the use of drugs (ciprofibrate) or by Cu^{2+} depletion (Reddy et al., 1984; Rao et al., 1985) [It is important to note the embryonic origin of part of the pancreas from liver and the possibility of trapped liver cells in the rudiment.]

2. Progressive restriction of developmental fates of the neural crest, wandering cells that will eventually form a plethora of structures, including pigment cells, neurons, and glial cells of the peripheral nervous system, the adrenal medulla, and head and face structures [This varied phenotypic expression is apparently brought about by the local milieu in which the neural crest cells, or a subpopulation, find themselves (Le Douarin, 1982; Weston, 1982). The provenance of this signal is unknown and acts on what can be considered originally as stem cells; it is unclear when neural crest derivatives are cultured (e.g., non-neuronal cells of sensory ganglia that will melanize), whether we are dealing with a homogeneous stem cell population or a determined subpopulation of these cells.]

3. Changes in their differentiated state of the epithelial cells (and some nerve cells) in Hydra during their constant mitotic division and change in location during growth (Bode and David, 1978)

4. Limb regeneration *in vivo* in newts, in which pluripotential fibroblasts of the dermis are converted into chondrocytes (Namenwirth, 1974)

Also included in this category of transdifferentiation is the elegant research of Schmid and Alder (1984) and Weber *et al.* (1987) on the transdifferentiation of striated muscle cells to smooth muscle cells in anthomedusae, marine jellyfish. In an earlier *in vitro* system, they demonstrated the only case in which cultured animal cells can transform into a functional organ (striated muscle into the sexual and feeding organ). In addition, fragments containing 20–50 striated muscle cells, after collagenase treatment to "destabilize," are cultured in artificial seawater. Transformation (only with this treatment) to new cell types (including smooth muscle cells) occurs. According to the authors, the cultured "piece" is not contaminated with other cells, and the striated muscle has an appropriate ultrastructural profile. The resulting smooth muscle cells exhibit *de novo* flagella and nonstriated contractile myofilaments. Inhibition of transcription and translation prevents this transdifferentiation. This work is by far the "cleanest" demonstration of transdifferentiation in invertebrates (and many vertebrates), but the phylogenetic niche of this organism must be considered. In contrast to what we know of vertebrate skeletal muscle, composed of multinucleated myotubes, terminally differentiated and stable in culture (Stockdale and Holtzer, 1961), medusa striated muscle cells are mononucleated and, after treatment, capable of change *in vitro*. Although such coelenterates exhibit specialized cell types, they generally have prodigious powers of regeneration. The spectrum of these types is limited, however (mesoderm is not present), and the versatility and scope of expression of a smaller number of individual cell types than that of vertebrates necessitate an inherently greater differentiative plasticity per cell type. Because of this less restrictive developmental birthright, the eventual information concerning gene expression may not be directly applicable. The system does provide us with a unique opportunity for elucidation of evolutionary links in metazoan differentiation mechanisms.

It can be seen that an obvious hierarchy exists in the expression of trans-differentiation in animals. The levels described in an attempt to discriminate among these examples is admittedly arbitrary, and overlap is possible. These inherent differences are valuable, however, in an effort to highlight those most stringent (primary) cases of transdifferentiation. An understanding of the experimental and biological parameters surrounding this spectrum will lead to the formulation of new and more rewarding interpretations and approaches to the problem of transdifferentiation, and ultimately phenotype expression.

5. Possible Mechanisms in Transdifferentiation

The most likely general candidates for roles in effecting transdifferentiation now appear to be alterations in the cell surface, growth factors, and elements affecting gene expression. A discussion of, and evidence for, these mechanisms, probably acting in concert rather than individually, follows. Most of the recent research that implicates them have used transdifferentiating eye tissues (usually of embryonic chick) as an experimental system.

5.1. Alterations in the Cell Surface

Transdifferentiation studies of embryonic chick, adult newt, and fetal human eye tissues have uniformly concluded that disruption of the tissue in some way is necessary for transdifferentiation to take place (e.g., Lopashov, 1977; Clayton, 1982; Yamada, 1982). The intact basement membrane of these tissues is thought to act as a stabilizer, to reduce the frequency of possible transformation *in situ*. Thus *in vitro* culture is usually necessary to elicit their transdifferentiation capability, and only under specified conditions, e.g., aggregate, rather than spreading, cultures will not transdifferentiate (Okada, 1983), nor will cultures with low-serum concentrations, adult serum, supplementary glucose, or F-12/M-199 media (Galli and dePomerai, 1984). Only newt eye tissue among adult vertebrates will undergo primary transdifferentiation *in situ*. Even then, removal of the normal lens or neural retinas to initiate regeneration from dorsal iris epithelium and pigmented retina, respectively, disrupts the intimate contact of the pupillary margin of the eccentrically placed lens with the dorsal iris or the contact of original neural retina with adjacent pigmented retinal epithelium. Yasuda (1979) reported that collagen substrates inhibit transdifferentiation *in vitro* of pigmented retina cells into lentoids. Indeed, Reh *et al.* (1987) demonstrated that laminin promotes *in vitro* transdifferentiation of pigmented retina cells into cells that bind a neuron-specific monoclonal antibody—apparently transdifferentiation to lens cells is suppressed. Lentectomy in adult newts stimulates $^{35}SO_4$ incorporation into newly synthesized glycosaminoglycans in newt iris, associated with de-differentiation (with loss of pigment) of iris epithelial cells prior to lens regeneration (Kulyk and Zalik, 1982). Hyaluronidase activity is also increased in the newt dorsal iris (the only lens regenerating sector *in situ*) compared with ventral iris after lentectomy (Kulyk *et al.*, 1987). Again, upon lentectomy in the newt, changes occur in the proliferating iris epithelial cells, resulting in discharge of melanosomes by exocytosis via elevation of intracellular $[Ca^{2+}]$ (Patmore and Yamada, 1982); this elevation is mediated by cyclic adenosine monophosphate (cAMP), whose level is highest at this time, and the cell configuration also becomes altered (Ortiz *et al.*, 1973; Velazquez and Ortiz, 1980). Zalik and Scott (1972, 1973) had earlier reported changes in electrophoretic mobilites of newt iris epithelial cells after lentectomy, indicating alterations in cell surface net charge. So, too, in the transdifferentiation of striated to smooth muscle in medusae (Weber *et al.*, 1987) discussed previously, collagenase treatment of the striated muscle cell fragment is necessary to "destabilize" the cells and permit formation of smooth muscle. Thus, we have increasing direct and indirect evidence of the importance of the cell surface in transdifferentiative events.

5.2. Growth Factors

The necessity of a neural retina factor necessary for newt lens regeneration *in situ* from dorsal iris has been noted. Earlier, lens regeneration was found to occur in intact irises *in vitro* if neural retina was provided (Yamada *et al.*, 1973) or if the iris was cultured in contact with pituitary glands (Connelly *et al.*, 1973). A mitogenic factor has been isolated from bovine retina (Barritault *et al.*, 1981; Courty *et al.*, 1985) and looms as a candidate for the as yet uncharacterized neural retina factor operative in newt lens regeneration. The extract has been fractionated into eye-derived growth factors (EDGF) I, II, and III. EDGF I and II have now been reported to be the retinal forms of basic and acidic Fibroblast Growth Factor from pituitary gland. Preliminary observations with an EDGF preparation (gift of Y. Courtois) injected into the eye of lentectomized newts whose dorsal iris had been removed, suggest a role in stimulating lens regeneration. Ventral iris, normally non-lens regenerating *in situ*, subsequently will frequently regenerate a lens (Borst, 1984). Using purified EDGF fractions, two laboratories recently reported success in eliciting lens regeneration from cultured newt irises. Cuny *et al.* (1986) report a stimulation of lens regeneration with EDGF II and III, while Connelly and Green (1987) find that only EDGF III has lens-regeneration stimulating activity. The discrepancy may be due to differing interpretations of baseline morphological lens regeneration. In any case, a growth factor is now available whose major characteristics (of retinal origin, and mitogenic, so as to increase the number of cell cycles the iris epithelial cells must undergo to form lens) fit the mold of the putative neural retina factor. Further reconciliation of the *in vitro* results, together with extension to the all-important *in vivo* situation (since transdifferentiation can occur so readily in eye tissues *in vitro* under varying conditions without exogenous growth factors) is eagerly awaited. Certainly a recurring problem will be the logistics of the tissue; because of the small amounts available (a single dorsal iris contains approximately 2000 epithelial cells), classic growth factor experiments (receptor affinity, with Scatchard plots, etc.) will always be difficult.

5.3. Elements Affecting Gene Expression

Transdifferentiation *in vitro* or *in vivo* indicates that previously inactive genes become activated, either by explantation to culture or, for example, by lentectomy. What permits these genes to express themselves? Knowledge of gene structure for the differentiated cell product may provide insight as to possible gene structural bases for the plasticity of gene expression. Some of the generalized features of the eukaryotic gene (and its flanking regions) and the processing of its mRNA transcript suggest themselves (reviewed by Breathnach and Chambon, 1981; Darnell, 1982; Marx, 1983). The RNA polymerase (II) molecule probably recognizes and binds to certain regions of DNA that are upstream (toward the 5' end) from the specific initiation site where transcription of DNA into RNA will begin. Certain consensus DNA sequences that have now been found appear to serve as promoter regions for RNA polymerase binding. One of these sites is centered 25–30 bases upstream from the initiation site and has the consensus sequence TATA (the TATA box, or Goldberg–Hogness box). Another region has the consensus sequence CCAAT and is located 70–80 nucleotides upstream from the initiation site. The deletion or alteration of these putative promoter regions is usually known to result in varying degrees of suppression of the levels of transcription, indicating that they are somehow involved in controlling the initiation of transcription. It is now believed that the TATA box is important in determining the precise initiation site and that the other regions upstream are involved in the efficiency of the transcription process. Enhancer elements (whose nucleotide sequence differs in different sources but seem to have a core sequence GTGGXXG, X=A, or T) were first reported in mammalian viruses. Enhancer sequences can stimulate transcription when placed either upstream or downstream of the 5' end of the gene, even when the orientation is inverted, and even from a considerable distance (several kilobases) from the beginning of the gene. They have been proposed as tissue-specific modulators of transcription in some mammalian genes (Gopal *et al.*, 1985; Garabedian *et al.*, 1986; Atchison and Perry, 1986).

There is also growing evidence for the importance of downstream control elements in eukaryotic genes. The results of Charnay *et al.* (1984) and Wright *et al.* (1984), using hybrid gene techniques, suggest that the β-globin gene, although it has functionally important upstream promoter elements, also contains an element within the structural gene necessary for expression in erythroid cells. These findings, together with other evidence of downstream regulatory sequences, e.g., *Xenopus* 5S ribosomal RNA gene (Schissel and Brown, 1984) and some mouse immunoglobulin (Ig) genes, with an enhancer element within an intron (Picard and Schaffner, 1984), illustrate the necessity to consider sequences both upstream and downstream of the transcription start site. It would be remiss not to mention, however, that alterations in the consensus CCAAT box sequence do not always result in reduced transcriptional efficiency, e.g., the herpes simplex virus TK gene injected into *Xenopus* oocytes (McKnight and Kingsbury, 1982), and that some eukaryotic genes do not contain a consensus CCAAT sequence, e.g., insulin and ovalbumin. It appears that these gene sequences are major pieces, but not the sole ones, in the eukaryotic gene regulatory puzzle.

It is only with the transdifferentiating systems (eye tissues) that produce newly synthesized crystallins as their end-cell product that relevant current research can be found. Evidence is beginning to accumulate that such gene structures as above may be associated with crystallin gene expression. Kondoh *et al.* (1983) reported that cloned δ-crystallin gene 1 injected into the nuclei of various mouse somatic cells was expressed poorly in all but lens epithelial cells, indicating that transcription of the genes is tissue specific (even though mouse lens would not normally produce δ-crystallin!). Of special interest is the expression of this gene when the chick DNA segment 5' of the δ-gene (including the promoter region) was replaced with a virus putative promoter sequence; the gene was no longer expressed preferentially in lens epithelial cells. This indicates the importance of the native promoter region in gene expression. A similar reciprocal experiment has demonstrated that a cloned mouse γ-crystallin promoter is active in lens explants derived from 14-day chick embryos but inactive in cells of nonlens origin (Lok *et al.*, 1985). γ-Crystallins are not normally present in the bird lens (McDevitt and Croft, 1977; Treton *et al.*, 1984). Using the pSVO–CAT expression vector in transfected chick embryonic lens epithelia, Borras *et al.* (1985) have found evidence for differential promoter activities for the two δ-crystallin genes. The δ1 gene contained a consensus CCAAT sequence and enhancer-like sequences. Previous reports have indicated little, if any, δ2-crystallin gene activity (Piatigorsky, 1984). Although post-transcriptional events that limit δ2 mRNA formation cannot be ruled out, these results suggest a promoter role in δ-crystallin gene expression. In addition, in a similar expression vector system, a DNA region upstream of the TATA box of the mouse αA crystallin gene has recently been implicated in *tissue-specific* expression. These regulatory sequences promoted CAT activity only in lens cells, while other, noncrystallin, promoter sequences were effective in both lens and nonlens cells (Chepelinsky *et al.*, 1985). Similar results on the mouse αA-crystallin promoter have been reported in the developing lens of transgenic mice containing the recombinant αA-crystallin-CAT gene (Overbeek *et al.*, 1985).

The interaction of these regions of the gene locus with "factors" has been increasingly suggested as a means of modulating gene expression. Treisman (1986), has demonstrated in the c-*fos* promoter region an element responding to serum growth factors; cell fusion experiments with heterokaryons have implicated positive and negative regulatory factors operating upon the rat prolactin promoter region (Lufkin and Bancroft, 1987); Artishevsky *et al.* (1987) report sequences in the 5'- and 3'-termini of a hamster histone gene involved in cell-cycle-related transcriptional regulation; and (more relevantly) Borrás *et al.* (1987), using deletion experiments, suggest the presence of both positive and negative regulatory domains in the promoter region of the δ1-crystallin gene. The DNA footprinting technique has been increasingly used to detect cellular factors that bind with eukaryotic genes, especially to the regulatory sequences previously described. Thus, Zinn and Maniatis (1986) employed it to demonstrate the binding of repressor molecules at two different sites, and a transcription factor at another, in the regulatory region of the human β-interferon (IFN$_\beta$) gene.

Such experiments suggest that this promising approach to the problem of transdifferentiation should have as longer-term objectives (using newt crystallin genes but applicable to other gene products of transdifferentiated cells):

1. Expression of isolated and characterized crystallin genes (e.g., McDevitt et al., 1986) in in vitro systems (e.g., Borrás et al., 1985; Chepelinsky et al., 1985)
2. Determination of the role, if any, of DNA methylation in the in vivo lens regeneration process [Hypomethylation of DNA is frequently associated with regulation of gene expression during development and cell phenotypic change (Jones et al., 1981; Yamada and McDevitt, 1984)]
3. Investigation into the existence of and possible role(s) for a "transcription complex" (as described and reviewed by Brown, 1984), as well as possible factors interacting with the gene regulatory regions (e.g., Zinn and Maniatis, 1986; Lufkin and Bancroft, 1987), as a basis for the differential crystallin gene activity seen in newt iris, before and after lentectomy
4. The use of newt crystallin cDNA probes as markers for differentiation in single and multiple (sequential) lens regeneration (cf. Hejtmancik et al., 1985, in chick lens development) and for chromosomal localization of the newt crystallin genes using in situ hybridization.

Only with such a multifaceted attack upon the cell and gene control mechanisms operative in the varying levels of transdifferentiation can we progress beyond the descriptive stage. Untouched upon in this treatment is the always unstated paradox or dilemma of explaining and reproducing the necessary integration of cells, resulting from transdifferentiation, into well-ordered tissue. In the transdifferentiation of neural retina, pigmented retina, and dorsal iris epithelium into "lens" in vitro, for example, the "lens" (actually lentoid) formed has no external layer/epithelium surrounding a fiber mass, as occurs with in situ lens regeneration. The tissue, cell, and molecular bases of transdifferentiation would provide a necessary framework for understanding the larger problem of differentiation in animals, per se.

ACKNOWLEDGMENTS. I thank my colleagues in the field for their discussions during the preparation of this chapter; the choice and treatment of the subject matter, however, are entirely mine. I am especially grateful to Dr. Randall Reyer, and the Biology Division Editorial Office, Oak Ridge National Laboratory, Oak Ridge, Tennessee, for providing the original photography for use in preparing Figs. 2 and 5 and Fig. 3, respectively. Most of the research cited herein was supported by grant EY-02534 from the National Eye Institute, National Institutes of Health.

References

Abe, S., and Eguchi, G., 1977, An analysis of differentiative capacity of pigmented epithelial cells of adult newt iris in clonal cell culture, *Dev. Growth Diff.* **19**:309–317.

Artishevsky, A., Wooden, S., Sherma, A., Resendez, E., and Lee, A. S., 1987, Cell-cycle regulatory sequences in a hamster histone promoter and their interactions with cellular factors, *Nature (Lond.)* **328**:823–827.

Atchison, M. L., and Perry, R. P., 1986, Tandem kappa immunoglobulin promoters are equally active in the presence of the kappa enhancer; implications for models of enhancer function, *Cell* **46**:253–262.

Barritault, D., Arruti, C., and Courtois, Y., 1981, Is there a ubiquitous growth factor in the eye?, *Differentiation* **18**:29–42.

Berardi, C. A., and McDevitt, D. S., 1982, Lens regeneration from the dorsal iris in *Eurycea bislineata*, the two-lined salamander, *Experientia* **38**:851–852.

Bloemendal, H., 1981, The Lens Proteins, in: *Molecular and Cellular Biology of the Eye Lens* (Bloemendal, ed.), pp. 1–47, Wiley, New York.

Bode, H. R., and David, C. N., 1978, Regulation of a multipotent stem cell, the interstitial cell of hydra, *Prog. Biophys. Mol. Biol.* **33**:189–206.

Borrás, T., Nickerson, J. M., Chepelinsky, A. B., and Piatigorsky, J., 1985, Structural and functional evidence for differential promoter activity of the two linked δ-crystallin genes in the chicken, *EMBO J.* **4**:445–452.

Borrás, T., Parker, D., Wawrousek, E. F., Peterson, C. A., and Piatigorsky, J., 1987, Regulation of the δ-crystallin gene family, *Invest. Ophthalmol. Vis. Sci.* **28**(suppl.):14.

Borst, D. E., and McDevitt, D. S., 1987, Eye lens regeneration and the crystallins in the adult newt, *Notophthalmus viridescens*, *Exp. Eye. Res.* **45**:419–441.

Borst, D. E., 1984, Eye lens regeneration and the crystallins in the adult newt, *Notophthalmus viridescens*, Doctoral thesis, University of Pennsylvania, Philadelphia.

Borst, D. E., DiRienzo, S. M., and McDevitt, D. S., 1981, Sequential differentiative decisions in newt iris epithelial cells, *J. Cell Biol.* **91**:28a.

Bours, J., 1971, Isoelectric focusing of lens crystallins in thin-layer polyacrylamide gels. A method for detection of soluble proteins in eye lens extract, *J. Chromatogr.* **60**:225–223.

Brahma, S. K., and McDevitt, D. S., 1974, Ontogeny and localization of the lens crystallins in *Xenopus laevis* lens regeneration, *J. Embryol. Exp. Morphol.* **32**:783–784.

Brahma, S. K., and van den Starre, H., 1976, Studies on biosynthesis of soluble lens crystallin antigens in the chick by isoelectric focusing in thin-layer polyacrylamide gel, *Exp. Cell Res.* **97**:175–183.

Breathnach, R., and Chambon, P., 1981, Organization and expression of eucaryotic split genes coding for proteins, *Annu. Rev. Biochem.* **50**:349–383.

Brown, D., 1984, The role of stable complexes that repress and activate eucaryotic genes, *Cell* **37**:359–365.

Charnay, P., Treissman, R., Mellon, P., Chao, M., Axel, A., and Maniatis, T., 1984, Differences in human α and β globin gene expression in mouse erythro-leukemic cells: The role of intragenic sequences, *Cell* **38**:251–263.

Chepelinsky, A. B., King, C. R., Zelenka, P. S., and Piatigorsky, J., 1985, Lens-specific expression of the chloramphenicol acetyltransferase gene promoted by 5′ flanking sequences of the murine αA-crystallin gene in explanted chicken lens epithelia, *Proc. Natl. Acad. Sci. USA* **82**:2334–2338.

Clayton, R. M., 1982, Cellular and molecular aspects of differentiation and transdifferentiation of ocular tissue *in vitro*, in: *Differentiation In Vitro* (Yeoman and Truman, eds.), pp. 83–120, Cambridge University Press, London.

Clayton, R. M., Jeanny, J.-C., Bower, D. J., and Errington, L. H., 1986, The presence of extralenticular crystallins and its relationship with transdifferentiation to lens, *Curr. Top. Dev. Biol.* **20**:137–151.

Collins, J. M., 1974, Structural changes in deoxyribonucleic acid during early stages of lens regeneration in *Triturus*, *J. Biol. Chem.* **249**:1839–1857.

Colucci, V. L., 1891, Sulla rigenerazione parziale dell'occhio nei tritoni. Istogenesi e sviluppo, *Mem. R. Accad Sci. Ist. Bologna* **5**:593–629.

Connelly, T. G., Ortiz, J. R., and Yamada, T., 1973, Influence of the pituitary on Wolffian lens regeneration, *Dev. Biol.* **31**:301–315.

Connelly, T. G., and Green, M. S., 1987, Influences of chromatographic fractions of extracts derived from bovine neural retina on newt (*Notophthalmus viridescens*) lens regeneration *in vitro*, *J. Exp. Zool.* **243**:233–243.

Coulombre, A. J., Steinberg, S. R., and Coulombre, J. L., 1963, The role of intraocular pressure in the development of the chick eye, *Invest. Ophthalmol. Vis. Sci.* **2**:83–89.

Courty, J., Loret, C., Moenner, M., Chevallier, B., Lagente, O., Courtois, Y., and Barritault, D., 1985, Bovine retina contains three growth factor activities with different affinity to heparin: Eye-derived growth factor I, II, III, *Biochimie* **67**:265–269.

Cronly-Dillon, J. R., 1968, Pattern of retinotectal connections after retinal regeneration, *J. Neurophysiol.* **31**:410–418.

Cuny, R., Jeanny, J.-C., and Courtois, Y., 1986, Lens regeneration from cultured newt irises stimulated by retina-derived growth factors, *Differentiation* **32**:221–229.

Dabagyan, N. V., and Oganesyan, R. O., 1971, An electrophysiological study of restoration of the function in the regenerating retina of adult newts, *Ontogen. Akad. Nauk. SSSR.* **2**:327–329.

Darnell, J. E., 1982, Variety in the level of gene control in eucaryotic cells, *Nature (Lond.)* **297**:365–371.

dePomerai, D. I., and Carr, A., 1987, Heat shock may relieve a postranscriptional block on δ crystallin synthesis in cultures of chick embryo neuroretinal cells, *Dev. Growth Diff.* **29**:37–46.

DiBerardino, M. A., Hoffner, N. J., and Etkin, L. D., 1984, Activation of dormant genes in specialized cells, *Science* **224**:946–952.

Dumont, J. N., and Yamada, T., 1972, Dedifferentiation of iris epithelial cells, *Dev. Biol.* **29**:385–401.

Dumont, J. N., Yamada, T., and Cone, M. V., 1970, Alteration of nucleolar ultrastructure in iris epithelial cells during initiation of Wolffian lens regeneration, *J. Exp. Zool.* **174**:184–204.

Eguchi, G., 1964, Electron microscopic studies on lens regeneration. II. Formation and growth of lens vesicle and differentiation of lens fibers, *Embryologia* **8**:247–287.

Eguchi, G., Abe, S., and Watanabe, K., 1974, Differentiation of lenslike structure from newt iris epithelial cells *in vitro*, *Proc. Natl. Acad. Sci. USA* **71**:5052–5056.

Eisenberg, S., and Yamada, R., 1966, A study of DNA synthesis during the transformation of the iris into lens in the lentectomized newt, *J. Exp. Zool.* **162**:353–368.

Errington, L. H., Bower, D. J., Cuthbert, J., and Clayton, R. M., 1985, The expression of chick α A₂-crystallin RNA during lens development and transdifferentiation, *Biol. Cell* **54**:101–108.

Filoni, S., Bosco, L., and Cioni, C., 1982, The role of neural retina in lens regeneration from cornea in larval *Xenopus laevis*, *Acta Embryol. Morphol. Exp.* **3**:15–28.

Freeman, G., 1963, Lens regeneration from cornea in *Xenopus laevis*, *J. Exp. Zool.* **154**:39–65.

Galli, M. A. H., and dePomerai, D. I., 1984, Differential effects of culture media on normal and foreign differentiation pathways followed by chick embryo neuroretinal cells *in vitro*, *Differentiation* **25**:238–246.

Garabedian, M. J., Shepherd, B. M., and Wensink, P. C., 1986, A tissue-specific transcription enhancer from the *Drosophila* Yolk Protein 1 gene, *Cell* **45**:859–867.

Gehring, W., 1967, Clonal analysis of determination dynamics in cultures of imaginal discs in *Drosophila melanogaster*, *Dev. Biol.* **16**:438–457.

Gehring, W., 1978, Imaginal discs: Determination, in: *The Genetics and Biology of Drosophila*, Vol. 2c (M. Ashburner and T. R. F. Wright, eds.), pp. 511–554, Academic, London.

Gopal, T. V., Shimada, T., Baur, A. W., and Nienhuis, A. W., 1985, Contribution of promoter to tissue-specific expression of the mouse immunoglobulin kappa gene, *Science* **229**:1102–1104.

Grigoryan, E. N., Mal'Chevskaya, I. E., Mitashov, V. I., Titov, M. I., Rubina, A. Yu., Vinogradov, V. A., 1987, Stimulation of lens regeneration from the ventral iris cells in newts, *Ontogenez* **18**: 605–613.

Hadorn, E., 1965, Problems of determination and transdetermination. *Brookhaven Symp. Biol.* **18**: 148–161.

Harding, J. J., and Dilley, K. G., 1976, Structural proteins of the mammalian lens: A review with emphasis on changes in development, aging and cataract, *Exp. Eye Res.* **22**:1–73.

Hejtmancik, J. F., Beebe, D. C., Ostrer, H., and Piatigorsky, J., 1985, δ- and β-crystallin mRNA levels in the embryonic and posthatched chicken lens: Temporal and spatial changes during development, *Dev. Biol.* **109**:72–81.

Jeanny, J.-C., Bower, D. J., Errington, L. H., Morris, S., and Clayton, R. M., 1985, Cellular heterogeneity in the expression of the δ-crystallin gene in non-lens tissues, *Dev. Biol.* **112**:94–99.

Jones, R. E., DeFeo, D., and Piatigorsky, J., 1981, Transcription and site-specific hypomethylation of the delta-crystallin genes in the embryonic chicken lens, *J. Biol.Chem.* **256**:8172–8176.

Karasaki, S., 1964, An electron microscopic study of Wolffian lens regeneration in the adult newt, *J. Ultrastruct. Res.* **11**:246–273.

Keefe, J. R., 1973, An analysis of urodelan retinal regeneration. I. Studies of the cellular source of retinal regeneration in *Notophthalmus viridescens* utilizing ^{3}H-thymidine and colchicine, *J. Exp. Zool.* **184**:185–206.

Kondoh, H., Yasuda, K., and Okada, T., 1983, Tissue-specific expression of a cloned chick δ crystallin gene in mouse cells, *Nature (Lond.)* **301**:440–442.

Kulyk, W. M., and Zalik, S. E., 1982, Synthesis of sulfated glycosaminoglycan in newt iris during lens regeneration, *Differentiation* **23**:29–35.

Kulyk, W. M., Zalik, S. E., and Dimitrov, E., 1987, Hyaluronic acid production and hyaluronidase activity in the newt iris during lens regeneration, *Exp. Cell Res.* **172**:180–191.

Le Douarin, N. M., 1982, *The Neural Crest*, Cambridge University Press, London.

Lindley, P., Narebor, M. E., Summers, L. J., and Wistow, G., 1985, The structure of the lens proteins, in: *The Ocular Lens* (H. Maisel, ed.), pp. 123–167, Dekker, New York.

Lok, S., Breitman, M. L., Chepelinsky, A. B., Piatigorsky, J., Gold, R. J. M., and Tsui, L.-C., 1985, Lens-specific promoter activity of a mouse γ-crystallin gene, *Mol. Cell. Biol.* **5**:2221–2230.

Lopashov, G. V., 1977, Levels in stabilization of cell differentiation and its experimental transformation, *Differentiation*, **9**:131–137.

Lufkin, L., and Bancroft, C., 1987, Identification by cell fusion of gene sequences that interact with positive trans acting factors, *Nature (Lond.)* **237**:283–286.

McDevitt, D. S., 1982, The crystallins of normal and regenerated newt eye lenses, *Adv. Exp. Biol. Med.* **158**:177–186.

McDevitt, D. S., and Brahma, S. K., 1973, Ontogeny and localization of the crystallins during embryonic lens development in *Xenopus laevis*, *J. Exp. Zool.* **186**:127–140.

McDevitt, D. S., and Brahma, S. K., 1981, Ontogeny and localization of the α, β, and γ crystallins in newt eye lens development, *Dev. Biol.* **84**:449–454.

McDevitt, D. S., and Brahma, S. K., 1982, α, β, and γ crystallins in the regenerating lens of *Notophthalmus viridescens*, *Exp. Eye Res.* **34**:587–594.

McDevitt, D. S., and Croft, L. R., 1977, On the existence of γ crystallin in the bird lens, *Exp. Eye Res.* **25**:473–481.

McDevitt, D. S., Hawkins, J. W., Jaworski, C. J., and Piatigorsky, J., 1986, Isolation and partial characterization of the human αA crystallin gene, *Exp. Eye Res.* **43**:285–291.

McGinnis, W., Garber, R. L., Wirz, J., Kuroiwa, A., and Gehring, W. J., 1984, A homologous protein-coding sequence in *Drosophila* homeotic genes and its conservation in metazoans, *Cell* **37**: 403–408.

McKnight, S. L., and Kingsbury, R., 1982, Transcriptional control signals of eukaryotic protein-coding gene, *Science* **217**:316–324.

Marx, J., 1983, Immunoglobulin genes have enhancers, *Science* **221**:735–737.

Mitashov, V. I., 1968, Autoradiographic investigations into the regeneration of the retina in pectinate newts (*Triturus cristatus*), *Dokl. Akad. Nauk. SSSR* **181**:411–415.

Mitashov, V. I., 1969, Characterization of mitotic cycles of pigmented epithelium and retinal rudiment cells during regeneration of the retina in adult newts (*Triturus cristatus* and *Triturus taeniatus*), *Dokl. Akad. Nauk. SSSR* **189**:831–833.

Moscona, A. A., 1983, On glutamine synthetase, carbonic anhydrase and Muller glia in the retina, in: *Progress in Retinal Research*, Vol. 2 (N. Osborne and G. Chader, eds.), pp. 111–135, Pergamon, Oxford.

Moscona, A. A., and Linser, P., 1983, Development and experimental changes in retinal glial cells: Cell interactions and control of phenotype expression and stability, *Curr. Top. Dev. Biol.* **18**:155–188.

Namenwirth, M., 1974, The inheritance of cell differentiation during limb regeneration in the axolotl, *Dev. Biol.* **41**:42–56.

Nathanson, M. A., and Hay, E. D., 1980, Analysis of cartilage differentiation from skeletal muscle grown on bone matrix. I. Ultrastructural aspects, *Dev. Biol.* **78**:301–331.

Nathanson, M. A., Hilfer, S. R., and Searls, R. L., 1978, Formation of cartilage by non-chondrogenic cell types, *Dev. Biol.* **64**:99–117.

Nogami, H., and Urist, M. R., 1974, Substrate prepared from bone matrix for chondrogenesis in tissue culture, *J. Cell Biol.* **62**:510–519.

Okada, T., 1983, Recent progress in studies of the transdifferentiation of eye tissue *in vitro*, *Cell Diff.* **13**:177–183.

Ortiz, J. R., Yamada, T., and Hsie, A. W., 1973, Induction of the stellate configuration in cultured iris epithelial cells by adenosine and compounds related to adenosine 3′:5′-cyclic monophosphate, *Proc. Natl. Acad. Sci. USA* **70**:2286–2290.

Overbeek, P. A., Chepelinsky, A. B., Khillan, J. S., Piatigorsky, J., and Westphal, H., 1985, Lens-specific expression and developmental regulation of the bacterial chloramphenicol acetyltransferase gene driven by the murine αA-crystallin promoter in transgenic mice, *Proc. Natl. Acad. Sci. USA* **82**:7815–7819.

Patmore, L., and Yamada, T., 1982, The role of calcium in depigmentation of iris epithelial cells during cell-type conversion, *Dev. Biol.* **92**:266–274.

Piatigorsky, J., 1984, Lens crystallins and their gene families. *Cell* **38**:620–621.

Picard, D., and Schaffner, W., 1984, A lymphocyte-specific enhancer in the mouse immunoglobulin κ gene, *Nature (Lond.)* **307**:80–82.

Rao, M. S., Subbarao, V., Scarpelli, D. G., and Reddy, J. K., 1985, Pancreatic hepatocytes in rats, *Toxicologist* **5**:160.

Reddy, J. K., Rao, M. S., Qureshi, S. A., Reddy, M. K., Scarpelli, D. G., and Lalwani, N. D., 1984, Induction and origin of hepatocytes in rat pancreas, *J. Cell Biol.* **98**:2082–2090.

Reese, D. H., 1973, *In vitro* initiation in the newt iris of some early molecular events of lens regeneration, *Exp. Eye Res.* **17**:435–444.

Reese, D. H., Puccia, E., and Yamada, T., 1969, Activation of ribosomal RNA synthesis in initiation of Wolffian lens regeneration, *J. Exp. Zool.* **170**:259–268.

Reyer, R. W., 1954, Regeneration of the lens in the amphibian eye, *Q. Rev. Biol.* **29**:1–46.

Reh, T. A., Nagy, T., and Gretton, H., 1987, Retinal pigmented epithelial cells induced to transdifferentiate to neurons by laminin, *Nature (Lond.)* **330**:68–71.

Reyer, R. W., 1971a, DNA synthesis and the incorporation of labelled iris cells into the lens during lens regeneration in adult newts, *Dev. Biol.* **24**:535–558.

Reyer, R. W., 1971b, The origins of the regenerating neural retina in two species of urodele, *Anat. Rec.* **169**:410–411.

Reyer, R. W., 1977, The amphibian eye: Development and regeneration, in: *Handbook of Sensory Physiology*. Vol. II. Part 5: *The Visual System in Vertebrates* (F. Crescitelli, ed.), pp. 309–390, Springer-Verlag, Berlin.

Sato, T., 1940, Vergleichende studien über die geschwindigkeit der Wolffschen linsenregeneration bei *Triton taeniatus* und bei *Diemyctylus pyrrhogaster*. *Arch. EntwMech. Org.* **140**:573–613.

Schlissel, M. S., and Brown, D. D., 1984, The transcriptional regulation of *Xenopus* 5S RNA genes in chromatin: The role of active stable transcription complexes and histone H1, *Cell* **37**:903–913.

Schmid, V., and Alder, H., 1984, Isolated, mononucleated, striated muscle can undergo pluripotent transdifferentiation and form a complex regenerate, *Cell* **38**:801–809.

Selman, K., and Kafatos, F. C., 1974, Transdifferentiation in the labial gland of silk moths: Is DNA required for cellular metamorphosis?, *Cell Diff.* **3**:81–94.

Steiner, E., Koller-Wiesinger, M., and Nothiger, R., 1981, Transdetermination in leg imaginal discs of *Drosophila melanogaster* and *Drosophila negromelanica*, *Dev. Biol.* **190**:156–160.

Stockdale, F. E., and Holtzer, H., 1961, DNA synthesis and myogenesis, *Exp. Cell Res.* **24**:508–520.

Stone, L. S., 1950, The role of retinal pigment cells in regenerating neural retinae of adult salamander eyes, *J. Exp. Zool.* **113**:9–31.

Stone, L. S., 1967, An investigation recording all salamanders which can and cannot regenerate a lens from the dorsal iris, *J. Exp. Zool.* **164**:87–104.

Stroeva, O. G., and Mitashov, V. I., 1983, Retinal pigment epithelium: Proliferation and differentiation during development and regeneration, *Int. Rev. Cytol.* **83**:221–293.

Thomson, I., dePomerai, D., Jackson, J. F., and Clayton, R. M., 1979, Lens-specific mRNA in cultures of embryonic chick neural retina and pigmented epithelium, *Exp. Cell Res.* **122**:73–81.

Thomson, I., Yasuda, K., dePomerai, D. I., Clayton, R. M., and Okada, T. S., 1981, The accumulation of lens-specific protein and mRNA in cultures of neural retina from 3½-day chick embryos, *Exp. Cell Res.* **135**:445–449.

Treisman, R., 1986, Identification of a protein-binding site that mediates transcriptional response of the c-*fos* gene to serum factors, *Cell* **46**:567–574.

Treton, J. A., Jones, R. E., King, C. R., and Piatigorsky, J., 1984, Evidence against γ-crystallin DNA or RNA sequences in the chicken, *Exp. Eye Res.* **39**:513–522.

Urist, M. R., 1965, Bone induction by autoinduction, *Science* **150**:893–895.

Velazquez, F. M., and Ortiz, J. R., 1980, Intracellular levels of adenosine 3′:5′-cyclic monophosphate in the dorsal iris of adult newt during lens regeneration, *Differentiation* **17**:117–120.

Waggoner, P. R., 1973, Lens differentiation from cornea following lens extirpation or cornea transplantation in *Xenopus laevis*, *J. Exp. Zool.* **186**:97–110.

Waggoner, P. R., and Reyer, R. W., 1975, DNA synthesis during lens regeneration in larval *Xenopus laevis*, *J. Exp. Zool.* **192**:65–72.

Weber, C., Alder, H., and Schmid, V., 1987, *In vitro* transdifferentiation of striated muscle to smooth muscle cells of a medusa, *Cell Diff.* **20**:103–115.

Weston, J. A., 1982, Mobile and social behaviour of neural crest cells, in: *Cell Behavior* (R. Bellairs, A. Curtis, and G. Dunn, eds.), pp. 429–470, Cambridge University Press, London.

Wolff, G., 1895, Entwicklungsphysiologische Studien. I. Die Regeneration der Urodelenlinse, *Roux Arch. Entwicklungsmech. Org.* **1**:380–390.

Wright, S., Rosenthal, A., Flavell, R., and Grosveld, R., 1984, DNA sequences required for regulated expression of β globin genes in murine erythroleukemia cells, *Cell* **38**:265–273.

Yamada, T., 1967, Cellular and subcellular events in Wolffian lens regeneration, *Curr. Top. Dev. Biol.* **2**:247–283.

Yamada, T., 1977, Control mechanisms in cell type conversion in newt lens regeneration, in: *Monographs in Developmental Biology*, Vol. 13, (A. Wolsky, ed.), pp. 1–126, S. Karger, Basel.

Yamada, T., 1982, Transdifferentiation of lens cells and its regulation, in: *Cell Biology of the Eye* (D. S. McDevitt, ed.), pp. 193–242, Academic, New York.

Yamada, T., and Beauchamp, J. J., 1978, The cell cycle of cultured iris epithelial cells: Its possible role in cell-type conversion, *Dev. Biol.* **66**:275–278.

Yamada, T,. and McDevitt, D. S., 1974, Direct evidence for transformation of differentiated iris epithelial cells into lens cells, *Dev. Biol.* **38**:104–118.

Yamada, T., and McDevitt, D. S., 1984, Conversion of iris epithelial cells as a model of differentiation control, *Differentiation* **27**:1–12.

Yamada, T., Reese, D. H., and McDevitt, D. S., 1973, Transformation of iris into lens *in vitro* and its dependency on neural retina, *Differentiation* **1**:65–82.

Yamada, T., and Roesel, M. E., 1971, Control of mitotic activity in Wolffian lens regeneration, *J. Exp. Zool.* **117**:119–128.

Yasuda, K., Okayama, K., and Okada, T., 1983, The accumulation of δ crystallin mRNA in transdetermination and transdifferentiation of neural retina cells into lens, *Cell Diff.* **12**:177–183.

Yasuda, K., 1979, Transdifferentiation of "lentoid" structures in cultures derived from pigmented epithelium was inhibited by collagen, *Dev. Biol.* **68:**618–623.

Zalik, S. E., and Scott, V., 1972, Cell surface changes during dedifferentiation in the metaplastic transformation of iris into lens, *J. Cell Biol.* **55:**134–146.

Zalik, S. E., and Scott, V., 1973, Sequential disappearance of cell surface components during dedifferentiation in lens regeneration, *Nature (New Biol.)* **244:**212–214.

Zinn, K., and Maniatis, T., 1986, Detection of factors that interact with the human β-Interferon regulatory region *in vivo* by DNAase 1 footprinting, *Cell* **45:**611–618.

Chapter 8

Genomic Activation in Differentiated Somatic Cells

MARIE A. DiBERARDINO

1. Introduction

The process of cellular differentiation in developing multicellular organisms culminates in the formation of specialized somatic cell types. Once cell types acquire a specific cellular phenotype for performing a specialized function, no further changes usually occur. We call this the stability of the differentiated state. It is now widely accepted that the process and stability of cell specialization are under genetic control. But the frequency and extent to which this genetic control involves irreversible genetic changes among eukaryotic animals remain unanswered.

Irreversible genetic changes involving DNA losses or whole chromosomes, so far as we know, occur as a regular event in only a few animal species (Hennig, 1986; Tobler 1986; Chapter 2, *this volume*). Examples of such losses are present in some protozoans, nematodes, crustaceans, insects, and mites of the lower animal phyla. Among vertebrates, DNA losses occur in the holocephalan fish, the mammalian Marsupialia, as well as in the genetic rearrangement of the immunoglobulin genes in chicken (Weill and Reynaud, 1987) and mammalian B lymphocytes (Alt *et al.*, 1987) and of certain receptor genes in mammalian T lymphocytes (Marrack and Kappler, 1987). Most of the evidence available today indicates that a large portion of the genome is retained in specialized somatic cells. For example, the activation of dormant genes in specialized cells is a relatively common phenomenon, consistently demonstrated in several experimental systems (DiBerardino *et al.* 1984), including cell cultures (Chapter 5, *this volume*), transdifferentiation (Chapter 7, *this volume*), heterokaryons (H. Harris, 1974; Ringertz and Savage, 1976; Blau *et al.*, 1985), cell hybrids (H. Harris, 1985; Chapter 5, *this volume*), cancer (Chapter 9, *this volume*) and nuclear transplantation into oocytes and eggs (DiBerardino, 1987). In these cases, the activation of dormant genes is accompanied by

MARIE A. DiBERARDINO • Department of Physiology and Biochemistry, The Medical College of Pennsylvania, Philadelphia, Pennsylvania 19129.

changes in the cell phenotype of specialized cells. Therefore, the evidence indicates that many silent genes are maintained in the genome in the absence of expression and that under favorable conditions they are stimulated to function. Nevertheless, a fundamental problem concerning the developmental process in multicellular organisms still remains unsolved, i.e., whether most differentiated somatic cell types retain the genomic totipotency of the zygote nucleus (DiBerardino, 1987, 1988).

The genomic potential of differentiated cells can best be evaluated by the transplantation of nuclei into oocytes and eggs, because this procedure has the potential to test the entire genome. The idea that the gene content of a cell might be tested by injecting a living nucleus into an enucleated egg was suggested by Hans Spemann (1938). The rationale was that the type of development that results is a reflection of the gene content of the introduced nucleus. This idea materialized when Briggs and King (1952, 1960) succeeded in obtaining metamorphosed frogs following transplantation of blastula and early gastrula nuclei into enucleated eggs of Rana pipiens. Later, McKinnell (1962), Gurdon (1962), and subsequently others obtained fertile adult frogs from embryonic nuclei, demonstrating, for the first time, the genomic totipotency of embryonic nuclei. These studies were extended to the insect Drosophila (Illmensee, 1973; Zalokar, 1973) and the teleost fish (Yan et al., 1984); their young embryonic nuclei were also shown to be totipotent. Recent studies in mice indicate that totipotency is restricted to early cleavage stage nuclei. Tsunoda et al. (1987) obtained several fertile mice from nuclei of the 8-cell stage, but no development ensued from nuclei of the inner cell mass of blastocysts; however, Illmensee and Hoppe (1981) reported fertile mice from these latter-stage nuclei. Among the larger mammals, three lambs (Willadsen, 1986) and two calves (Prather et al., 1987) have resulted from 8-cell and 9- to 15-cell-stage nuclei, respectively. So far, the success of nuclear transfers in mammals is limited to stages that are much earlier than that in amphibians (see Section 2); however, mammalian nuclear transfer studies are still in their infancy, and we await future developments in this area.

This chapter deals mainly with nuclear transfers from terminally differentiated somatic cells, i.e., cells in which the genome is imprinted to specify a specific and final cellular phenotype. So far, successful nuclear transfer studies from terminally differentiated somatic cells have been performed in frogs. The results of these studies summarized below demonstrate that the genomes of several differentiated somatic cell types can (1) undergo widespread activation, (2) direct the formation of a multiplicity of cell types, and (3) display genetic multipotentiality, but (4) evidence for genomic totipotency of differentiated somatic cells is still lacking.

2. Nuclear Transfers from Differentiated Somatic Tissues

The critical experiment posed by Spemann (1938) concerned the genetic repertoire of differentiated somatic cells. The term differentiated cell has undergone multiple uses. For example, the unfertilized egg is considered by some to be a highly differentiated cell (gamete), yet at fertilization it becomes totipotent. The critical test of nuclear equivalence can best be assayed on genomes that have been imprinted to specify a specific and final cellular phenotype, i.e., genomes of cells that normally have no other potential in the organism. Here we consider such cell types as definitive tests of the genomic potential of differentiated cells.

Nuclear transfer experiments that assess the genetic potential of differentiated cells must include evidence for two conditions: (1) the donor cells are differentiated, and (2) the nuclear transplants are derived from the donor nucleus, and not from the egg nucleus. We shall consider two series of experiments that differ in the certainty that the donor cells were derived from differentiated cells. The first series (Table I; Section 2.1) comprises the initial attempts to work with differentiated cells at a time when the techniques usually precluded distinction between stem cells and differentiated cells in the tissues employed. In the second series (Table II; Section 2.2), the specialized properties of the donor cells were documented. In both series only those nuclear transplants that developed to postneurula stages, larval (tadpole) stages, and adults will be considered. Postneurula embryos have developed primitive organ systems, which in the latter stages display muscle and heart function. During the early larval stages, the major cell types, tissue, and organ systems have differentiated and are functional. Thus, transplanted nuclei capable of programming eggs to develop to these stages are considered genetically multipotent. Nuclei that direct the formation of fertile frogs are interpreted to be totipotent.

Verification that the nuclear transplants developed from the transplanted test nucleus and not from the egg nucleus that remains due to faulty enucleation has been performed in the following ways. In *Xenopus*, the egg nucleus is inactivated by ultraviolet (UV) irradiation (Gurdon, 1960a) but, since this procedure is successful in only 61–92% of cases (DuPasquier and Wabl, 1977), a genetic marker is required. The most used marker in *Xenopus* has been the 1-nu (nucleolar) mutation (Elsdale *et al.*, 1960) that can be seen in cells from the gastrula stage and beyond. However, as pointed out by DuPasquier and Wabl (1977), it is necessary to verify that one nucleolus is accompanied by a diploid set of chromosomes and does not result from haploidy or aneuploidy. In *Rana pipiens*, triploidy has served as a convenient cytogenetic marker (McKinnell *et al.*, 1969) because the triploid set of chromosomes can be identified throughout the life cycle of the frog. Another verification procedure has also been used in *Rana pipiens* because microsurgical removal of the egg nucleus can be accomplished in 98–100% of cases. The exovate formed at the time of enucleation adheres to the vitelline membrane outside of the egg. This membrane is later removed, sectioned, and stained with the Feulgen procedure that specifically stains DNA and exhibits the presence of the egg nucleus in the exovate. This information together with the determination of chromosome number of the nuclear transplant and the time of first cleavage insures that development ensued from the test nucleus (DiBerardino and Hoffner, 1971; Orr *et al.*, 1986).

Table I. Nuclear Transfers from Differentiated Somatic Tissues[a]

Donor		Identification of donor cells	Total nuclei tested	Total transfers[b] at stage								Verification	Reference
Stage	Tissue			Postneurula		Larva		Metamorphosis		Adult			
				N	%	N	%	N	%	N	%		
Larva	Intestine[x]	Size	726	36	5.0	31	4.3	—	—	42F	0.6	1-nu	Gurdon and Uehlinger (1966)
	Intestine[x]	Size	522	8	2.0	8	2.0	1	0.2	—	—	98% haploids[c]	Marshall and Dixon (1977)
	Epidermis[x]	Nonciliated	440	2	0.4	2	0.4	—	—	1F	0.2	1-nu	Kobel et al. (1973)
	Minced tadpoles[x]	Cell culture	3686	23	0.6	9	0.2	3	0.1	—	—	1-nu[d]	Gurdon and Laskey (1970)
Adult	Liver[x]	Cell line	365	2	0.6	2	0.6	—	—	—	—	—	Kobel et al. (1973)
	Skin, lung, kidney[x]	Cell culture	2322	26	1.1	7	0.3	—	—	—	—	1-nu[d]	Laskey and Gurdon (1970)
	Intestine[x]	Size	1112	10	0.9	6	0.5	—	—	—	—	95% haploids[c]	McAvoy et al. (1975)

[a]Superscript letters: X, Xenopus laevis; F, fertile.
[b]Includes the results of serial transfers.
[c]Denotes efficiency of UV irradiation in controls.
[d]Chromosome number determined in nuclear transplants. 1-nu, one nucleolus.

Table II. Nuclear Transfers from Differentiated Somatic Cells[a]

Donor		Differentiated state of donor cells	Total nuclei tested	Total transfers[b] at stage						Verification	Reference
Stage	Tissue/cell			Postneurula		Tadpole					
						Prefeeding		Feeding[c]			
				N	%	N	%	N	%		
Larva	Melanophores cell culture[X]	Pigmented	257	2	0.8		—	—	—	1-nu	Kobel et al. (1973)
Juvenile frog	Erythrocytes[R]	Shape, hemoglobin	51	4	8.0	4	8.0	3	5.9	Triploid[d]	DiBerardino et al. (1986)
Adult	Erythrocytes[R]	Shape, hemoglobin	130	11	8.5	6	4.6	—	—	Exovate[d]	DiBerardino and Hoffner (1983)
	Erythroblasts[X]	Shape, hemoglobin	442	8	2.0	8	2.0	—	—	1-nu[d]	Brun (1978)
	Skin, cell culture[X]	Keratin antibody	129	6	4.6	4	3.1	—	—	1-nu[d]	Gurdon et al. (1975)
	Spleen[X]	Immunogen antibody	100	6	6.0	6	6.0	—	—	1-nu[d]	Wabl et al. (1975)

[a] Superscript letters: X, Xenopus laevis; R, Rana pipiens.
[b] Includes the results of serial transfers.
[c] Stages of hindlimb bud.
[d] Chromosome number determined in the nuclear transplants.

2.1. Differentiated Somatic Tissues

The advanced development of nuclear transplants obtained from differentiated somatic tissues of *Xenopus* larvae and adults is summarized in Table I. Five adult frogs were obtained from five original larval nuclei, four from nuclei of the larval intestine (Gurdon and Uehlinger, 1966), and one from a nucleus of larval epidermis (Kobel *et al.*, 1973). Three of these adult frogs were fertile—two from intestinal nuclei, and one from a nonciliated cell of the epidermis. The donor cells from the intestine were selected on the basis of large size. The epidermal cell from the experiments of Kobel *et al.* (1973) was identified as nonmotile (nonciliated). Nuclei from epidermal cells that were motile (ciliated), and therefore specialized, did not give advanced development. The authors concluded that the single case showing totipotency was derived from a nonspecialized cell. In addition to the studies tabulated in Table I, endodermal nuclei of the primitive gut taken from postneurula embryos and initial larval stages before tissue differentiation directed *Xenopus* eggs to develop into 20 fertile frogs (~3%), most of which bore the 1-nu genetic marker of the donor nucleus (Gurdon, 1962, 1986).

At least four metamorphosed frogs developed from eggs injected with larval nuclei—one from an intestinal cell selected on the basis of large size (Marshall and Dixon, 1977), and three from cell cultures (Gurdon and Laskey, 1970). The latter three were obtained from cell cultures, initiated by mincing whole tadpoles that were in the stage of beginning circulation in the gills, but it is not known if all the donor cells were differentiated. The above tissues (intestine, epidermis, and cell cultures of tadpoles) also contained nuclei capable of directing eggs to develop into 69 postneurula embryos, and 50 of these proceeded into larval stages.

Nuclear transfers of adult somatic nuclei into *Xenopus* eggs resulted in 38 postneurula embryos, and 15 of these developed into larvae. The somatic nuclei were derived from a cell line established from liver (Kobel *et al.*, 1973), from cell cultures of skin, lung, and kidney (Laskey and Gurdon, 1970), and from *in vivo* intestinal cells selected on the basis of size (McAvoy *et al.*, 1975).

Several conclusions can be made from the above studies. First, cells from both larval and adult tissues contain some nuclei that can direct the formation of the diverse cell types found in postneurula embryos and tadpoles. Therefore, these nuclei display genetic multipotentiality. Second, genetic totipotency has been shown for a few nuclei from advanced embryonic and young larval stages, since these nuclei directed the formation of fertile frogs. Third, the interpretation of these results remains equivocal, because with the exception of the nonciliated epidermal cells, no valid criteria were used to distinguish between stem cells and differentiated cells in the tissues employed.

2.2. Differentiated Somatic Cells

In this section we consider the advanced development obtained from nuclei of donor cells that were unequivocally differentiated. Among the cell types listed in Table II, melanophores, erythrocytes, and erythroblasts were directly identified at the time of nuclear transfer. For tests of melanophore nuclei, short term monolayer cultures from prefeeding *Xenopus* larvae were used, and the melanophore donor cells were identified on the basis of pigment inclusions (Kobel *et al.*, 1973). Their nuclei promoted enucleated eggs to develop into postneurula embryos (0.8%) that contained the 1-nu genetic marker of the donor cells.

Nucleated erythroid cells of amphibians are particularly ideal for nuclear transfer experiments: (1) they are unequivocally differentiated; (2) they can be conveniently isolated from the peripheral blood by low speed centrifugation; (3) the cells exist singly in the plasma, and therefore, no chemicals are required to achieve cell dissociation as is the case when using donor cells from solid tissue; and, most importantly, (4) individual erythroid cells can be directly identified with the stereomicroscope at the time of nuclear transfer because of their oval shape and presence of hemoglobin. Erythrocyte nuclei from juvenile *Rana* frogs promoted the highest yield of tadpoles and expressed the greatest genomic potential among the differentiated somatic cell types tested: 8% of the original nuclear population promoted the formation of prefeeding tadpoles and 5.9% directed the development of *feeding* tadpoles that advanced to hindlimb bud stages, and in the best cases survived up to a month (DiBerardino *et al.*, 1986). All the tadpoles bore the triploid marker of the donor nucleus. Nuclear transfer studies performed on adult erythrocyte nuclei from *Rana* resulted in 4.6% prefeeding tadpoles (DiBerardino and Hoffner, 1983). The origin of the latter tadpoles was verified by recovery of the egg nucleus in the exovate and determination of chromosome number. Another study of adult erythroid nuclei was performed in *Xenopus* (Brun, 1978). In the latter study, erythroblast nuclei (2.0%) promoted development as far as the first tadpole stage at which hatching is initiated, and all these nuclear transplants bore the 1-nu genetic marker.

Two other studies examined the genomic potential of adult skin and spleen cells. The donor cells were derived from a nearly homogeneous population. In the case of the skin cells, correlative studies showed that over 99% of the skin cells from short term culture were producing immunoreactive keratin (Gurdon *et al.*, 1975). Nuclear transfers from similarly cultured skin cells resulted in 3.1% prefeeding tadpoles that carried the 1-nu genetic marker. In the case of the spleen cells, their nuclei directed the development of prefeeding tadpoles (6.0%) that bore the 1-nu genetic marker of the donor cells of which 96.1–98.7% were judged to be producing immunoglobins (Ig) (Wabl *et al.*, 1975).

In summary, nuclei from five documented differentiated somatic cell types promoted the formation of 37 postneurula embryos. With the exception of cultured melanophores, nuclei from the other 4 cell types promoted the formation of 28 tadpoles. Thus, there is evidence for genomic multipotentiality of differentiated somatic cells from the larva, juvenile frog and adult. *However, to date, no nucleus of a documented differentiated somatic cell nor of any adult somatic cell (stem or specialized) has yet been shown to be totipotent.*

3. Nuclear Transfers from Germ Cells

The genomic totipotency of germ cell nuclei certainly does not require formal proof via the nuclear transplantation assay, because the germ cells of an individual retain the repertoire of zygotic genes, except for an occasional mutation. Nevertheless, nuclear transfer tests of germ cells have been informative because they showed that their genome after transplantation into eggs expresses different degrees of developmental potential depending on the stage of differentiation of the donor cells (Table III).

Smith (1965) tested nuclei of primordial germ cells taken from *Rana pipiens* tadpoles at the initial feeding stage. The primordial germ cells are large and yolky and can be easily distinguished from the small somatic cells of the genital ridge. The investigator's efficiency in microsurgical removal of the host nucleus was 99.5%, as judged by the production of androgenetic haploids. Among the 31 normal tadpoles that developed from 7.6% of the injected eggs, 10 were reared for periods of 1–3 months, were then autopsied and found to be normal. Although the precise larval stages were not reported, a photograph of one metamorphosing tadpole was illustrated.

Lesimple *et al.* (1987) tested germ cell nuclei of the salamander *Pleurodeles waltl* taken from progressive stages of larval development up to metamorphosis. The larger germ cells were distinguished from the smaller somatic cells. Among the total injected eggs, 12 (0.4%) attained the feeding larval stage, 9 (0.3%) metamorphosed and 6 (0.2%) developed into adults. Two of the adults were mated and produced normal progeny. The adults arose from experiments in which the donor nuclei were not genetically marked. The authors argue that it is unlikely that the adults arose from the egg nuclei, because in *Pleurodeles* gynogenesis results in 50% ZZ males and 50% WW females, but they obtained 3 ZZ males and 3 ZW females and no WW females. Nevertheless, proof of this interpretation is lacking.

One study examined the developmental potential of adult germ cell nuclei, since their nuclei are derived from differentiated germ cells (DiBerardino and Hoffner, 1971). The cell suspension of donor cells from adult testes was enriched for spermatogonia, and the donor cells were selected on the basis of size. Enucleated *Rana pipiens* eggs injected with these nuclei yielded 4 (3.5%) postneurulae and 1 (0.9%) feeding tadpole. Recovery of the egg nucleus in the exovate indicated that development was directed by the transferred nucleus. Failure to reveal the genomic totipotency of differentiated germ cells in nuclear transplantation experiments indicates that their genome and/or chromatin undergo modifications required for specifying the phenotype of a differentiated cell. Under the experimental conditions, the modifications are stable and prevent the full expression of their genomic potential, but certainly they are not irreversible (DiBerardino and Hoffner, 1971). Lesimple *et al.* (1987) reported a decrease in the percentage of hatching larvae when the germinal nuclei were taken after the mid-larval stages. They suggested that differentiation of germinal stem cells into spermatogonia or oogonia might account for the

Table III. Nuclear Transfers from Germ Cells[a]

	Donor		Total nuclei tested	Total transfers at stage									Reference
				Postneurula		Feeding larva		Metamorphosis		Adult			
Stage	Gonad	Cell		N	%	N	%	N	%	N	%		
Larva	Genital ridge	Primordial germ cells[R]	410	35	8.5	31	7.6[b]	—	—	—	—	Smith (1965)
Larva to metamorphosis	Genital ridge, undifferentiated and differentiated gonads	Primordial germ cells, gonocytes[P]	2986	46	1.5	12	0.4	9	0.3	6	0.2[F]	Lesimple et al. (1987)
Adult	Testis	Enriched for spermatogonia[R]	116	4	3.5	1	0.9	—	—	—	—	DiBerardino and Hoffner (1971)

[a]Superscript letters: R, *Rana pipiens*; P, *Pleurodeles waltl*; 2F, two adults were fertile.
[b]Ten larvae raised, final stage not indicated, one in the climax of metamorphosis; see text.

decrease in successful nuclear transplantations, while the retention of a few stem cells (gonocytes) might account for the successful nuclear transplants.

4. Imprinting of the Genome

Recent nuclear transfer studies in the mouse have shown that normal development requires the presence of both the female and male pronuclei. Embryos with two female or two male pronuclei arrest early in embryogenesis (reviewed in Chapter 4, *this volume*). Those embryos with solely maternal or paternal chromosomes exhibited different phenotypes. Embryos with a diploid set of maternal chromosomes developed into well formed but small 25-somite embryos; however, the extraembryonic tissues were underdeveloped. By contrast, embryos with a diploid set of paternal chromosomes differentiated poorer embryonic structures but better extraembryonic tissues than the embryos with maternal chromosomes. These results indicate that the maternal and paternal genomic contributions to embryogenesis are functionally different and that both are essential form normal development. The investigators have concluded that a differential imprinting of the genome occurs in the male and female gametes during gametogenesis and suggest that different subsets of genes are inactivated in the parental genomes but in a complementary pattern. This imprinting of the genome is reversible, however, because homozygous uniparental mouse embryos in chimeric combination with normal embryos can develop into fertile adults that produce normal gametes derived from the uniparental genome (Anderegg and Markert, 1986). These latter investigators interpret the cases of totipotency to stem from reimprinting the genome during gametogenesis.

In amphibians, gynogenetic diploids produced by suppression of the second polar body can develop into fertile adults. These cases include *Rana pipiens* (Richards and Nace, 1978) and *Xenopus laevis* (Reinschmidt et al., 1985). Thus, if differential imprinting of amphibian parental genomes occurs during gametogenesis, such imprinting must be easily reversed in amphibian embryos.

Amphibian embryos also differ from mouse embryos in their onset of zygotic RNA transcription. In *Xenopus*, the first detectable RNA transcription occurs at the sixth cleavage stage (Nakakura et al., 1987), whereas in murine embryos RNA synthesis is initiated during the first cleavage stage (Clegg and Piko, 1983). The later stage of onset of RNA transcription in frog embryos would permit the genome of a transplanted frog nucleus additional cell cycles to adjust to a new program of gene expression compared with the one cell cycle in the mouse. This difference in the onset of RNA synthesis might account for successful nuclear transplantations from advanced stages of frog cells, whereas successful nuclear transfers in mammals are so far limited to cleavage stage nuclei. Whether the temporal program of gene expression, parental imprinting of genomes or other factors account for the difference in nuclear reprogramming between mammals and frogs remains to be elucidated.

Although the successful cases of amphibian nuclear transplants are those that are frequently emphasized, many frog nuclei from determined regions of embryos before overt tissue differentiation displayed developmental restrictions after transfer to enucleated eggs. Somatic nuclei tested from different embryonic stages, germ layers and primitive organs from various anuran and urodelen amphibians (*Ambystoma, Bufo, Pleurodeles, Rana,* and *Xenopus*) showed a progressive decrease in the percentage of normal nuclear transplants when the donor nuclei were taken from progressively more advanced stages of embryogenesis (reviewed by: Briggs and King, 1959; Gallien, 1966; King, 1966; DiBeradino and Hoffner, 1970; Gurdon, 1974, 1986; McKinnell, 1978; Briggs, 1979). Other main points derived from these studies were (1) a few of nuclei still displayed genetic totipotency, and this was especially evident in the endodermal nuclei of hatched larvae of *Xenopus* (Gurdon, 1962); (2) but most nuclei did not promote normal development of the test eggs; (3) the developmental restrictions analyzed in endodermal (King and Briggs, 1956; Gurdon, 1960b; Subtelny, 1965) and neural nuclear transplants (DiBerardino and King, 1967) were stable, since they could not be reversed through serial transplantation; (4) furthermore, the restrictions expressed by endodermal nuclei were intrinsic since they were not corrected by parabiosing nuclear transplant embryos with normal ones (Briggs *et al.*, 1961), nor could they be corrected when a haploid set of egg chromosomes was combined with the diploid set from the endodermal nucleus (Subtelny, 1965).

Among the *Rana* nuclear transplants derived from endodermal and neural nuclei, a small number displayed a pattern of developmental abnormalities consistent with the origin of their nuclei. During postneurula stages some nuclear transplants derived from late gastrula endodermal nuclei exhibited deficiencies and degenerative changes in ectodermal and mesodermal organs, but not in endodermal tissues (Briggs and King, 1957). This phenotype was not reversed by serial cloning (King and Briggs, 1956), alleviated by parabiosis with normal embryos (Briggs *et al.*, 1961), or associated with obvious chromosomal abnormalities (Briggs *et al.*, 1961; Subtelny, 1965). Similar studies of nuclei from the presumptive medullary plate regions of late gastrula and the definitive medullary plate of early neurula and mid-neurula embryos showed a small group with a phenotype different from the endodermal nuclear transplants. The neural nuclear transplant group (13%) displayed during larval stages cellular deficiencies mainly in mesodermal and endodermal derivatives but showed good differentiation in the organs and tissues of ectodermal origin and did not present detectable chromosomal abnormalities (DiBerardino and King, 1967). These distinct phenotypes suggested that by late gastrulation some endodermal and neural nuclei have acquired stable properties for specific pathways of cell differentiation. Although these phenotypes were described more than 20 years ago, the mechanism probably is analogous to the imprinting of the genome recently described in murine nuclear transplants. In the mouse differential imprinting of parental genomes, presumably during gametogenesis, leads to two distinct phenotypes, depending on whether the conceptus developed from two male or two female pronuclei. In some of the amphibian nuclear transplants, we probably observed the effect of genomic imprinting initiated during embryogenesis. The limitation of these studies is that the data are not extensive. Perhaps, an analysis of the pattern of DNA methylation in the endoderm and neural nuclear transplants would clarify this interesting problem. Presumably, the normal frog nuclear transplants were derived from stem cells present among the other differentiating donor cells and would differ from the abnormal nuclear transplants in their pattern of DNA methylation.

5. Reversal of Genomic Imprinting

The genome of differentiated cells has been imprinted by molecular mechanisms to specify a particular cellular phenotype. Evidence available indicates that this imprinting results from the interaction of chromosomal proteins with the genome and modification of its DNA by methylation. Here we consider the extent to which the genomic imprinting in differentiated somatic frog cells can be erased under differing cytoplasmic environments.

5.1. Nuclear Transfer into First Meiotic Metaphase Oocytes

Nuclei of *Rana* erythrocytes have expressed the greatest genomic potential among the genomes of differentiated somatic cells so far tested (see Section 2.2). The main difference in the experimental procedure was that the erythrocyte nuclei were initially injected into oocytes, whereas the other nuclear types were injected directly into eggs. The standard host for nuclear transplantation has been the egg, the host originally proposed by Spemann (1938). However, we hypothesized that the expression of the genetic potential of differentiated somatic cells might be enhanced if the donor nuclei were first exposed to the molecular components in the oocyte that normally prepare the oocyte chromosomes to participate in fertilization (DiBerardino, 1980). Initially, we tested whether somatic nuclei of embryonic cells would, in fact, function properly in the egg after residing in oocyte cytoplasm (Hoffner and DiBerardino, 1980). Oocytes at the stage of first meiotic metaphase were used for hosts. Such oocytes display in the animal pole the first black dot, indicative of the first meiotic metaphase. A single broken embryonic cell, together with its nucleus, was injected into the animal hemisphere near the equator region. At this time, the oocytes are not activatable, but become so about 24 hr, later when they have matured *in vitro* into eggs (18°C). The matured oocytes (eggs) were activated by penetration of a glass microneedle. Within 10 min, the second black dot, indicative of the second meiotic metaphase of the egg, appeared and was removed microsurgically with a glass microneedle. We found that embryonic somatic nuclei exposed to cytoplasm directing meiotic events could still support development through embryogenesis. Cytological studies performed on a parallel series of nuclear transplants showed that transplanted embryonic nuclei formed metaphase chromosomes on newly induced spindles in concert with the behavior of the oocyte nucleus during first meiotic metaphase. When the matured oocytes were activated, the injected nuclei transformed into pronuclei. Thus, somatic nuclei can reversibly respond to the molecular cues directing meiotic and mitotic events in a sequential manner and still direct the host to develop through embryogenesis. These findings indicated that nuclei of differentiated cells could now be tested in oocytes to determine whether the imprinting of their genome could be erased by exposure to the molecular components of maturing oocytes.

5.2. Genomic Potential of Erythrocytes

The erythrocyte satisfies the most stringent criterion of a differentiated cell (see Section 2.2); it is a noncycling and terminally differentiated cell, one that is almost transcriptionally quiescent. Its oval shape and the presence of hemoglobin (Hb) permit its direct identification under the stereomicroscope ($\times$ 100 magnification) at nuclear transfer. Thus, any development ensuing from the transplantation of an erythrocyte nucleus would unequivocally result from a differentiated genome.

The developmental potential of the adult erythrocyte genome was tested in two ways (DiBerardino and Hoffner, 1983). In one series, the nuclei were injected into enucleated eggs; in another series, the nuclei were injected into oocytes at the stage of first meiotic metaphase (Fig. 1). Some of the blastulae derived from the hosts of each series were used to provide donor nuclei for a second transplant generation. The blastula nuclei were injected singly into enucleated eggs and nuclear clones were produced. None of the transplants from the egg series developed beyond the early gastrula stage. However, among the 12 clones derived from the original oocyte series, eight clones (67%) contained members that developed into postneurula embryos, and six clones (50%) had members that differentiated into prefeeding tadpoles. When erythrocyte nuclei from juvenile frogs (Fig. 2) were tested under the same conditions, they directed the development of feeding tadpoles that formed hind limb buds in 75% of the four nuclear clones (DiBerardino et al., 1986). These tadpoles survived for up to 1 month and are the longest surviving and most advanced nuclear transplants derived from documented differentiated somatic cells (see Table II). The interval from zygote to metamorphosis is approximately 3 months. Thus, the month old tadpoles developed and survived for about one third of this interval.

Several conclusions resulted from the nuclear transfer studies of the noncycling and terminally differentiated erythrocytes of *Rana*. First, the erythrocyte genome retained the genes for directing the differentiation of the cell types, tissues, and organ systems of tadpoles. In fact, the specification of the various cell types in these tadpoles requires the activation of dormant genes that never functioned in the cell lineage of an erythrocyte. Second, exposure of the erythrocyte genome to the cytoplasm of maturing oocytes induced the most widespread activation of the erythrocyte genome ever obtained in an experimental system. In this regard, the imprinting of the genome that formerly spec-

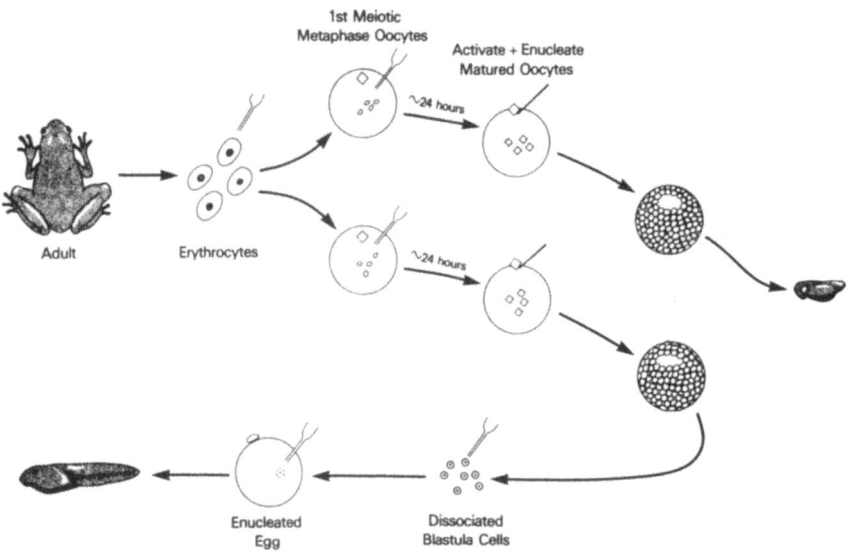

Figure 1. Erythrocytes obtained by intracardiac puncture of adult *Rana pipiens* frogs were broken in distilled water by osmotic shock and microinjected into oocytes near the equator at first meiotic metaphase. Approximately 24 hr later (at 18°C), when the oocyte matured, the matured oocyte (egg) was activated by pricking with a glass needle, and the egg nucleus was removed microsurgically. (Top) Original transplant generation; prehatching tadpoles resulted. In some cases, nuclear transplant blastulae were dissected and their animal hemisphere nuclei were transplanted singly into activated–enucleated eggs. (Bottom) First retransfer generation; swimming tadpoles resulted. Artwork by Betti Goren. (From DiBerardino *et al.,* 1984.) (Copyright 1984 by the AAAS.)

ified the phenotype of an erythrocyte, was erased to the extent of specifying the cell phenotypes of tadpoles. Third, serial transplantation of erythrocyte nuclei through eight transplant generations showed that once the erythrocyte genome is activated by its progression through the cytoplasm of the oocyte and activated egg, it maintains its potential not only for replication, but also for directing the complex functions of organogenesis in excess of 100 cell cycles (Orr *et al.,* 1986). In addition, there was no evidence that the mitotic progeny of the erythrocyte nuclei lost their ability to replicate their genomes and continue cell cycling. Fourth, some tadpoles fed and therefore functioned as independent organisms (Fig. 2). Thus, the developmental block previously exhibited in prefeeding tadpoles from differentiated somatic nuclei (see Table II) was surmounted. The progression of tadpoles to the feeding stage is a critical developmental event, because at this stage maternal (egg) components are largely used. Thereafter, continued development, further differentiation, and survival of the tadpoles are dependent on an external source of nourishment.

An important factor that must be considered in evaluating the formation of nuclear transplants is the contribution of maternal RNAs and proteins to their development. It is certain that the transplanted nucleus contributes a signifi-

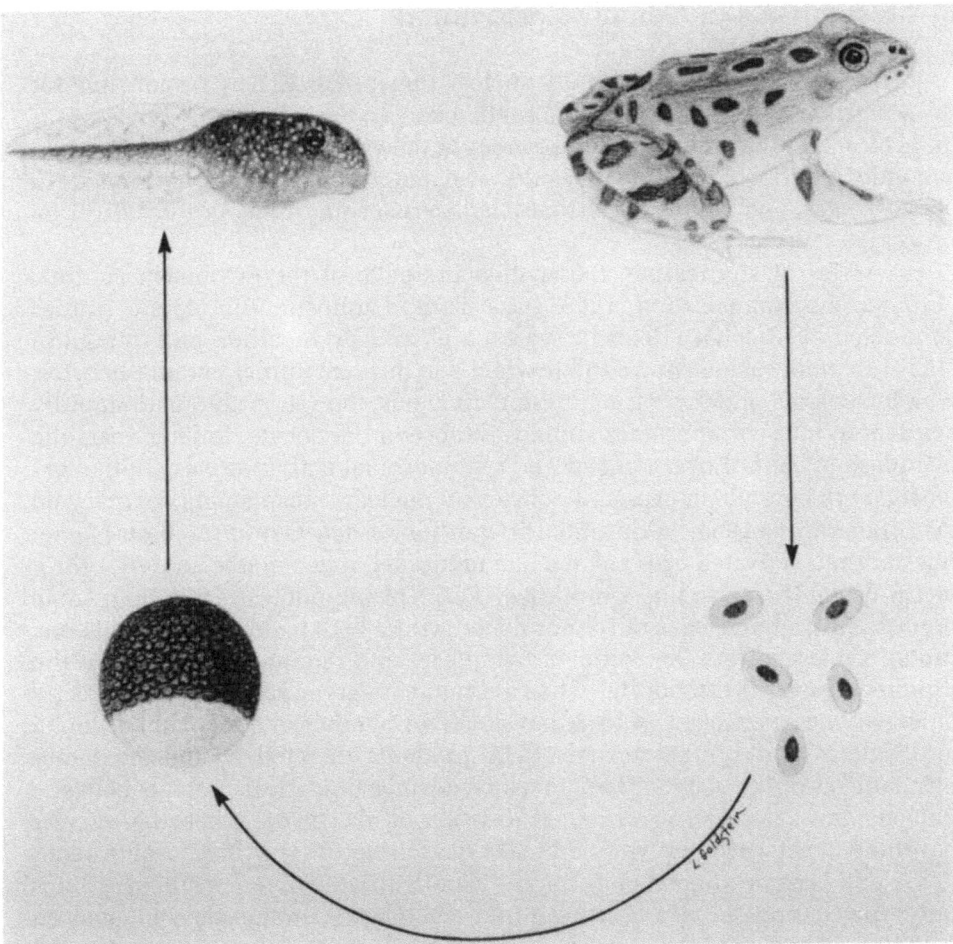

Figure 2. Erythrocyte nuclei derived from juvenile *Rana pipiens* frogs were injected into oocytes at the stage of first meiotic metaphase and blastulae developed. Nuclei from these blastulae were injected singly into enucleated eggs that developed into feeding tadpoles and advanced to hind limb bud stages. See legend of Fig. 1 for details of the procedure.

cant role to the development of nuclear transplants, because eggs lacking a functional nucleus but containing a functional centriole develop at best only into partially and abnormally cleaved blastulae (Briggs *et al.*, 1951). Nevertheless, there is not enough information to evaluate exactly the relative contribution of maternal RNAs and proteins to the development of nuclear transplants. For this reason, the survival of the nuclear transplants as independent organisms is crucial, because most maternal products are presumably used and degraded within a restricted early period. Therefore, the most critical evaluation of the genomic potential of differentiated somatic nuclei should be made on those nuclear transplants that function as independent organisms.

6. Mechanisms of Genomic Activation

The primary mechanism(s) in nuclear transplants that is responsible for the activation of the genome of differentiated cells is not known. However, it is quite clear that the molecular components in the cytoplasm control nuclear and genomic function. Some key results are cited from studies performed in oocytes, eggs, and cell cultures that characterize some steps in the activation process.

A series of studies has shown that the state of the cytoplasm controls chromosome condensation, DNA sythesis, and mitosis. The classic studies conducted by Brachet (1922) in sea urchins and by Bataillon and Tchou-Su (1934) in amphibians showed that when sperm prematurely entered oocytes, which were still undergoing maturation divisions, the sperm chromatin rapidly condensed into chromosomes similar to those of the oocyte. In later years the cytoplasmic control over nuclear and chromosomal activities was studied extensively in amphibian eggs and oocytes by nuclear transplantation (reviewed by DiBerardino, 1980; Etkin and DiBerardino, 1983; Gurdon, 1986). Nuclei injected into activated eggs behave like pronuclei; they enlarge, undergo chromatin decondensation and synthesize DNA. During nuclear activation, some nonhistone proteins leave the donor nuclei, while both histone and nonhistone proteins migrate from the recipient cytoplasm into the injected nuclei. At the diplotene stage of the oocyte, when its nucleus (germinal vesicle) is greatly enlarged and is engaged in RNA but not DNA synthesis, transplanted nuclei also enlarge and synthesize new RNA products directed by the regulatory mechanisms of the oocyte. Here, too, an exchange of nuclear proteins between nucleus and cytoplasm is observed (Gurdon *et al.*, 1976). When the nuclear envelope of the germinal vesicle breaks down, the oocyte chromosomes condense and become aligned on the spindle of the first meiotic metaphase. Similarly, injected nuclei also transform into metaphase chromosomes aligned on spindles. The cytoplasmic control of nuclear activity also operates in the mouse oocyte (Clarke and Masui, 1986) as well as in oocytes of other species. The same phenomena were reported in cultured mammalian cells when cells from different phases of the cell cycle were fused together. Cells in S phase induced cells in G1 phase to undergo DNA synthesis, whereas mitotic cells fused with G1, G2, or S cells caused premature chromosome condensation (reviewed by Adlakha and Rao, 1986). The main conclusion drawn from the above studies is that a new program of genomic expression is induced in test nuclei by a specific set of molecular factors residing in the oocyte, egg and different phases of the cell cycle, respectively. The availability of cytoplasmic extracts from activated eggs that can induce sperm to transform into pronuclei and mitotic chromosomes *in vitro* will ultimately permit the biochemical analysis of the factors in eggs that induce DNA synthesis and chromosomal condensation (Lohka and Masui, 1983). With respect to extracts from oocytes, Korn *et al.* (1982) identified a positive regulator that activates dormant 5S RNA genes of erythrocytes. The activating factor is sensitive to protease and heat treatment suggesting that protein(s) are involved.

Heterokaryons have also provided clues into the process of gene activation. An outstanding example is the activation of certain dormant genes in the hen erythrocyte. The classic experiment by H. Harris (1974) involved the fusion of hen erythrocytes with HeLa cells. In the heterokaryons, the nuclei of the parent cells remain separate, and each nucleus can be identified and monitored. The erythrocyte nuclei enlarged, their highly condensed chromatin dispersed, they resumed synthesis of RNA and DNA, and later, a series of chick proteins were synthesized. During the activation process, there is an uptake of mammalian nuclear proteins by the chick erythrocyte nucleus and a loss of histone 5 (H5) from the chromatin of the erythrocyte (Ringertz et al., 1985). Blau et al. (1985) fused mouse muscle cells with a variety of nonmuscle cell types of human origin. Such heterokaryons containing muscle cells do not undergo cell division, and therefore, the heterokaryons are stable and retain the complement of chromosomes. In these heterokaryons, various specific mRNAs and proteins, normally made by muscle cells were also made by the nonmuscle genomes. In addition to extending previous studies on heterokaryons, the authors suggest that their system will be amenable to isolating the trans-acting regulators that induce muscle gene activation.

In the case of *Rana* erythrocyte nuclei, we have shown that exposure of erythrocyte nuclei to the components in the cytoplasm from the stage of first meiotic metaphase and subsequent maturation stages leads to widespread activation of the genome (DiBerardino and Hoffner, 1983). This activation involves in some manner the ability of noncycling and nonreplicating erythrocyte nuclei to undergo significant DNA replication in activated eggs. Autoradiographic studies by Leonard et al. (1982) demonstrated that only 24% of adult erythrocyte nuclei transplanted to eggs synthesized DNA and, in these cases, only a small portion of the genome engaged in DNA synthesis. If, however, adult erythrocyte nuclei were first injected into oocytes that later matured and were activated, more then 75% synthesized DNA, and more than one half of these nuclei synthesized DNA in amounts similar to those of the egg pronucleus. Thus, exposure of erythrocyte nuclei to oocyte cytoplasm leads not only to induction of DNA synthesis in more nuclei, but also to significant enhancement of the extent of replication of the genome. We have proposed that a remodeling of the proteins in the erythrocyte chromatin occurs in the oocyte cytoplasm that renders the chromatin capable of responding more normally to the signals in activated egg cytoplasm that induce the synthesis of DNA (Leonard et al., 1982; DiBerardino and Hoffner, 1983; DiBerardino, 1987, 1988). Such remodeling of chromosomal proteins could lead to changes in chromatin structure and in patterns of DNA methylation that eventually permit the genome to respond later to the appropriate transcription factors and synthesize the appropriate RNAs during embryogenesis. Other autoradiographic studies examined the nucleocytoplasmic exchange of nonhistone proteins when endodermal nuclei were transplanted to eggs. These studies showed that during the first cell cycle there is an accumulation of cytoplasmic egg nonhistone proteins by transplanted nuclei and also a major loss of nonhistone proteins from the transplanted nuclei into the cytoplasm (DiBerardino and Hoffner, 1975;

Hoffner and DiBerardino, 1977). Perhaps a similar but more effective nucleocytoplasmic exchange occurs in erythrocyte nuclei placed in the oocyte, not only because of a longer exposure to the cytoplasm, but also because the nuclei are exposed to additional factors not present in the egg. With respect to chromosomal proteins there is evidence that changes occur in the proteins of chromatin in germ cells during spermatogenesis (Dixon, 1972), oogenesis (Masui et al., 1979), and after fertilization (Poccia, 1986). Thus, there are precedents that substantiate our hypothesis that remodeling of chromosomal proteins may have occurred in erythrocyte nuclei exposed to oocyte cytoplasm. This hypothesis can be tested by examining appropriate chromosomal proteins, as well as the state of DNA methylation and the chromatin structure of specific genes in test nuclei before and after exposure to oocyte cytoplasm. Such studies, together with the identification of specific extracts of oocyte cytoplasm that control molecular changes in the chromatin of test nuclei, might ultimately indicate the primary mechanisms involved in the activation of an entire genome of differentiated somatic cells. This information might also indicate how the germ cell chromatin of oocytes is prepared for participation in fertilization and embryogenesis.

In addition to the remodeling of chromosomal proteins, an essential replication factor may be required. For example, Blow and Laskey (1988) have shown that sperm chromatin can rereplicate in a cell-free system extracted from *Xenopus* eggs, if the nuclear membrane is permeabilized or breaks down. They suggest that an essential replication factor gains access to DNA when the nuclear envelope breaks down during mitosis and that one role of the nuclear envelope is to limit DNA to one replication per cycle. In our transfers of nuclei into oocytes at first meiotic metaphase, the nuclear membrane breaks down and the interphase chromatin is transformed into metaphase chromosomes aligned on a newly induced spindle (Hoffner and DiBerardino, 1980; Leonard et al., 1982). These events occur by 1 hr or earlier after transfer to the oocytes. It is likely that noncycling erythrocyte nuclei lack such an essential replication factor or have negligible amounts, but acquire it from the oocyte cytoplasm after nuclear membrane breakdown. Nuclei injected initially into eggs retain their membranes until after the period of DNA synthesis and would therefore be deficient in the replication factor. The above speculation is consistent with our findings that adult erythrocyte nuclei incubated in oocyte cytoplasm undergo significant DNA synthesis in activated eggs and support tadpole development, whereas those injected in eggs undergo little DNA synthesis and support development only to the early gastrula stage (Leonard et al., 1982; DiBerardino and Hoffner, 1983).

7. Conclusions and Perspectives

Amphibian nuclear transplantations from several differentiated somatic cell types into oocytes and eggs have shown that their nuclei still contain the genes required for the development of prefeeding tadpoles. In addition, erythrocyte nuclei of *Rana* have directed the formation of independent organisms (feeding tadpoles) that advanced to stages of hindlimb bud and survived up to 1 month. These results demonstrate that the genomes of several differentiated somatic cell types can (1) undergo widespread activation, (2) direct the formation of a multiplicity of cell types in prefeeding and feeding tadpoles, and (3) display genetic multipotentiality. However, evidence for genomic totipotency of differentiated somatic nuclei is still lacking. With regard to this unsolved problem, the results from amphibian nuclear transplantation studies permit two interpretations: either (1) irreversible genetic changes have occurred during somatic cell specialization that prevent feeding tadpoles from maturing into fertile frogs, or (2) the nuclear transplantation procedure is still inadequate to show the complete genetic repertoire of differentiated somatic nuclei. Although information is not available at this time to choose between these alternatives, we shall examine each one and propose the most likely interpretation.

The evidence at hand shows that irreversible genetic changes concomitant with cell differentiation are *not a general* occurrence among the phyla (see Section 1). In the lymphocyte genes and the genes coding for the antigenic proteins in the cell membrane of the protozoan *Trypanosoma*, the mechanism responsible for the changes in the primary structure of DNA involve genomic sequence rearrangement and nucleotide losses, while gene conversion operates at the mating type locus of the yeast plant. These genetic changes appear to be a prerequisite for gene activation in these presumably special cases. Translocations of the size that occur in the Ig genes do not occur in a number of genes that have been analyzed, nor are they required for expression, e.g., the chicken ovalbumin gene, chicken transferrin and lysozyme genes, *Drosophila* salivary gland glue protein genes, mouse gene for salivary and liver α-amylase, and class I mouse transplantation antigen genes. However, these studies would not have detected mutational changes involving a few nucleotides, small deletions, inversions, insertions, or conversions (reviewed by Davidson, 1986).

In contrast to the special cases cited, there is overwhelming evidence that many silent genes in differentiated cells are retained and can be activated under appropriate conditions, such as transdifferentiation, transfection experiments, cell fusion, cancer, and nuclear transplantation. It is likely that genomic totipotency of some differentiated somatic cell types is still a tenable hypothesis. However, since genomic totipotency of animal genomes of differentiated somatic cells has not yet been demonstrated, we cannot formally exclude the possibility of subtle irreversible changes in the DNA of certain genes and/or in their *cis*- and *trans*-controlling regions. In fact, nucleotide changes in a master switch gene(s) that controls development from the tadpole stage to metamorphosis could limit genomic totipotency.

The alternate interpretation of nuclear transplantation studies is that in most cases the complete gene content of the zygote nucleus is retained in differentiated somatic cells, but the current procedure fails to indicate totipotency. This is a plausible interpretation, since several modifications in the procedure have led to enhanced expression of the genetic potential of somatic nuclei (reviewed in DiBerardino, 1987). For example, residence of erythrocyte nuclei in the cytoplasm of maturing oocytes resulted in tadpoles, but development proceeded only to early gastrulae when the nuclei were injected initially into eggs (DiBerardino and Hoffner, 1983). It should be noted that the nuclei of adult differentiated germ cells have not promoted complete development of injected eggs (DiBerardino and Hoffner, 1971). Successful nuclear transfers of germ cells according to LeSimple *et al.* (1987) appear to be restricted to larval stem cells (see Section 3). The most plausible explanation for failure to show genomic totipotency of differentiated somatic and germ cells is that we have not yet discovered the proper conditions for completely erasing the imprinting of the genome. The chromatin of differentiated somatic and germ cells consists of its own set of chromosomal proteins required for its specific phenotype. When its nucleus is transplanted into the cytoplasm of an oocyte or egg, its chromatin must undergo remodeling of its proteins and structure suitable for rapid DNA replication during early cleavage stages. If such changes are not complete, incomplete replicates of the genome would be expected, followed by deletions in chromosomes that have been observed in nuclear transplants that cease development during embryogenesis and early larval stages (DiBerardino and Hoffner, 1970; DiBerardino, 1979, 1987). Although no detectable aberrations in the metaphase chromosomes of feeding tadpoles were detected (Hoffner, Orr, and DiBerardino, unpublished observations), we cannot exclude the presence of submicroscopic changes.

The main conclusion that may be drawn from the amphibian nuclear transplantation experiments is that the genes required for embryogenesis, and for the morphogenesis and cellular differentiation of prefeeding tadpoles and of the independent organisms of feeding tadpoles (see Table II) have not been altered irreversibly in those differentiated somatic cell types tested. Failure to obtain fertile frogs from differentiated somatic nuclei transplanted into oocytes and eggs is not evidence for irreversible genetic changes. Only losses or changes in DNA sequences of the genome would be irreversible, and such cases are still not a general phenomenon during cell differentiation of eukaryotic animals. The lymphocyte is probably unique, for the DNA rearrangements lead to antibody and receptor diversity. But other types of somatic cell specialization that do not require phenotypic diversity in their terminal state of differentiation probably achieve their specialized phenotype by mechanisms of epigenetic imprinting of the genome (e.g., DNA methylation–hypomethylation, chromatin structure, and DNA–protein interactions) that control specific gene expression without an irreversible change in the nuclear genome (DiBerardino and Hoffner, 1970; DiBerardino *et al*, 1984; DiBerardino, 1987, 1988).

Two principal problems regarding genomic activation of differentiated somatic cells are likely to be pursued. First, the unsolved problem of totipotency of differentiated somatic nuclei from larvae and adults should continue to be studied in amphibians by means of nuclear transfer into oocytes and eggs and extended to other nonhuman forms. Attempts should focus on finding ways to transform the chromatin of differentiated cells into that similar to the chromatin of the egg nucleus. An alternate approach for studying the question of nuclear equivalence is available, i.e., DNA sequencing of an entire genome.

Even if this herculean task is accomplished and the normal organization of genetic linkage is discovered, this information in itself still would not result in the production of a living and integrated organism. The second principal problem to be pursued is the molecular analysis of the factors involved in activating nearly an entire genome. The feasible approach is to study the molecular changes in chromatin exposed to extracts of oocyte cytoplasm. Pursuit of problems one and two will eventually reveal the normal mechanisms that imprint the genome with restrictions and control cell specialization. The long-range outcome of these studies may settle the century old problem of whether or not the nuclei of somatic cell types generally retain the genetic repertoire of the zygote nucleus. If totipotency can be demonstrated, we would then be capable of controlling the activation of an entire genome. In the meantime, efforts should focus on the mechanisms leading to the activation of nearly an entire genome—a feat already accomplished. Such knowledge could be extended to controlling genomic activation in various cellular disorders.

ACKNOWLEDGMENTS. The author thanks Laurence D. Etkin and Robert G. McKinnell for constructive comments on the manuscript. This research was supported by grant GM23635 from the National Institutes of Health.

References

Adlakha, R. C., and Rao, P. N., 1986, Molecular mechanisms of the chromosome condensation and decondensation cycle in mammalian cells, *BioEssays* **5**:100–105.

Alt, F. W., Blackwell, T. K., and Yancopoulos, G. D., 1987, Development of the primary antibody repertoire, *Science* **238**:1079–1087.

Anderegg, C., and Markert, C. L., 1986, Successful rescue of microsurgically produced homozygous uniparental mouse embryos via production of aggregation chimeras, *Proc. Natl. Acad. Sci. USA* **83**:6509–6513.

Bataillon, E., and Tchou-Su, 1934, L'analyse expérimentale de la fécondation et sa definition par les processus cinétiques, *Ann. Sci. Natl. Zool. Biol. Anim.* **17**:9–36.

Blau, H. M., Pavlath, G. K., Hardeman, E. C., Chiu, C-P., Silberstein, L., Webster, S. G., Miller, S. C., and Webster, C., 1985, Plasticity of the differentiated state, *Science* **230**:758–766.

Blow, J. J., and Laskey, R. A., 1988, A role for the nuclear envelope in controlling DNA replication within the cell cycle, *Nature (Lond.)* **332**:546–548.

Brachet, A., 1922, Recherches sur la fécondation prématurée de l'oeuf d'oursin (*Paracentrotus lividus*), *Arch. Biol.* **32**:205–244.

Briggs, R., 1979, Genetics of cell type determination, *Int. Rev. Cytol.* (Suppl. **9**):107–127.

Briggs, R., and King, T. J., 1952, Transplantation of living nuclei from blastula cells into enucleated frogs' eggs, *Proc. Natl. Acad. Sci. USA* **38**:455–463.

Briggs, R., and King, T. J., 1957, Changes in the nuclei of differentiating endoderm cells as revealed by nuclear differentiation, *J. Morphol.* **100**:269–312.

Briggs, R., and King, T. J., 1959, Nucleocytoplasmic interactions in eggs and embryos, in: *The Cell*, Vol. I (J. Brachet and A. E. Mirsky, eds.), pp. 537–617, Academic, New York.

Briggs, R., and King, T. J., 1960, Nuclear transplantation studies on the early gastrula (*Rana pipiens*), *Dev. Biol.* **2**:252–270.

Briggs, R., Green Ufford, E., and King, T. J., 1951, An investigation of the capacity for cleavage and differentiation in *Rana pipiens* eggs lacking "functional" chromosomes, *J. Exp. Zool.* **116**:455–500.

Briggs, R., King, T. J., and DiBerardino, M. A., 1961, Development of nuclear transplant embryos of known chromosome complement following parabiosis with normal embryos, in: *Symposium on the Germ Cells and Earliest Stages of Development* (S. Ranzi, ed.), pp. 441–477, Fond. A. Baselli, Instituto Lombardo, Milan.

Brun, R. B., 1978, Developmental capacities of *Xenopus* eggs, provided with erythrocyte or erythroblast nuclei from adults, *Dev. Biol.* **65:**271–284.

Clarke, H. J., and Masui, Y., 1986, Transformation of sperm nuclei to metaphase chromosomes in the cytoplasm of maturing oocytes of the mouse, *J. Cell Biol.* **102:**1039–1046.

Clegg, K. B., and Pikó, L., 1983, Quantitative aspects of RNA synthesis and polyadenylation in 1-cell and 2-cell mouse embryos, *J. Embryol. Exp. Morphol.* **74:**169–182.

Davidson, E. H., 1986, *Gene Activity in Early Development*, 3rd ed., Academic, Orlando, Florida.

DiBerardino, M. A., 1979, Nuclear and chromosomal behavior in amphibian nuclear transplants, *Int. Rev. Cytol.* (Suppl. **9**):129–160.

DiBerardino, M. A., 1980, Genetic stability and modulation of metazoan nuclei transplanted into eggs and oocytes, *Differentiation* **17:**17–30.

DiBerardino, M. A., 1987, Genomic potential of differentiated cells analyzed by nuclear transplantation, *Am. Zool.* **27:**623–644.

DiBerardino, M. A., 1988, Genomic multipotentiality of differentiated somatic cells, *Cell Diff. Develop.* **25**(suppl.):129–136.

DiBerardino, M. A., and Hoffner, N., 1970, Origin of chromosomal abnormalities in nuclear transplants—A reevaluation of nuclear differentiation and nuclear equivalence in amphibians, *Dev. Biol.* **23:**185–209.

DiBerardino, M. A., and Hoffner, N., 1971, Development and chromosomal constitution of nuclear transplants derived from male germ cells, *J. Exp. Zool.* **176:**61–72.

DiBerardino, M. A., and Hoffner, N. J., 1975, Nucleocytoplasmic exchange of nonhistone proteins in amphibian embryos, *Exp. Cell Res.* **94:**235–252.

DiBerardino, M. A., and Hoffner, N. J., 1983, Gene reactivation in erythrocytes: Nuclear transplantation in oocytes and eggs of *Rana*, *Science* **219:**862–864.

DiBerardino, M. A., and King, T. J., 1967, Development and cellular differentiation of neural nuclear-transplants of known karyotype, *Dev. Biol.* **15:**102–128.

DiBerardino, M. A., Hoffner, N. J., and Etkin, L. D., 1984, Activation of dormant genes in specialized cells, *Science* **244:**946–952.

DiBerardino, M. A., Orr Hoffner, N., and McKinnell, R. G., 1986, Feeding tadpoles cloned from *Rana* erythrocyte nuclei, *Proc. Natl. Acad. Sci. USA* **83:**8231–8234.

Dixon, G. H., 1972, The basic proteins of trout testis chromatin: Aspects of their synthesis, postsynthetic modifications and binding to DNA, *Acta Endocrinol.* (*Copenh.*) **168:**130–154.

DuPasquier, L., and Wabl, M. R., 1977, Transplantation of nuclei from lymphocytes of adult frogs into enucleated eggs. Special focus on technical parameters, *Differentiation* **8:**9–19.

Elsdale, T. R., Gurdon, J. B., and Fischberg, M., 1960, A description of the technique for nuclear transplantation in *Xenopus laevis*, *J. Embryol. Exp. Morphol.* **8:**437–444.

Etkin, L. D., and DiBerardino, M. A., 1983, Expression of nuclei and purified genes microinjected into oocytes and eggs, in: *Eukaryotic Genes* (N. Maclean, S. P. Gregory, and R. A. Flavell, eds.), pp. 127–156, Butterworths, London.

Gallien, L., 1966, La greffe nucléaire chez les amphibiens, *Ann. Biol.* **5–6:**241–269.

Gurdon, J. B., 1960a, The effects of ultraviolet irradiation on uncleaved eggs of *Xenopus laevis*, *Quart. J. Microsc. Sci.* **101:**299–311.

Gurdon, J. B., 1960b, The developmental capacity of nuclei taken from differentiating endoderm cells of *Xenopus laevis*, *J. Embryol. Exp. Morphol.* **8:**505–526.

Gurdon, J. B., 1962, Adult frogs derived from the nuclei of single somatic cells, *Dev. Biol.* **4:**256–273.

Gurdon, J. B., 1974, The genome in specialized cells, as revealed by nuclear transplantation in amphibia, in: *The Cell Nucleus*, Vol. I (H. Busch, ed.), pp. 471–489, Academic, Orlando, Florida.

Gurdon, J. B., 1986, Nuclear transplantation in eggs and oocytes, *J. Cell Sci.* **4**(suppl.):287–318.

Gurdon, J. B., and Laskey, R. A., 1970, The transplantation of nuclei from single cultured cells into enucleate frogs' eggs, *J. Embryol. Exp. Morphol.* **24:**227–248.

Gurdon, J. B., and Uehlinger, V., 1966, "Fertile" intestine nuclei, *Nature (Lond.)* **210:**1240–1241.

Gurdon, J. B., Laskey, R. A., and Reeves, O. R., 1975, The developmental capacity of nuclei transplanted from keratinized skin cells of adult frogs, *J. Embryol. Exp. Morphol.* **34:**93–112.

Gurdon, J. B., Partington, G. A., and DeRobertis, E. M., 1976, Injected nuclei in frog oocytes: RNA synthesis and protein exchange, *J. Embryol. Exp. Morphol.* **36:**541–553.

Harris, H., 1974, *Nucleus and Cytoplasm*, Clarendon, Oxford.

Harris, H., 1985, Suppression of malignancy in hybrid cells: The mechanism, *J. Cell Sci.* **79:**83–94.

Hennig, W. (ed.), 1986, *Results and Problems in Cell Differentiation*, Vol. 13: *Germ Line-Soma Differentiation* (W. Hennig and J. Reinert, eds.), Springer-Verlag, Berlin.

Hoffner, N. J., and DiBerardino, M. A., 1977, The acquisition of egg cytoplasmic nonhistone proteins by nuclei during nuclear reprogramming, *Exp. Cell Res.* **108:**421–427.

Hoffner, N. J., and DiBerardino, M. A., 1980, Developmental potential of somatic nuclei transplanted into meiotic oocytes of *Rana pipiens, Science* **209:**517–519.

Illmensee, K., 1973, The potentialities of transplanted early gastrula nuclei of *Drosophila melanogaster.* Production of their imago descendants by germline transplantation, *Roux Arch.* **171:** 331–343.

Illmensee, K., and Hoppe, P. C., 1981, Nuclear transplantation in *Mus musculus:* Developmental potential of nuclei from preimplantation embryos, *Cell* **23:**9–18.

King, T. J., 1966, Nuclear transplantation in amphibia, *Methods Cell Physiol.* **2:**1–36.

King, T. J., and Briggs, R., 1956, Serial transplantation of embryonic nuclei, *Cold Spring Harbor Symp. Quant. Biol.* **21:**271–290.

Kobel, H. R., Brun, R. B., and Fischberg, M., 1973, Nuclear transplantation with melanophores, ciliated epidermal cells, and the established cell-line A-8 in *Xenopus laevis, J. Embryol. Exp. Morphol.* **29:**539–547.

Korn, L. J., Gurdon, J. B., and Price, J., 1982, Oocyte extracts reactivate developmentally inert *Xenopus* 5s genes in somatic nuclei, *Nature (Lond.)* **300:** 354–355.

Laskey, R. A., and Gurdon, J. B., 1970, Genetic content of adult somatic cells tested by nuclear transplantation from cultured cells, *Nature (Lond.)* **228:**1332–1334.

Leonard, R. A., Hoffner, N. J., and DiBerardino, M. A., 1982, Induction of DNA synthesis in amphibian erythroid nuclei in *Rana* eggs following conditioning in meiotic oocytes, *Dev. Biol.* **92:**343–355.

Lesimple, M., Dournon, C., Labrousse, M., and Houillon, C., 1987, Production of fertile salamanders by transfer of germ cell nuclei into eggs, *Development* **100:**471–477.

Lohka, M. J., and Masui, Y., 1983, Formation in vitro of sperm pronuclei and mitotic chromosomes induced by amphibian ooplasmic components, *Science* **220:**719–721.

Marrack, P., and Kappler, J., 1987, The T cell receptor, *Science* **238:**1073–1079.

Marshall, J. A., and Dixon, K. E., 1977, Nuclear transplantation from intestinal epithelial cells of early and late *Xenopus laevis* tadpoles, *J. Embryol. Exp. Morphol.* **40:**167–174.

Masui, Y., Meyerhof, P. G., and Ziegler, D. H., 1979, Control of chromosome behavior during progesterone induced maturation of amphibian oocytes, *J. Steroid Biochem.* **11:**715–722.

McAvoy, J. W., Dixon, K. E., and Marshall, J. A., 1975, Effects of differences in mitotic activity, stage of cell cycle, and degree of specialization of donor cells on nuclear transplantation in *Xenopus laevis, Dev. Biol.* **45:**330–339.

McKinnell, R. G., 1962, Intraspecific nuclear transplantation in frogs, *J. Hered.* **53:**199–207.

McKinnell, R. G., 1978, *Cloning, Nuclear Transplantation in Amphibia*, University of Minnesota Press, Minneapolis.

McKinnell, R. G., Deggins, B. A., and Labat, D. D., 1969, Transplantation of pluripotential nuclei from triploid frog tumors, *Science* **163:**394–396.

Nakakura, N., Miura, T., Yamana, K., Ito, A., and Shiokawa, K., 1987, Synthesis of heterogeneous mRNA-like RNA and low-molecular-weight RNA before the midblastula transition in embryos of *Xenopus laevis, Dev. Biol.* **123:**421–429.

Orr Hoffner, N., DiBerardino, M. A., and McKinnell, R. G., 1986, The genome of frog erythrocytes displays centuplicate replications, *Proc. Natl. Acad. Sci. USA* **83:**1369–1373.

Poccia, D., 1986, Remodeling of nucleoproteins during gametogenesis, fertilization, and early development, *Int. Rev. Cytol.* **105:**1–65.

Prather, R. S., Barnes, F. L., Sims, M. M., Robl, J. M., Eyestone, W. H., and First, N. L., 1987, Nuclear transplantation in the bovine embryo: Assessment of donor nuclei and recipient oocyte, *Biol. Reprod.* **37**:859–866.

Reinschmidt, D., Friedman, J., Hauth, J., Ratner, E., Cohen, M., Miller, M., Krotoski, D., and Tompkins, R., 1985, Gene-centromere mapping in *Xenopus laevis, J. Hered.* **76**:345–347.

Richards, C. M., and Nace, G. W., 1978, Gynogenetic and hormonal sex reversal used in tests of the XX-XY hypothesis of sex determination in *Rana pipiens, Growth* **42**:319–331.

Ringertz, N. R., and Savage, R. E., 1976, *Cell Hybrids,* Academic, Orlando, Florida.

Ringertz, N. R., Nyman, U., and Bergman, M., 1985, DNA replication and H5 histone exchange during reactivation of chick erythrocyte nuclei in heterokaryons, *Chromosoma* **91**:391–396.

Smith, L. D., 1965, Transplantation of the nuclei of primordial germ cells into enucleated eggs of *Rana pipiens, Proc. Natl. Acad. Sci. USA* **54**:101–107.

Spemann, H., 1938, *Embryonic Development and Induction,* repr. 1967, Hafner, New York.

Subtelny, S., 1965, On the nature of the restricted differentiation-promoting ability of transplanted *Rana pipiens* nuclei from differentiating endoderm cells, *J. Exp. Zool.* **159**:59–92.

Tobler, H., 1986, The differentiation of germ and somatic cell lines in nematodes, in: *Results and Problems in Cell Differentiation,* Vol. 13: *Germ Line-Soma Differentiation* (W. Hennig and J. Reinert, eds.), pp. 1–69, Springer-Verlag, Berlin.

Tsunoda, Y., Yasui, T., Shioda, Y., Nakamura, K., Uchida, T., and Sugie, T., 1987, Full-term development of mouse blastomere nuclei transplanted into enucleated two-cell embryos, *J. Exp. Zool.* **242**:147–151.

Wabl, M. R., Brun, R. B., and DuPasquier, L., 1975, Lymphocytes of the toad *Xenopus laevis* have the gene set for promoting tadpole development, *Science* **190**:1310–1312.

Weill, J-C., and Reynaud, C-A., 1987, The chicken B cell compartment, *Science* **238**:1094–1098.

Willadsen, S. M., 1986, Nuclear transplantation in sheep embryos, *Nature (Lond.)* **320**:63–65.

Yan, S., Lu, D., Du, M., Li, G., Lin, L., Jin, G., Wang, H., Yang, Y., Xia, D., Liu, A., Zhu, Z., Yi, Y., and Chen, H., 1984, Nuclear transplantation in teleosts, *Sci. Sin. [B]* **27**:1029–1034.

Zalokar, M., 1973, Transplantation of nuclei into the polar plasm of *Drosophila* eggs, *Dev. Biol.* **32**:189–193.

Chapter 9

Neoplastic Cells
Modulation of the Differentiated State

ROBERT GILMORE McKINNELL

1. Introduction: The Concept of Cancer Cell Differentiation

For at least a century and a half, biologists have been aware that, relative to normal cells, many cancer cells appear morphologically undifferentiated. While some have concluded that cancer cells must arise by the dedifferentiation of normal cells, others consider it more likely that many cancer cells develop from normal cells that never fully differentiated. Since these undifferentiated cancer cells can be shown to share many fundamental characteristics with normal cells, it is not altogether surprising that some neoplasms can be induced to differentiate and become postmitotic benign cells. If only one or a few malignant tumors had the potentiality for differentiation, the phenomenon would be of interest primarily because of its rarity. However, many types of malignant neoplasms, such as plant teratomas, fish tumors, amphibian cancers, and malignant neoplasms of all three embryonic germ layers of mammals, can be induced to differentiate. This indicates that the phenomenon is not rare, and it suggests the possibility that the control of differentiation could be exploited in the treatment of cancer.

It is the purpose of this chapter to consider the diversity of cancers that differentiate and the multiplicity of inducers that elicit these differentiations. It would be impractical to list all the literally hundreds of agents known to induce differentiation of cancer cells, but a sufficient number will be mentioned to indicate the heterogeneous nature of agents with this property. Moreover, inasmuch as induced embryonic differentiation and induced cancer cell differentiation share some commonalities, both phenomena are considered.

Although research into the control of cancer cell differentiation is a timely endeavor (Marx, 1987; Potter, 1987; McKinnell, 1989; Pierce and Speers, 1988), it is interesting to note that speculation and experimentation in this area have their roots in nineteenth and early twentieth century biology. A brief chronicle

ROBERT GILMORE McKINNELL • Department of Genetics and Cell Biology, University of Minnesota, Saint Paul, Minnesota 55108-1095.

is included in this chapter to trace the evolution of current notions concerning cancer cell differentiation.

There is ample evidence that terminal differentiation abrogates malignancy. That abrogation of malignancy is a principal rationale for continued study of cancer cell differentiation.

2. Cancer and Stem Cells

It is not inappropriate to compare normal cells and cancer cells. This is because studies of cell anatomy (Bernhard, 1969) and cell function (Pierce *et al.*, 1978) show that cancer cells could not survive *in vivo* or *in vitro* if they were not provided with the usual array of nucleic acids, enzymes, and physiologically essential molecules in an architectural order that allows for survival and proliferation. Thus, it may be stated that cancer cells share many attributes with normal cells. However, compared with normal differentiated adult cells, malignant neoplastic cells appear incompletely or improperly differentiated (Siminovitch and Axelrad, 1963). It will be argued here that the appearance of neoplastic cells is due to the fact that many of them are transformed stem cells frozen in that state by an oncogenic event.

This is not to suggest that the only difference between a cancer cell and its normal cell of origin is that the former is a mitotically active stem cell precluded from maturation because of transformation. Carcinogenesis by chemicals (Pitot, 1985), viruses (Gross, 1983), and radiation (Fry, 1985) obviously causes subtle changes in cancer cells that ultimately are manifested by altered cell function. However, even with significant changes in DNA, genetic defects can be bypassed in some cancer cells, and the cells can be induced to differentiate with appropriate treatment despite the genetic lesion.

Stem cells in normal adult tissue are cells that divide and give rise to daughter cells with "unlimited growth potentiality" in a process of self-renewal (Siminovitch and Axelrad, 1963). Adult tissues that depend on a stem cell population include the lining of the gut, skin, and blood-forming cells in the bone marrow. The stem cells also provide mitotic progeny, which differentiate and replenish cells lost to wear or senescence. Differentiated cells generally do not transform into cells of another type, and—when terminally differentiated—they are usually postmitotic.

If the notion that cancer cells are similar to normal cells in many respects is accepted, it becomes useful to consider the possibility that the undifferentiated appearance of many cancer cells is not due to dedifferentiation (an uncommon event in animals) but is due to the overproduction of mitotically active transformed stem cells that retain morphological similarity, not to fully differentiated postmitotic cells, but to normally regulated stem cells. Evidence will be presented that at least some cancer cells not only resemble normal stem cells morphologically, but (as their normal stem cell compatriots) also retain the capacity for differentiation. If enough is learned about the phenomenon of cancer cell differentiation, it may be possible to induce the less than fully differentiated neoplastic cells to differentiate and become postmitotic without the usual need to kill either the cancer cells intentionally or normal cells inadvertently.

3. Neoplasms That Differentiate

3.1. Crown Gall Tumor of Tobacco

All multicellular organisms that undergo differentiation may become afflicted with neoplasms. This generalization applies as much to plants as it does to animals (Levin and Levine, 1920; Lippincott and Lippincott, 1981; Kalil and Hildebrandt, 1981; Smith, 1988). A non-self-limiting neoplastic disease of dicotyledonous plants known as crown gall (Braun, 1947, 1968, 1972; Meins, 1974) is initiated when the T_i plasmid (Watson *et al.*, 1975; Schilperoot *et al.*, 1978; Ream and Gordon, 1982; Thomashow *et al.*, 1986) is transmitted to a wound on a susceptible plant by a specific bacterium, *Agrobacterium tumefaciens* (Smith and Townsend, 1907).

Normal cells and tissues from many mature plants can be cultured *in vitro* (Torrey, 1985). Some somatic cells and protoplasts, when cultured *in vitro*, appear totipotent by virtue of their capacity to grow and differentiate all the cell types of an entire plant (Muir *et al.*, 1958; Steward *et al.*, 1964, 1970; Earle and Torrey, 1965). Tobacco (*Nicotiana tabacum*) cells exhibit such totipotency (Takebe *et al.*, 1971; Vasil and Hildebrandt, 1965).

The crown gall with a known etiology and well-characterized mode of transfection, coupled with advanced plant cell culture technique, was an ideal subject for characterization of tumor cell differentiative potential.

Tumors in the tobacco plant induced by the T_i plasmid may grow as typical crown gall or as teratomas, depending on culture conditions (Braun, 1953). Teratomas grow as a chaotic arrangement of tissues and organs, including abnormal shoots and leaves. The abnormal shoots and leaves may be grafted serially (i.e., tumor tissue may be grafted to several hosts successively) giving rise to functional leaves, stems, diploid flower parts such as petals and filaments, and all other specialized tissue types (Braun, 1951; Braun and Wood, 1976; Turgeon *et al.*, 1976; Binns *et al.*, 1981).

Were these abnormal shoots and leaves that developed on the surface of the teratomas produced by inappropriate cell differentiation of the teratoma cells, or were they produced from normal cells swept along from normal portions of the tobacco plant by the rapidly growing tumor? To answer that question, tumors were produced by cloning single teratoma cells. The cloned cells developed as teratomas and produced the expected chaotic array of cells plus abnormal shoots and leaves. The latter were thus shown to be derived from teratoma cells. The cloned teratoma structures were then grafted serially to normal hosts where the grafts differentiated into normal plants that had the capacity to flower and set seed (Braun, 1959, 1968, 1981; Meins, 1974; Binns *et al.*, 1981). These experiments demonstrate unequivocally that the transformed condition can be reversed with plant tumor cells giving rise to the entire spectrum of normal cell types. Therefore, neoplasia in the crown gall teratoma is not caused by a deletion or other irreversible alteration in the genome of the tobacco cell, but rather results from inappropriate gene activity. That inappropriate gene activity can be normalized with serial grafting, altering the tumor environment sufficiently to permit normal cell differentiation and morphogenesis.

3.2. Pigment Cell Tumor of the Goldfish

Many species of aquatic vertebrates are vulnerable to spontaneous tumors (Schlumberger and Lucké, 1948; Harshbarger et al., 1981). Invasive neoplasms of the goldfish, *Carassius auratus*, provide one example (Ishikawa et al., 1978). Spontaneous erythrophoromas (red pigment cell neoplasms) of these fish are the source of cell lines (Matsumoto et al., 1980) that may proliferate indefinitely as stem cells or may differentiate as a result of treatment with chemicals such as dimethylsulfoxide (DMSO), 12-O-tetradecanoyl-phorbol 13-acetate (TPA) or autologous serum. These goldfish tumor-derived cell lines differentiate dermal bone-like structures, scales, fin rays, teeth, neuronlike cells, and lentoid bodies (Matsumoto et al., 1981, 1983, 1985; Akiyama et al., 1986).

3.3. Neoplasms of Amphibians

3.3.1. Lucké Renal Adenocarcinoma

A herpesvirus-induced renal adenocarcinoma afflicts some members of the northern leopard frog species, *Rana pipiens* (Lucké, 1934; Naegele et al., 1974; McKinnell 1973a, 1984; McKinnell and Duplantier, 1970; Asashima et al., 1987). The neoplasm is spontaneously metastatic (Lucké and Schlumberger, 1949; McKinnell and Cunningham, 1982; McKinnell and Tarin, 1984), and its invasive properties have been studied *in vitro* (Mareel et al., 1985; McKinnell et al., 1986, 1988a).

The technique of nuclear transplantation, which is extremely useful in revealing differentiative potential, has been used to study the R. *pipiens* renal adenocarcinoma. The nuclear transplantation procedure was originally developed in the northern leopard frog for the purpose of revealing the developmental potential of embryonic somatic nuclei (Briggs and King, 1952; DiBerardino, 1980, 1987; DiBerardino and Hoffner, 1980; McKinnell, 1978, 1985). In the early studies, transplantation of blastula nuclei into enucleated mature ova resulted in embryos that developed into frogs capable of reaching sexual maturity. Transplantation of older nuclei, however, resulted in progressively fewer embryos that were able to reach maturity. No enucleated ovum receiving an adult nucleus has ever developed to maturity. The results of testing with the nuclear transplantation procedure suggest that, in the course of embryonic development, nuclei progressively lose their ability to execute the diverse differentiative programs required for normal development.

Although nuclear transfer techniques available at the present time suggest that adult nuclei are unable to program for all aspects of normal development, even terminally differentiated nuclei have been shown to be competent to program for many cell types when tested by nuclear transplantation. DiBerardino and colleagues (Chapter 8, *this volume*) have demonstrated that red blood cell (RBC) nuclei, when pretreated by transfer to oocyte cytoplasm (DiBerardino and Hoffner, 1983) followed by retransfer to mature egg cytoplasm, are competent to program for all tissue types of swimming larvae that feed and develop limb buds (DiBerardino *et al.*, 1986; Orr *et al.*, 1986). The most advanced developed obtained with transplantation of adult nuclei, thus far, is of these oocyte pretreated erythrocyte nuclei. Ordinarily, terminal differentiation is associated with a loss of proliferative potential (Scott *et al.*, 1988). The remarkable experiments of DiBerardino and colleagues have reawakened interest in the mitotic and developmental potential of terminally differentiated cells.

Nuclear transplantation studies were begun with nuclei of the Lucké renal adenocarcinoma (Fig. 1) a number of years ago (King and McKinnell, 1960; King and DiBerardino, 1965; McKinnell, 1972, 1973b, 1979). Swimming larvae ensued in some experiments (Fig. 2). No pretreatment of nuclei with oocyte cytoplasm was used to obtain the reported results. The tumor nuclear transplant larvae differentiated the full gamut of tissues appropriate for larvae which include brain and spinal cord, notochord, skin, muscle, connective tissue, heart, blood vessels, and gills. The tadpoles had the capability of swimming, indicating not only that appropriately formed muscles and nerves were present but that their function was fully coordinated.

Lucké renal adenocarcinomas were then produced that varied in chromosome number from ordinary frog cells (diploid number: 26; DiBerardino *et al.*, 1963) to serve as cytogenetically labeled nuclear donors. This was done because the yield of larvae programmed by the renal adenocarcinoma nuclei was relatively low; in one experiment, 7.4% (12 of 161) developed beyond the gastrula stage (DiBerardino *et al.*, 1961). Cytogenetically labeled donor nuclei would provide the means to ensure that the larvae were truly descended from the transplanted renal adenocarcinoma nucleus and not from an inadvertently surviving egg nucleus. Triploid tumors were produced by injecting Lucké tumor herpesvirus (LTHV) (Fig. 3) into triploid (Fig. 4) embryos (McKinnell and Tweedell, 1970). Seven larvae developed from the transplantation of triploid renal adenocarcinoma nuclei. All seven larvae were triploid (McKinnell *et al.*, 1969). The triploidy of the tumor nuclear transplant larvae excluded the remote possibility that the larvae developed by parthenogenesis and thereby provided direct genetic evidence that nuclei from the Lucké renal adenocarcinoma were reprogrammed to differentiate all tissue types of prefeeding swimming larvae.

A carcinoma is composed of malignant epithelium and of nonmalignant components such as connective tissue. Are tumor nuclear transplant larvae descended from mitotic progeny of carcinoma cells or from nonmalignant connective tissue cells of the tumor? To answer this question, dissociated cells of the Lucké renal adenocarcinoma were studied by their fluorescence in the ultraviolet (UV) range after staining with acridine orange (Tweedell, 1965;

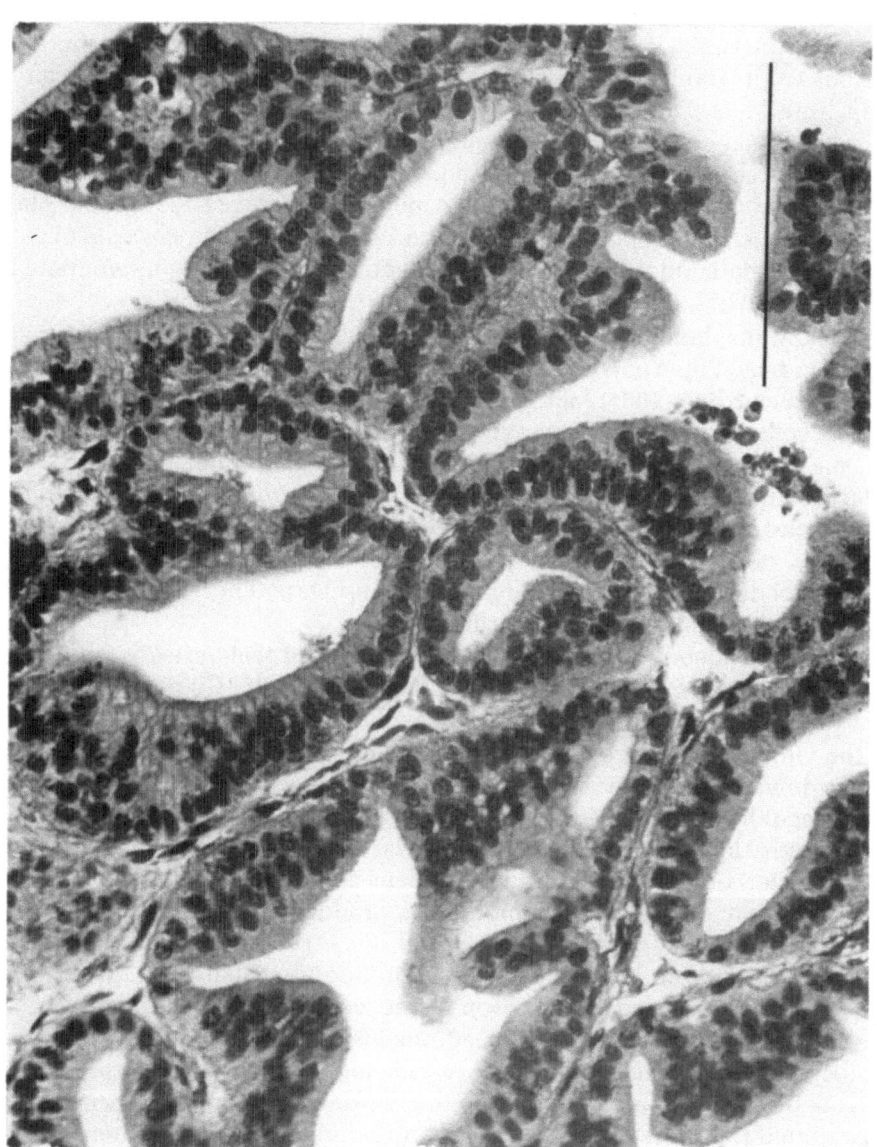

Figure 1. Histological section of a Lucké renal adenocarcinoma. Bar = 100 μm. (From McKinnell and Duplantier, 1970.)

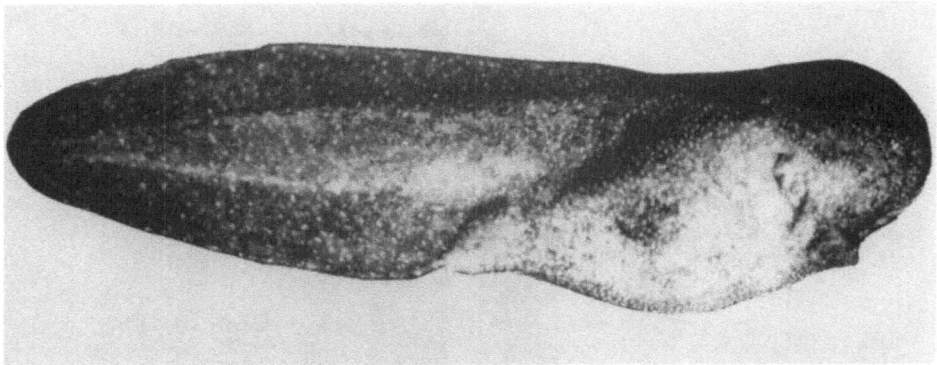

Figure 2. *Rana pipiens* larva resulting from the transplantation of a Lucké renal adenocarcinoma nucleus into an enucleated mature egg. (From McKinnell, 1972.)

Harrison *et al.*, 1975). Nuclei stain yellow green in both carcinoma and connective tissue. However, the cytoplasm of tumor cells is red, in contrast to the deep green cytoplasm of connective tissue. Hence, carcinoma cells are easily distinguished from connective tissue cells by their fluorescence in the UV range. Most (98.5%) of the dissociated cells were carcinoma (McKinnell *et al.*, 1976; Seppanen *et al.*, 1984), strongly supporting the view that the larvae previously reported could only have developed in response to the insertion of a renal adenocarcinoma nucleus into an enucleated ovum.

Additional evidence supporting the view that malignant carcinoma nuclei program for the nuclear transplant larvae was obtained by transplantation of renal adenocarcinoma nuclei whose cells were cultured *in vitro*. The cell cultures were epithelial in appearance, as would be expected from cultures of a carcinoma. Larvae were produced from transplantation of the cultured nuclei, corroborating and strengthening the earlier studies (DiBerardino *et al.*, 1983). The inescapable conclusion is that mitotic descendants of the Lucké renal adenocarcinoma nuclei are competent to differentiate into all the tissues characteristic of a swimming larva. The molecular mechanisms controlling differentiation are yet to be explained.

As is true of most tumors (Mintz, 1978; Freshney, 1985; Pierce and Speers, 1988), the cell type of origin of the Lucké renal adenocarcinoma is unknown. However, several lines of evidence argue compellingly that the target cell for neoplastic change is either a stem cell or a less than fully differentiated precursor cell of the frog kidney. One line of evidence concerns the developmental stages that are vulnerable to Lucké tumor herpesvirus (LTHV) transformation. Only eggs, embryos, and young larvae are susceptible to the oncogenic action of LTHV (Tweedell, 1967; McKinnell and Tweedell, 1970). A pronephric cell (or perhaps an antecedent stem cell of the pronephros) can be transformed by the virus, resulting in pronephric carcinomas at metamorphosis (Tweedell, 1978; Ogilvie *et al.*, 1984). Since the pronephros exists only in immature stages, the pronephric renal tumors support the view that an embryonic cell, not a fully

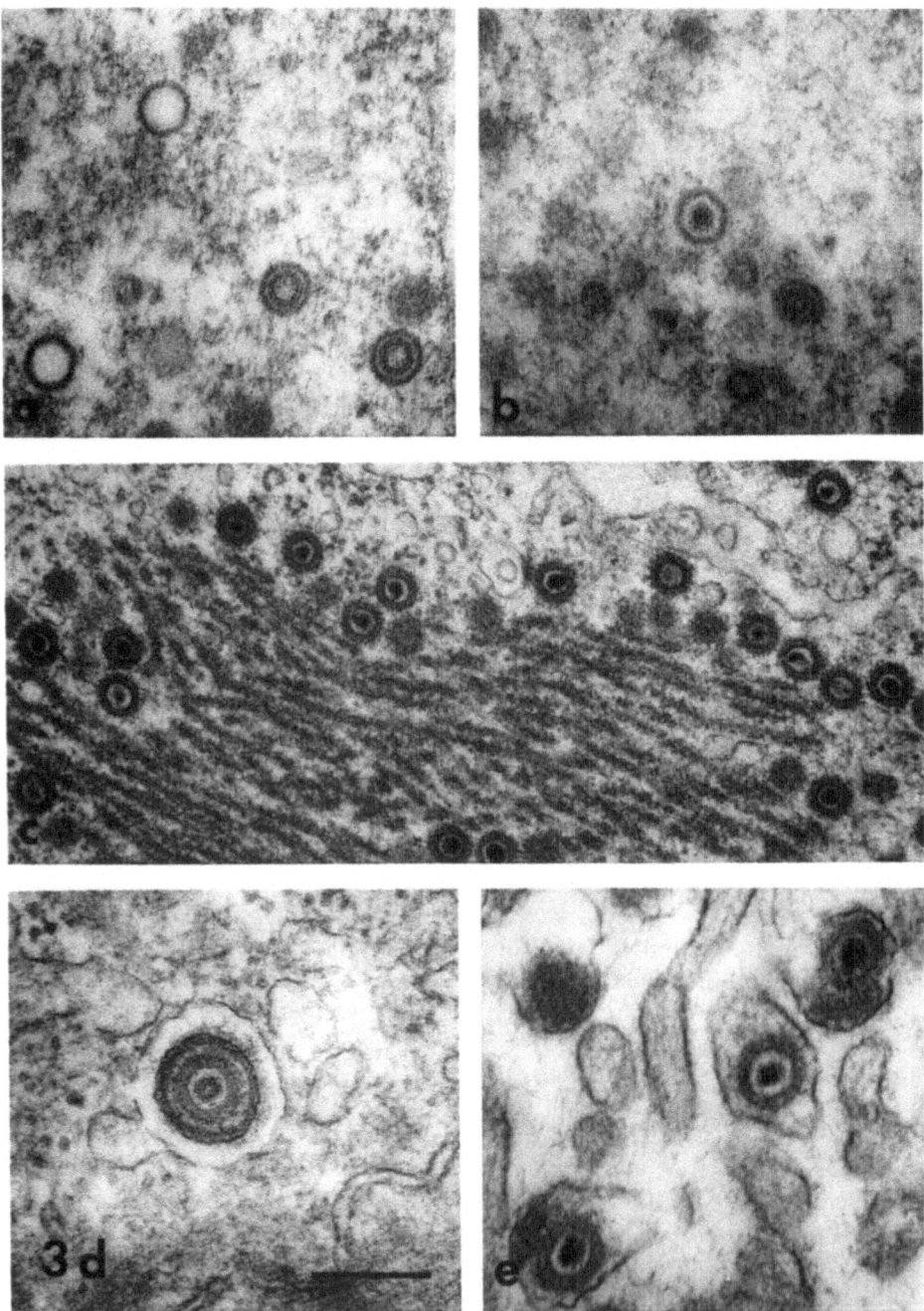

Figure 3. Electron micrographs of Lucké tumor herpesviruses. (a) Single- and double membrane capsids. (b) Single-membrane capsid with DNA core. (c) Double-membrane capsids with DNA cores. (d) Mature virion in cytoplasm; bar = 0.1 μm. (e) Extracellular mature virion with loose envelope. (From McKinnell, 1973a.)

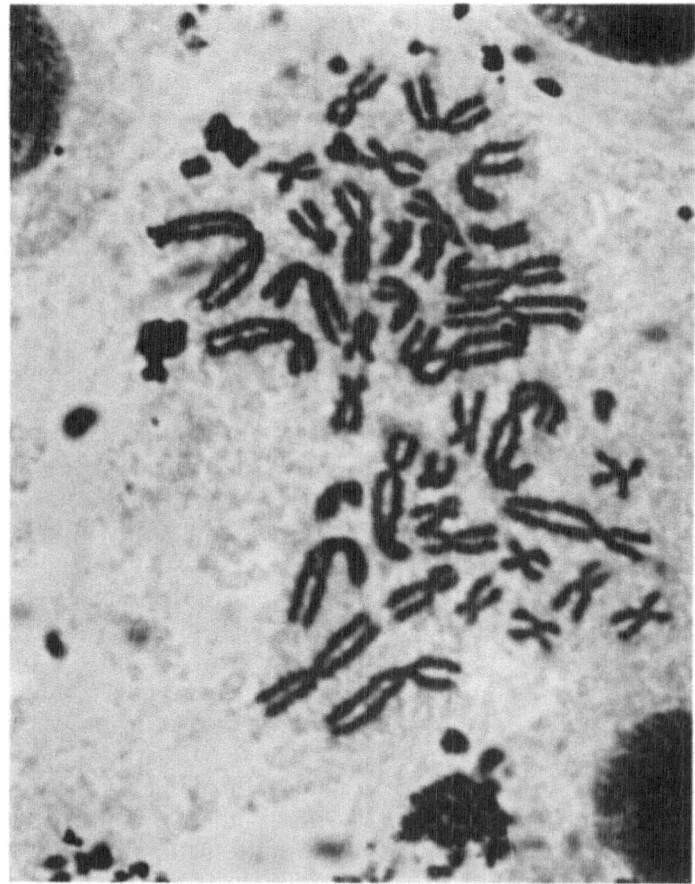

Figure 4. Aceto-orcein metaphase chromosome preparation from a triploid larva (3n=39). (Courtesy of Cynthia Nicholson.)

differentiated adult cell, is the target for LTHV transformation. In addition to the relatively rare induced pronephric tumors, many mesonephric tumors are induced by injection of LTHV. Nephric blastemas become separated from lateral plate mesoderm in recently hatched frog larvae (Shumway, 1940, stage 21), and differentiation of mesonephric nephrons from the nephric blastema does not begin in larvae until the operculum is completed (Shumway, 1940, stage 25) (Witschi, 1956). Since eggs, embryos, and young larvae, bereft of differentiated renal cells, are highly vulnerable to the oncogenic effect of LTHV, it is reasonable to suspect that the target of LTHV is not a mature renal cell but a less than fully differentiated nephric precursor cell. Induction of tumors in adult frogs by exposure of early embryonic frogs to LTHV has interesting counterparts in mammalian tumor induction studies where viruses and chemical carcinogens have greatest efficacy when administered to neonatal subjects (Gross, 1951; Pietra *et al.*, 1959; Dawe and Law, 1959; Baluda and Jamieson, 1961).

Other evidence relating to the embryonic origin of the Lucké renal adenocarcinoma is derived from the transplantation experiments. Normal development, or extensive embryonic development in nuclear transfer embryos, is
characteristic primarily of transplanted young embryonic nuclei. Nuclei from
older embryos and adults generally require pretreatment, either with low temperature and a polycationic amine (Hennen, 1970) or by pretreatment by exposure of nuclei to oocyte cytoplasm (DiBerardino et al., 1986). Yet no pretreatment of any kind is required to condition renal adenocarcinoma nuclei to
participate in embryonic development after nuclear transplantation. In other
words, the Lucké renal adenocarcinoma nuclei behave as if they were embryonic nuclei. That might be, as is suggested here, because the Lucké renal adenocarcinoma cells are embryonic stem cells frozen in that stage by exposure to
LTHV.

An additional line of evidence suggesting that an immature cell is target for
transformation by LTHV is temperature sensitivity of the cytoplasmic microtubule complex. Cells cultured from metamorphic larvae (i.e., larvae with large
posterior limbs and fully formed anterior limbs enclosed within the gill cavity
just before transformation into juvenile frogs) display cytoplasmic microtubules at both 7°C and 28°C, in contrast to cultured Lucké renal adenocarcinoma cells, which have well-developed cytoplasmic microtubules at 28°C but
have profoundly disorganized microtubules at 7°C (McKinnell et al., 1984). The
sensitivity to loss of organization of the cytoplasmic microtubule complex at
the reduced temperature of these tumor cells is similar to, and characteristic of,
early embryonic cells (McKinnell et al., 1988b). These lines of evidence support the view that the Lucké renal adenocarcinoma may have an origin from
less than fully differentiated precursor cells as other tumors are thought to have
(Mintz, 1978; Pierce et al., 1978; Paul, 1978; Prescott and Flexer, 1986; Sachs,
1986).

One goal of differentiation therapy is to induce the genome of neoplastic
cells to yield cell progeny that are terminally differentiated and postmitotic. It
may be asked if cell differentiation can suppress expression of the neoplastic
phenotype without concomitant loss of proliferative potential (Scott et al.,
1988). Differentiation of the full spectrum of larval tissues has been demonstrated in these studies of a neoplastic genome. Cytogenetic analysis of metaphase chromosomes of the swimming tadpoles is reported above. Thus, it is
clear because of the presence of the metaphase chromosomes that neoplastic
suppression is compatible with retained mitotic potential.

The experiments resulting in differentiated mitotic progeny of the Lucké
renal adenocarcinoma would be of limited value if they had no relevance to
tumors of higher organisms. However, it has been reported that a human congenital nephroma, which grew in vitro as undifferentiated cells totally devoid
of renal structures, could be cocultured with fetal mouse dorsal spinal cord
with the result that the nephroma differentiated fetal nephrons complete with
Bowman's capsules, glomeruli, and proximal tubules (Crocker and Vernier,
1972). The human nephroma was thus shown to differentiate when confronted
with a normal kidney tubule inducer (Grobstein, 1955; Unsworth and Grob-

stein, 1970). Therefore, both the frog and the human renal neoplasms retain genetic potentiality, as do many other malignant neoplasms, for differentiation of cell types not expressed in the neoplastic condition.

3.3.2. Epithelial Tumor of *Triturus cristatus*

Skin tumors have been induced in the European newt, *Triturus cristatus*, by subcutaneous injection of a mixture of dibenzanthracene and benzpyrene. The tumors became invasive and in a number of instances metastatic nodules were detected at autopsy. Without any treatment, the tumors underwent what was termed spontaneous regression but was in fact differentiation of the neoplastic cells into keratinized skin cells and epithelial pearls, pigment cells, and mucous cells (Seilern-Aspang and Kratochwil, 1962).

3.4. Ectodermal Neoplasms of Mammals

3.4.1. Squamous Cell Carcinoma

Skin neoplasms of rats include a transplantable squamous cell carcinoma with islands of well differentiated epithelial pearls. The pearls are composed of differentiated squamous cells lacking mitoses (How and Snell, 1967). Rats with such tumors were pulse-labeled with [³H]thymidine in order to detect the origin of the epithelial pearls. While undifferentiated squamous cell carcinoma nuclei were labeled after 2 hr with this treatment, epithelial pearls remained unlabeled. Tumors examined at 96 hr postlabeling, however, contained epithelial pearls that were labeled (Pierce and Wallace, 1971). This study indicates that the well-differentiated pearls, comprised of postmitotic cells, were derived from the malignant squamous cell carcinoma (Fig. 5). The designations squamous cell carcinoma and epithelial pearl are morphological and do not indicate the capacity of either to form malignant growths. Therefore, fragments of carcinoma and dissected pearls were grafted to appropriate rat hosts. The epithelial pearls failed to form tumors, in contrast to many of the carcinoma fragments, which did. It was concluded that the squamous cell carcinoma stem cells differentiated into the nonmalignant squamous cell pearls which were incompetent to form neoplastic growths (Pierce and Wallace, 1971).

Frequently, the concept of differentiation therapy is used in juxtaposition to cytotoxic therapy. This is in contrast to a therapy that seeks control of neoplasia without the need for cell death with another therapy, which by design, is lethal to cells. Squamous cells normally differentiate and die. Thus, in the special case of squamous cell carcinoma, differentiation therapy may lead to cell death. Cell death was indeed reported after squamous cell carcinoma, cultured as three-dimensional tumoroids, was treated with IFN$_\gamma$. The treated cells became near-normal morphologically and contained more of a 67-kDa keratin (the 67 kDa keratin is thought to be differentiation specific because of its presence in epidermis and because it is found in reduced amounts in cell

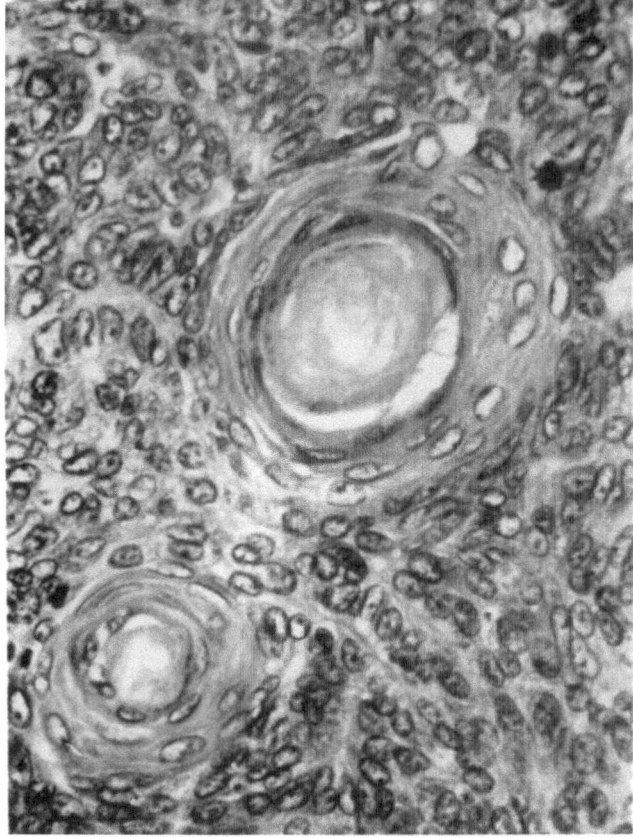

Figure 5. Rat squamous cell carcinoma with undifferentiated neoplastic cells and two differentiated squamous cell pearls with keratotic centers. (From Pierce and Wallace, 1971.)

cultures and in poorly differentiated neoplasms, Cooper *et al.*, 1985; see also Yuspa and Roop, 1988) than the control cells (Chang *et al.*, 1988).

3.4.2. Neuroblastoma

The capacity of human neuroblastomas to differentiate was reported more than 40 years ago, but the significance of the observations to the idea of cancer cell differentiation may have escaped the investigators. In an attempt to develop a procedure that would distinguish neuroblastomas from lymphosarcomas and small cell carcinomas, eight human neuroblastomas were cultured *in vitro*. Dendritic processes and axons (neurites) were a conspicuous trait of the cultured neuroblastoma cells (Murray and Stout, 1947). Since that pioneering study, other investigators have reported similar results, i.e., neuroblastomas, which typically resemble masses of small lymphocytes, will, when cultured, mature and differentiate (Goldstein *et al.*, 1964; Prasad, 1982; Tsokos *et al.*,

1987). Differentiation may occur spontaneously *in vitro* (Murray and Stout, 1947; Goldstein *et al.*, 1964) or result from treatment with a diversity of chemical substances (see Section 5) (Schubert *et al.*, 1969, 1971; Seeds *et al.*, 1970; Spiegelman *et al.*, 1979; Goldstein and Plurad, 1980; Sidell, 1981; Mattsson *et al.*, 1984; Prasad, 1982; Tsokos *et al.*, 1987; Gotti *et al.*, 1987; Sugimoto *et al.*, 1988). Differentiated neuroblastoma cells, when transplanted into the brains of nonhuman primates, survive at least 9 months. The neuroblastoma cells do not revert to a malignant state during that time (Kordower *et al.*, 1987).

Just as the mammalian blastocyst has the capacity to induce differentiation of teratocarcinoma cells (see Section 3.6.2), there may be an analogous differentiating environment that exerts biological control over neuroblastoma cells. Cells from neuroblastomas have been injected into the neural crest migratory route (Podesta *et al.*, 1984) and into the embryonic adrenal gland at the time when neural crest cells arrive at that organ (Wells and Miotto, 1986). While the fate of the injected neuroblastoma cells is not known, it was observed that tumors failed to occur in expected numbers suggesting the possibility of neuroblastoma cell differentiation.

3.5. Mesodermal Tumors of Mammals

3.5.1. Rhabdomyosarcoma

Striated muscle neoplasms of rats (Corbeil, 1967) and of humans (Kroll *et al.*, 1963) contain both undifferentiated mononucleated cells and multinucleated striated muscle fibers. The mononucleated cells are mitotically active, incorporate [³H]thymidine, and have DNA values typical of dividing cells. By contrast, the postmitotic multinucleated striated fibers fail to be labeled with [³H]thymidine after a 2-hr exposure. If the cells are incubated an additional 16 hr after unincorporated radioactive precursors have been removed, the differentiated cells become labeled, indicating that the origin of the muscle fibers is from the mononucleated stem cells (Nameroff *et al.*, 1970).

Undifferentiated mouse striated muscle tumor cells become differentiated when treated with sodium butyrate (Dexter *et al.*, 1981) or with N,N-dimethylformamide (Dexter, 1977). Animals injected with tumor cells that have differentiated develop fewer tumors than do animals receiving undifferentiated rhabdomyosarcoma cells (Dexter *et al.*, 1978).

Rat rhabdomyosarcoma cells will differentiate in response to hydrocortisone, which in turn inhibits metastases to other tissues. Metastasis is a fundamental characteristic of malignancy. Differentiated postmitotic cells should, and apparently do, fail to metastasize (Becker *et al.*, 1985).

3.5.2. Myeloid Leukemia

Factors that control growth, viability, and differentiation of granulocyte and macrophage stem cells have been purified and characterized as glycoproteins produced by normal tissues. Differentiation factors induce mouse and rat myeloid leukemic cells to differentiate in vitro (Metcalf, 1985; Jimenez and Yunis, 1987). The human myeloid leukemic cell lines ML-1 and HL-60 (Fig. 6) can be induced to differentiate in vitro by treatment with a number of different agents (Takeda et al., 1982; Mendelsohn et al., 1980; Matzner et al., 1987; Yunis et al., 1987; Langdon and Hickman, 1987a; Samara et al., 1987). Terminal differentiation of murine and human leukemic cell lines has been induced with human colony-stimulating factors produced with recombinant DNA techniques (Metcalf, 1986; Souza et al., 1986). A combination of interferon- and retinoic acid-induced differentiation in vitro of leukemic cells obtained from patients with acute myelogenous leukemia (Gallagher et al., 1987). Curiously, HL-60 cells differentiate in culture medium deprived of a single essential amino acid, despite the failure of protein synthesis inhibitors to induce differentiation (Pilz et al., 1987).

Some myeloid tumor cell lines have the competence to grow but fail to differentiate properly. Some of these cells, designated D^+ for differentiation positive, respond to an exogenous supply of differentiation inducer by producing cells that are mitotically inactive and that are morphologically and physiologically indistinguishable from normal macrophages and granulocytes. The D^+ cells respond not only to the natural inducer but by differentiation to a relatively long list of chemical agents as well (Sachs, 1978). Other leukemic cells, designated D^- for differentiation negative, fail to respond to natural differentiation inducers and represent a different stage of neoplastic progression than do the D^+ type cells. The D^- myeloid leukemic cells may respond to other inducers, or combinations of inducers, together with normal differentiation factors, to produce differentiated, postmitotic myeloid blood cells. The D^- cells may respond to steroid hormones, X-rays, some vitamins, insulin, lectins, some phorbol esters, and low doses of conventional chemotherapeutic agents (Sachs, 1986, 1987a). These studies demonstrate that both the D^+ and the D^- cells have retained the genes necessary for differentiation, even though their chromosomes may vary from the normal euploid karyotype (Azumi and Sachs, 1977). It is interesting to note that while there may be a genetic basis for the malignant behavior of leukemic cells, the genetic alteration can be bypassed with appropriate treatment (Sachs, 1978; Lotem and Sachs, 1978, 1981, 1984).

3.5.3. Murine Erythroleukemia

Murine erythroleukemia cell (MELC) lines, derived from mice with Friend virus-induced leukemia (Friend et al., 1966; Friend, 1978, 1980), grow in vitro as pleomorphic populations of undifferentiated cells and cells at various levels of erythroid differentiation. Changes in MELC that accompany differentiation include greatly increased levels of globin mRNA, α- and β-globin synthesis, appearance of erythrocyte membrane antigens, synthesis of heme and hemoglobin, decreased cell volume and cell water, and a limited capacity for cell renewal (Friend et al., 1971; Friend, 1980; Ellis et al., 1987).

Chemicals known to induce differentiation of MELC in vitro include dimethylsulfoxide (DMSO), hexamethylene bisacetamide (HMBA), purines and

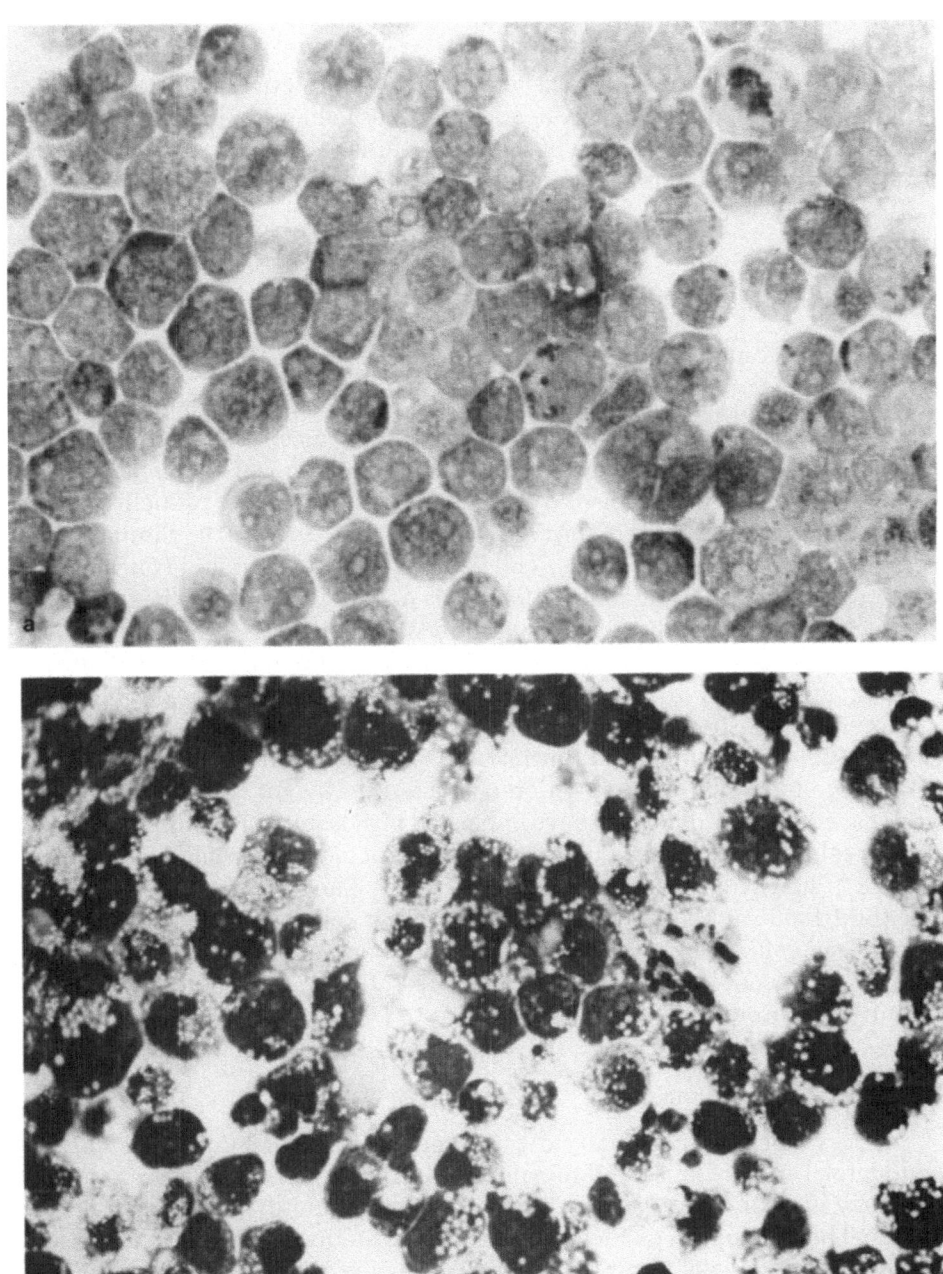

Figure 6. Phagocytosis by differentiated HL-60 cells induced with human lung conditioned medium (HLCM). Latex beads (10⁸/ml) were added to both control and HLCM treated cells. Subsequently, the beads were removed and the cells were washed. (a) Control cells with no intracellular latex beads, (b) HLCM differentiated phagocytic cells with plentiful intracellular latex beads. (From Yunis *et al.*, 1987.)

purine analogues, short-chain fatty acids, hemin, cardiac glycosides, and metabolic inhibitors (reviewed by Friend, 1980; Marks and Rifkind, 1978; Marks *et
al.*, 1987; Waxman *et al.*, 1986). Recently, the antibiotic trichostatins and a
protein differentiation factor were added to those chemicals that have the capacity to induce differentiation of MELC (Yoshida *et al.*, 1987; Eto *et al.*, 1987)
and thus may be added to the more than 300 known inducers of differentiation.

3.6. Endodermal Tumors of Mammals

3.6.1. Human Colon Cancer

Human colon carcinoma cell lines grow in soft agar and form tumors when
injected subcutaneously in mice. Treatment of those cell lines with the polar
solvent *N,N*-dimethylformamide induces differentiation of the cells and the
loss of clonogenicity in semisolid medium as well as a major reduction in the
tumor forming capabilities of the cells (Dexter *et al.*, 1978, 1979). Similarly, the
growth of grafted colon carcinoma cell lines was inhibited in nude mice treated
with polar solvents (Dexter *et al.*, 1982). It is not known whether the inhibition
of growth in treated nude mice is caused by differentiation of the grafted colon
carcinoma cells or by some effect of the host against the inoculum. Be that as it
may, an induction of differentiation with a concurrent reduction of tumorigenicity were demonstrated with the human colon carcinoma cells. More recently,
a human colon carcinoma cell line (HT-29), cultured in the absence of glucose,
was shown to exhibit differentiation characterized by gut cells that are polar-
ized, have brush borders, and have enhanced glycosaminoglycans synthesis as
well as cells which secrete mucous (Zweibaum *et al.*, 1985; Simon-Assmann *et
al.*, 1987; Huet *et al.*, 1987; Dudouet *et al.*, 1987). Another colon carcinoma cell
line (LIM 1863) similarly differentiates columnar cells and goblet cells *in vitro*
(Whitehead *et al.*, 1987). Colon cancer cell differentiation in response to chemicals, culture conditions (confluent or postconfluent), and other factors was
recently reviewed (Kim, 1988).

3.6.2. Teratocarcinoma

It may not be appropriate to place teratocarcinomas under the caption
endodermal tumors, but the rationale for placing them here relates to the fact
that they are considered germ cell tumors and primordial germ cells in the
mammal have an origin on the posterior yolk sac of the embryo (Witschi, 1948).

Teratocarcinomas occur spontaneously in the gonads of some strains of
mice (Stevens, 1973) or are produced by grafting embryos to ectopic sites (Stevens, 1970a,b; Solter *et al.*, 1970; Damjanov *et al.*, 1987). Mouse teratocarcinomas are composed of a wide diversity of tissues in various stages of differentiation, including central nervous system (CNS), cartilage, bone, stratified
squamous epithelium, striated muscle fibers, respiratory and alimentary
mucosae, and glands (Askanazy, 1907; Fekete and Ferrigno, 1952; Pierce *et al.*,

1959; Willis, 1962; Stevens, 1981) in a hodgepodge arrangement described by Needham (1942) as an approximation to chaos. Sometimes structures appear that resemble early mammalian embryos; these are called embryoid bodies (Dixon and Moore, 1953). The internal cells of the embryoid bodies were thought to be multipotential and were believed to give rise to the diversity of cell types by differentiation (Pierce and Dixon, 1959). Grafts to the anterior eye chamber (Stevens, 1960) or subcutaneously in mice (Pierce et al., 1960) established experimentally the multipotentiality of the cells. Kleinsmith and Pierce (1964) transplanted single teratocarcinoma cells isolated in capillary tubes into mice and obtained experimental tumors composed of embryonal carcinoma and many well-differentiated somatic tissues.

Brinster (1974) first demonstrated that teratocarcinoma cells, transferred to a mouse blastocyst, will participate in normal development. He illustrated his study with a random bred Swiss albino mouse displaying a stripe of agouti fur. The teratocarcinoma cells, carrying the genotype for agouti color, had differentiated into fur-producing cells in the presence of the normally differentiating embryonic host cells. Brinster's studies were extended by others (Mintz and Illmensee, 1975; Illmensee and Mintz, 1976; Dewey et al., 1977; Papaioannou et al., 1975, 1978; Stewart and Mintz, 1981), who injected genetically marked teratocarcinoma cells into blastocysts that were subsequently placed in foster female mothers (the methodology for this procedure is described by Bradley, 1987). Normal live mosaic mice were born. The teratocarcinoma cells differentiated all tissue types found in normal mouse development.

A teratocarcinoma cell line, E Ca 247, has the competence to differentiate trophectoderm after treatment in the blastocyst. If inner cell mass cells are the normal counterpart of malignant teratocarcinoma, one might suspect that some inner cell mass cells could differentiate trophectoderm also. Some do. In this interesting study, knowledge of cancer cell biology has led to a new appreciation of the differentiative potential of normal embryonic cells (Pierce et al., 1988).

Teratocarcinoma-derived cell lines differentiate in vitro (Rudnicki and McBurney, 1987). Retinoic acid will induce such differentiation (Strickland and Mahdavi, 1978), and varying the concentration of retinoic acid used will induce different cell types. A subclone of P19 teratocarcinoma cells exposed to retinoic acid at 10^{-9} M differentiates into cardiac cells, whereas retinoic acid at 10^{-8} M induces skeletal cells, and retinoic acid at 10^{-5} to 10^{-7} M causes the differentiation of neurons and astroglia (Fig. 7) (Edwards and McBurney, 1983). Retinoic acid treatment of teratocarcinoma cells affects DNA methylation (Razin et al., 1984), expression of homeobox genes (Colberg-Poley et al., 1985; Hauser et al., 1985; Deschamps et al., 1987) and X-chromosome inactivation (Paterno and McBurney, 1985) and will decrease poly(ADP-ribose) synthesis (Ohashi et al., 1984). Cell-surface glycolipid antigens change as an embryonal carcinoma cell line (NTERA-2) differentiates in response to retinoic acid (Fenderson et al., 1987).

Not only will murine teratocarcinoma cells differentiate but so too will the human counterpart, both in vitro (Webb et al., 1986; Levine and Flynn, 1986;

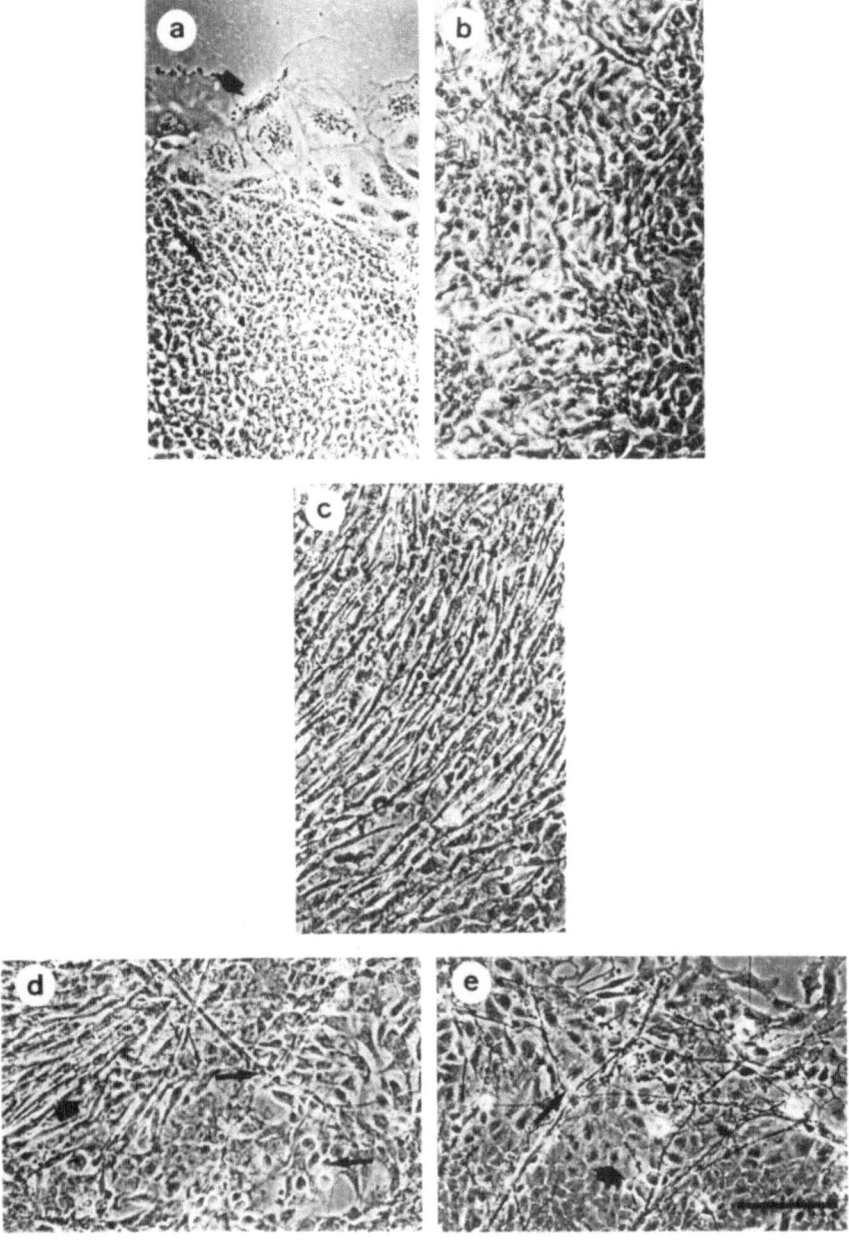

Figure 7. The concentration of retinoic acid (RA) determines the differentiated cell type formed by an embryonal carcinoma (EC) cell line in culture. (a) Undifferentiated EC cells (thin arrow) and extraembryonic endodermal cells (thick arrow) with 0 M RA. (b) Cardiac muscle cells that contract rhythmically with 10^{-9} M RA. (c) Bipolar myoblasts and a few myotubes with 10^{-8} M RA. (d) Both myotubes (thick arrow) and neurons (thin arrows) with 10^{-8} M RA. (e) Neuron (thin arrow) growing above an area with astroglia cells (thick arrow) with 10^{-7} M RA. Magnification is the same in all panels, length of index line in (e): 100 μm. (From Edwards and McBurney, 1983.)

Fenderson *et al.*, 1987; Wartiovaara and Rechardt, 1985) as well as *in vivo* (Smithers, 1969; Carr *et al.*, 1981).

4. Cancer Cell Differentiation: A Brief Chronicle

The possibility of exploiting induced cancer cell differentiation as a new mode of cancer therapy is becoming increasingly attractive (see references cited, last paragraph, Section 5). One might view the biological research leading to contemporary clinical trials as existing only in the recent literature. It may therefore come as a mild surprise that notions concerning cancer cell differentiation have their roots in early nineteenth century biology.

As early as 1829, Recamier (see Oberling, 1944) envisioned neoplastic cells as embryonic cells that persisted during development and into adulthood. The embryonic cells were thought to become neoplastic because of differentiation in an inappropriate anatomical site or at an improper time in ontogeny. This view was shared by Remak in 1854 (Rather, 1978), Cohnheim (1875, 1889) and Beard (1902) (reviewed by Triolo, 1965). These early workers invoked the need to consider the origin of cancer cells as it relates to morphology (i.e., some tumors appear to be undifferentiated and embryonic compared with adult tissue) and mitotic potential (i.e., embryos and neoplasms frequently share similar growth rates that contrast with the reduced or absent growth of many normal adult tissues).

The maturation and differentiation of leukemic cells were reported early in this century by Awrorow and Timofejewsky (1914) and Timofejewsky and Benewolenskaya (1929), followed by Bessis (1954), Friend *et al.* (1966), and Ichikawa (1969).

Needham (1936, 1942) was aware of the Lucké renal adenocarcinoma (see Section 3.3.1) and proposed an experiment that he hoped would "master" the "wildly growing material." Needham suggested grafting Lucké renal adenocarcinoma fragments to the limbs of urodeles (which regenerate) (Rose 1970; Hay 1974). After the graft had grown, he proposed that the limb be amputated with the thought that the intense biological activity of the regeneration blastema might force the frog neoplasm to differentiate. Rose and Wallingford (1948) and Rose (1949) attempted the Needham experiment and reported preliminary success; others failed in their attempt to repeat the Needham experiment (Ruben, 1955, 1956; Ruben and Balls, 1964; Sheremetieva, 1965; Tsonis, 1984).

It is worth reconsidering the Needham/Rose experiment in light of current concepts. One would speculate that, in view of the stability of cell-type specialization of most tissues, it would be unlikely that renal adenocarcinoma would differentiate limb tissues. If grafted Lucké renal adenocarcinoma in the blastema differentiated normal cells, it would probably differentiate normal kidney cells. However, limb blastema would seem an inappropriate environment for the induction of kidney cell differentiation. Consider the fact that teratocarcinoma cells differentiate in the blastocyst (see Section 3.6.2) and that neuroblastoma cells may differentiate (see Section 3.4.2) in the neurula. These examples would appear to suggest that cancer cells differentiate in an environment appropriate for the differentiation of normal stem cell counterparts, pointing to the embryonic kidney as an appropriate site for differentiation of the Lucké renal adenocarcinoma. No one, thus far, has studied the capability of differentiating embryonic kidney to induce differentiation of the grafted Lucké renal adenocarcinoma. The Needham/Rose experiment is considered here because in the evolution of ideas it proposed a mode of inducing differentiation, it was ahead of its time, and contemplation of the experiment helps clarify the need for special environments for induced cancer cell differentiation.

Cohnheim and Needham were not alone. Other pioneers who believed that with an understanding of normal differentiation might emerge greater understanding of cancer were Waddington (1935), Berrill (1943), Pierce (1961), Markert (1968), and Potter (1969).

5. Control of Differentiation of Embryo Cells and Cancer Cells: Are There Commonalities?

Earlier in this chapter (see Section 2) it was stated that the anatomy and function of normal cells and cancer cells have much in common. Following that discussion, a number of cancers that undergo maturation and differentiation were reviewed. Inasmuch as embryo cells also undergo maturation and differentiation, one could ask if there are common mechanisms that lead these two categories of cells to the differentiated state. The question is unanswered because primary embryonic induction remains an unsolved problem after over a half century of study (Saxén et al., 1978). In fact, Gurdon (1987) stated that the "molecular basis of induction and the inductive response remains almost totally obscure." That statement is as applicable to cancer cell differentiation as it is to embryo cell differentiation.

While the molecular mechanisms of embryonic induction are admittedly obscure, much is known about the operational aspects. For example, it is recognized that primary induction of the embryonic axis and neural plate occurs through an interaction of the ectoderm with underlying chordamesoderm (Spemann and Mangold, 1924; Spemann, 1938; Saxén and Toivonen, 1962; Tarin, 1973; Sanchez and Barbieri, 1988). Induction of neural tissue can occur experimentally in heterotopic sites in vivo or as a result of the interaction of inducer and responding tissue in sandwiches in vitro. Curiously, living inducer is not required for induction, making a forceful argument that the reaction is a chemical, not biological, event (Bautzmann et al., 1932; Saxén and Toivonen, 1962). Following the realization that a chemical substance may trigger the events leading to differentiation, there was frenzied activity to identify chemicals critical to this event. Within a short time, many unrelated and frequently bizarre chemicals were identified that have the capacity to induce neural differentiation of ectoderm fragments in vitro (Okada, 1938; Barth, 1939; Ranzi and Tamini, 1939; Brachet and Rapkine, 1939; Beatty et al., 1939; Shen 1939). The understanding of induced differentiation was further complicated when it was found that neural differentiation *in vitro* would occur with no inducer at all (Barth, 1941). It would seem that there are too many inducers and the inducers appear to be unrelated chemically. This observation, coupled with the reports that induction occurs with no inducer at all, led to the notion that perhaps some (many?) of the chemicals were not inducers themselves but were substances that act to liberate inducers intrinsic to the reacting cells. It was postulated that they did this by injury to the plasma membrane resulting in a sublethal cytolysis that in turn activated substances leading to neural differentiation (Holtfreter 1947, 1948). Curiously, the concept of cell injury, which provided a common denominator linking diverse inductive reactions, reappeared recently in a discussion of induced cancer cell differentiation (Hickman and Friedman, 1988). While prospective neural plate (both *in vivo* and *in vitro*) seems poised and ready to differentiate in response to a diversity of seemingly unrelated stimuli, it should be remembered that in the intact embryo embryonic ectoderm differentiates as epidermis in the absence of chordamesoderm.

There are operational similarities in the differentiation of cancer cells to that described for the differentiation of normal embryonic cells. Spontaneously, squamous cell carcinoma (see Section 3.4.1) differentiates in intact animals and humans. So too does neuroblastoma (see Section 3.4.2). Thus, cancer cell differentiation occurs *in vivo*, as it is known to occur in intact embryos. Tumor cells such as teratocarcinoma (see Section 3.6.2) and neuroblastoma (see Section 3.4.2) can be placed in a new embryonic environment in which differentiation occurs, perhaps analogous to heterotopic differentiation in experimental embryos. And cancer cells will differentiate in response to a diversity of substances (which might be described as unrelated and even bizarre) *in vitro* much as normal embryonic differentiation can be induced *in vitro*.

Biological agents that induce cancer cell differentiation encompass a substantial literature. Biological response modifiers discussed with myeloid leukemia (see Section 3.5.2) are the macrophage granulocyte inducers (MGI), also known as granulocyte macrophage colony stimulation factor (GM-CFS) (Sachs, 1987a,b; Metcalf, 1985). A protein, the erythroid differentiation factor (EDF), has been isolated from a human leukemic cell line that induces differentiation of murine erythroleukemia cells (Eto *et al.*, 1987) (see Section 3.5.3). Glial maturation factor (GMF) is an acidic protein that promotes differentiation of cultured astroblasts and cultured human glioma cell lines (Lim *et al.*, 1986). Interferon alone has no effect on differentiation of HL-60 cells, but retinoic acid induced-differentiation is enhanced when combined with interferon treatment (Kohlhepp *et al.*, 1987). Similarly, tumor necrosis factor and retinoic acid potentiate the differentiative capacity of each other (Trinchieri *et al.*, 1987). Other biological agents that effect differentiation of cancer cells include factors in the cytoplasm of the amphibian egg (see Section 3.3.1), which inhibit expression of those genes characteristic of the Lucké renal adenocarcinoma and awake from dormancy the genes associated with normal differentiation of amphibian larvae, and the interior of the murine blastocyst, which permits differentiation of teratocarcinoma cells (see Section 3.6.2).

Cancer cells also respond to a wide spectrum of relatively simple chemical substances. The induced differentiation of teratocarcinoma cells occurs by a variety of chemical substances: retinoic acid (Moore *et al.*, 1986), retinal, retinol, 5-bromouracil 2'-deoxyribose (BUdR), 5-iodouracil 2'-deoxyribose (IUdR), hexamethylene bisacetamide (HMBA), dimethylacetamide (DMA), dimethylsulfoxide (DMSO) (Andrews *et al.*, 1986), and ouabain (Zimmerman and Speers, 1988). Promyelocytic leukemia cells (HL-60) are induced to differentiate by retinoic acid, 12-*O*-tetradecanoyl-phorbol-13-acetate (TPA), phytohemagglutinin-leukocyte-conditioned medium (Kaplinsky *et al.*, 1986; Paukovits *et al.*, 1986; Maddox and Haddox, 1988), cyclopentenyl and cyclopentyl analogues of cytidine (Glazer *et al.*, 1986), dimethylsulfoxide (DMSO), and dimethylformamide (DMF) (Mukherjee *et al.*, 1985; Matzner *et al.*, 1987), and tunicamycin (Kos *et al.*, 1987), as well as recombinant human leukocyte interferon in combination with other inducers (Grant *et al.*, 1985). This is only a partial listing of differentiation agents. Hickman and Friedman (1988) report that 73 compounds are known that induce the terminal differentiation of HL-60 promyelocytic leukemia cells. The HL-60 list seems small compared with the 300 agents known to induce differentiation in mouse erythroleukemia cells (Waxman *et al.*, 1986). Thus, it is beyond the scope of this chapter either to list all the agents or explain their function. Perhaps the diversity of chemical differentiation inducers is more impressive than the sheer numbers of these substances. That diversity is reminiscent of the range of chemicals that induce CNS development in vertebrate embryos.

One might observe here, as was noted in the discussion of CNS differentiation: "there are too many inducers." Hickman and Friedman (1988) suggest that an understanding of how the various agents act may be simplified a bit if it is recognized that there may be a limited repertoire of cellular responses to a multiplicity of chemically induced changes or lesions. These investigators note that many of the agents that have efficacy in inducing cancer cell differentiation are cytotoxic drugs used at "marginally toxic concentrations." More specifically, it may be noted that the concentrations of polar compounds (e.g., *N,N*-dimethylformamide, *N*-methylformamide, and related ureas and acetamides) required to induce differentiation of the promyelocytic leukemia cell line HL-60 were marginally lower than those that were cytostatic or cytotoxic, which suggests that "a toxic threat to the cells was sufficient to induce differentiation (Langdon and Hickman, 1987*b*)." HL-60 cells differentiate in response to an enormous diversity of chemical gents; Langdon and Hickman (1987*b*) note that these agents include not only the polar solvents they studied, but chemical compounds as different as anthracycline antibiotics (Schwartz *et al.*, 1983), methotrexate (Bodner *et al.*, 1981), and tunicamycin (Nakayasu *et al.*, 1980) as well. A property of all of these agents is that they are either cytotoxic or cytostatic, or both. It could be that the more recent studies of cancer cell differentiation will provide new insight into the old, and almost forgotten, sublethal cytolysis of differentiating embryonic CNS proposed a generation ago by Holtfreter. Furthermore, it is at least within the realm of possibility that the adjective noncytotoxic will have to be deleted from at least some forms of differentiation therapy. A useful catalogue that sorts agents into five categories is that of Gabrilove (1986), in which inducers of cancer cell differentiation are divided into chemotherapeutic agents, polar compounds, phorbol esters, vitamin analogs, and physiologically important cytokines. Similar listings of agents that induce cancer cell differentiation are found in Freshney (1985) and Waxman and Takaku (1988).

The multiplicity of chemical substances that induce cancer cell differentiation provides no insight into molecular mechanisms responsible for that differentiation. However, two areas of research are converging that have the potentiality of providing new understanding of cancer cell differentiation. Those two areas involve research into growth factors and the study of oncogenes.

Growth factors are peptides that cause an increase of cell proliferation by binding to specific cell membrane receptors. Some neoplastic cells exhibit an escape from growth factor control by displaying a decreased or absent requirement for specific growth factors (Sporn et al., 1985; Goustin et al., 1986; Burgess, 1987).

There is a direct connection between several important growth factors and cellular oncogenes. Examples of the relationship of specific peptide growth factors and specific oncogenes include platelet-derived growth factor (PDGF), which is structurally related to the protein product of the oncogene v-sis (Doolittle et al., 1983; Waterfield et al., 1983). Production of PDGF may lead to the expression of a second oncogene, c-myc, which in turn is believed to be related to cell proliferation (Kelly et al., 1983). This series of sequentially related events (i.e., an oncogene that codes for a growth factor, the production of which in turn causes the expression of another oncogene that is related to cell proliferation) has been described as a hierarchical control of cell proliferation (Sporn et al., 1985).

Growth factor receptors may be encoded by certain oncogenes. For example, c-erb B codes for a protein with close similarity to the epidermal growth factor (EGF) receptor (Downward et al., 1984) and c-fms gene product is similar to colony-stimulating factor-1 (CSF-1) receptor (Sherr et al., 1985). The p21 gene product of ras has been termed an obligatory intermediate in the transduction of a growth factor signal (Mulcahy et al., 1985; Goustin et al., 1986). Further linking of growth factors with oncogenes is evidenced by growth factors that increase transcription of certain oncogenes (Kelly et al., 1983; Greenberg and Ziff, 1984; Muller et al., 1984; Kruijer et al., 1984). Obviously, then, growth factors and oncogenes are intimately interconnected. That leads to the question of the relationship of oncogene expression to cancer cell differentiation.

If cellular oncogenes are involved in malignancy, and there is ample evidence that they are (Aaronson et al., 1986; Barbacid, 1986; Bishop, 1983, 1987; Kline and Kline, 1986; Pimentel, 1986), the cancer cell differentiation described in this chapter would seem an appropriately rich arena for establishing the role of various oncogenes (and growth factors). This is because it should be possible to ascertain the level of c-onc mRNA in a cell population while the cells are phenotypically malignant, during the time that they are differentiating, and after they display a benign phenotype. It is difficult to imagine a direct role in oncogenesis for a particular c-*onc* if transcriptional activity of that gene remains invariant with phenotypic change. However, if the level of mRNA of the c-*onc* is directly related to the state of differentiation, a role for that gene can be envisioned. As would be expected, there is an emerging literature concerning this question. The literature remains somewhat difficult to interpret.

An example is the proto-oncogene c-*myc*. The c-*myc* mRNA level is reported to be cell proliferation dependent (Kelly *et al.*, 1983). It is expressed in normally growing cells as well as neoplastic cells (Bishop, 1983; Müller and Verma, 1984). Ordinarily, with growth arrest, c-*myc* expression is greatly reduced (as expected). c-*myc* mRNA reappears when cells are stimulated to resume cell division. The oncogene is expressed especially high in the human promyelocytic leukemia cell line, HL-60 (Rowley and Skuse, 1987). As anticipated, there is an almost total inhibition of c-*myc* transcription after induced differentiation of HL-60 cells with either Me_2SO or retinoic acid (Westin *et al.*, 1982). Similarly, c-*myc* mRNA levels are decreased with differentiation induced by the dihydroxymetabolite of vitamin D_3. Removal of the vitamin D metabolite after the onset of differentiation results in the resumption of elevated levels of c-*myc* (Reitsma *et al.*, 1983). Likewise, decreased levels of c-*myc* mRNA with differentiation were reported by Einat *et al.* (1985), Grosso ad Pitot (1984), and Slungaard *et al.* (1987). Contrast those studies with chronic lymphocytic leukemic cells that were induced to differentiate by the action of a phorbol ester. Transcription of c-*myc* was increased and remained high during the differentiation process. In this instance, it is clear that downregulation of c-*myc* is not a prerequisite for differentiation (Larsson *et al.*, 1987). In another study, HL-60 cells were treated with DMSO with or without growth inhibition. The kinetics of c-*myc* expression parallel closely the increased proportion of differentiated myeloid cells but not by changes in the proportion of proliferating cells (Filmus and Buick, 1985). Finally, with Friend erythroleukemia cells treated with DMSO, there is a marked decrease in c-*myc* mRNA for 12 hr, followed by an increase in c-*myc* mRNA to the pretreatment level by 18 hr, followed by a gradual decline (Lachman and Skoultchi, 1984). One would expect variant expression of c-*myc* in different neoplastic cell lines. Such variant expression is reported. However, it is difficult to generalize at the present time from these studies whether to expect c-*myc* expression to increase, decrease, or follow a pattern of alternating levels with cell maturation. Hickman and Friedman (1988) commented that much of the work reported concerning transcriptional activity of oncogenes has been without parallel measurements of the protein gene products. These workers suggest that post-translationally regulated changes may be as important as transcriptional modulation. They state that "at this moment in time, it cannot be claimed that a clear role for oncogene expression in differentiation has emerged" (Hickman and Friedman, 1988).

The caption of this section asks the rhetorical question as to whether there are commonalities in the differentiation of embryonic and cancer cells. While a definitive answer to that question is lacking because of a paucity of understanding concerning the molecular mechanisms of either, it seems likely that heightened insight of the one will augment understanding of the other—with the possibility of a practical event emerging from this greater enlightenment. That event would be the continued and enhanced development of nontoxic differentiative cancer therapy (Pierce, 1961, 1974; Paul, 1978; Sachs, 1980, 1986, 1987a,b; Sporn and Roberts, 1984; Freshney, 1985; Gabrilove, 1986; Metcalf, 1986; Daenen et al., 1986; Barrett, 1987; Fujii et al., 1987; Jimenez and Yunis, 1987; Kasukabe et al., 1987; Kohlhepp et al., 1987; Langdon and Hickman, 1987a,b; Marks, 1985; Marks et al., 1987; Marx, 1987; Nicola, 1987; Samara et al., 1987; Trinchieri et al., 1987; Yunis et al., 1987; Breitman and Sherman, 1988; Francis and Pinsky, 1988; McKinnell, 1989; Pierce and Speers, 1988; Waxman and Takaku, 1988; Wiemann et al., 1988). The multiplicity of recent papers concerning cancer cell differentiation, only some of which are cited here, witnesses the increasingly widely held view that differentiation therapy has potential value in the treatment of malignancy.

ACKNOWLEDGMENTS.   The author thanks Dr. Marion Namenwirth, for her critical reading of the manuscript, and Dr. G. Barry Pierce, for helpful discussions during the writing of this paper. Debra L. Kane and Beverly W. Kerr detected errors and made useful comments concerning the manuscript. Tina M. Lorsung typed this paper. The current work of the author's laboratory is supported by grants RD-248 and IN-Y-8 from the American Cancer Society and by NATO grant 94/0438.

References

Aaronson, S. A., Bishop, J. M., Sugimura, T., Terada, M., Toyoshima, K., and Vogt, P. K. (eds.), 1986, *Oncogenes and Cancer*, Japan Scientific Societies Press, Tokyo.

Akiyama, T., Matsumoto, J., Ishikawa, T., and Eguchi, G., 1986, Production of crystallins and lens-like structures in differentiation-induced neoplastic pigment cells (goldfish erythrophoroma cells) *in vitro*, *Differentiation* **33**:34–44.

Andrews, P. W., Gonczol, E., Plotkin, S. A., Dignazio, M., and Oosterhuis, J. M., 1986, Differentiation of TERA-2 human embryonal carcinoma cells into neurons and HCMV permissive cells. Induction by agents other than retinoic acid, *Differentiation* **31**:119–126.

Asashima, M., Oinuma, T., and Meyer-Rochow, V. B., 1987, Tumors in amphibia, *Zool. Sci.* **4**:411–425.

Askanazy, M., 1907, Teratome nach ihrem Bau, ihrem Verlauf, ihrer Genese und im Vergleich zum experimentellen Teratoid, *Verhandl. Dtsch. Pathol. Ges.* **11**:39–82.

Awrorow, P., and Timofejewsky, A. D., 1914, Kultivierungsversuche von leukamischen Blute, *Virchows Arch. Pathol. Anat. Physiol.* **216**:184–214.

Azumi, J-I., and Sachs, L., 1977, Chromosome mapping of the genes that control differentiation and malignancy in myeloid leukemic cells, *Proc. Natl. Acad. Sci. USA* **74**:253–257.

Baluda, M. A., and Jamieson, P. P., 1961, *In vivo* infectivity studies with avian myeloblastosis, virus, *Virology* **14**:33–45.

Barbacid, M., 1986, Oncogenes and human cancer: Cause or consequence?, *Carcinogenesis* **7**:1037–1042.

Barrett, J. C., 1987, Cellular and molecular mechanisms for suppression and version of tumorigenicity—A chemical pathology study section workshop, *Cancer Res.* **47**:2514–2520.

Barth, L. G., 1939, The chemical nature of the amphibian organizer. III. Stimulation of the presumptive epidermis of Ambystoma by means of cell extracts and chemical substances, *Physiol. Zool.* **12:**22–29.

Barth, L. G., 1941, Neural differentiation without organizer, *J. Exp. Zool.* **87:**371–383.

Bautzmann, H., Holtfreter, J., Spemann, H., and Mangold, O., 1932, Versuche zur Analyse der Induktionsmittel in der Embryonalentwicklung, *Naturwissenschaften* **20:**971–974.

Beard, J., 1902, Embryological aspects and etiology of carcinoma, *Lancet* **1:**1758–1761.

Beatty, R. A., de Jong, S., and Zielinski, M. A., 1939, Experiments on the effect of dyes on induction and respiration in the amphibian gastrula, *J. Exp. Biol.* **16:**150–154.

Becker, M., Moczar, E., Korach, S., Lascaux, V., and Poupon, M. F., 1985, Relationship between *in vitro* effects and *in vivo* control of metastasis induced by hydrocortisone in a rat rhab-domyosarcoma, in: *Treatment of Metastasis: Problems and Prospects* (K. Hellmann and S. A. Eccles, eds.), pp. 183–186, Taylor and Francis, London.

Bernhard, W., 1969, Ultrastructure of the cancer cell, in: *Handbook of Molecular Cytology* (A. Lima-de-Faria, ed.), pp. 687–715, North-Holland, Amsterdam.

Berrill, N. J., 1943, Malignancy in relation to organization and differentiation, *Physiol. Rev.* **23:** 101–123.

Bessis, M., 1954, La différenciation et la maturation des cellules leucémiques. Considérations cytologiques et cliniques, *Rev. Hematol.* **9:**745–758.

Binns, A. N., Wood, H. N., and Braun, A. C., 1981, Suppression of the tumorous state in crown gall teratomas of tobacco: A clonal analysis, *Differentiation* **19:**97–102.

Bishop, J. M., 1983, Cellular oncogenes and retroviruses, *Annu. Rev. Biochem.* **52:**301–354.

Bishop, J. M., 1987, The molecular genetics of cancer, *Science* **235:**305–311.

Bodner, A. J., Ting, R. C., and Gallo, R. C., 1981, Induction of differentiation of human prom-yelocytic leukemia cells (HL-60) by nucleosides and methotrexate, *J. Natl. Cancer Inst.* **67:** 1025–1030.

Brachet, J., and Rapkine, L., 1939, Oxydation et réduction d'explantats dorsaux et ventraux de gastrulas (Amphibiens), *C. R. Soc. Biol.* **131:**789–791.

Bradley, A., 1987, Production and analysis of chimaeric mice, in: *Teratocarcinomas and Embryonic Stem Cells: A Practical Approach* (E. J. Robertson, ed.), pp. 113–151, IRL Press Ltd., Oxford.

Braun, A. C., 1947, Thermal studies on the factors responsible for tumor initiation in crown-gall, *Am. J. Bot.* **34:**234–240.

Braun, A. C., 1951, Recovery of crown-gall tumor cells, *Cancer Res.* **11:**839–844.

Braun, A. C., 1953, Bacterial and host factors concerned in determining tumor morphology in crown gall, *Bot. Gaz.* **114:**363–371.

Braun, A. C., 1959, A demonstration of the recovery of the crown-gall tumor cell with the use of complex tumors of single-cell origin, *Proc. Natl. Acad. Sci. USA* **45:**932–938.

Braun, A. C., 1968, The multipotential cell and the tumor problem, in: *The Stability of the Differentiated State* (H. Ursprung, ed.), pp. 128–135, Springer-Verlag, New York.

Braun, A. C., 1972, The usefulness of plant tumor systems for studying the basic cellular mechanisms that underlie neoplastic growth generally, in: *Cell Differentiation* (R. Harris, P. Allin, and D. Viza, eds.), pp. 115–118, Munksgaard, Copenhagen.

Braun, A. C., 1981, An epigenetic model for the origin of cancer, *Q. Rev. Biol.* **56:**33–60.

Braun, A. C., and Wood, H. N., 1976, Suppression of the neoplastic state with the acquisition of specialized functions in cells, tissues, and organs of crown gall teratomas of tobacco, *Proc. Natl. Acad. Sci. USA* **73:**496–500.

Breitman, T. R., and Sherman, M. I., 1988, *In vivo* systems for differentiation therapy of leukemia and solid tumors, in: *The Status of Differentiation Therapy of Cancer* (S. Waxman, G. B. Rossi, and F. Takaku, eds.), pp. 263–275, Raven, New York.

Briggs, R., and King, T. J., 1952, Transplantation of living nuclei from blastula cells into enucleated frogs' eggs, *Proc. Natl. Acad. Sci. USA* **38:**455–463.

Brinster, R. L., 1974, The effect of cells transferred into the mouse blastocyst on subsequent development, *J. Exp. Med.* **140:**1049–1056.

Burgess, A., 1987, Growth factors and oncogenes, in: *Oncogenes and Growth Factors* (R. A. Bradshaw and S. Prentis, eds.), pp. 123–134, Elsevier, Amsterdam.

Carr, B. I., Gilchrist, K. W., and Carbone, P. P., 1981, The variable transformation in metastases from testicular germ cell tumors: The need for selective biopsy, *J. Urol.* **126:**52–54.

Chang, E. H., Ridge, J., Yu, Z., Richtsmeier, W. J., Hartford, J. B., and Black, R., 1988, Induction of altered oncogene expression and differentiation of squamous cell carcinoma in vitro, in: *The Status of Differentiation Therapy of Cancer* (S. Waxman, G. B. Rossi, and F. Takaku, eds.), pp. 63–77, Raven, New York.

Colberg-Poley, A. M., Voss, S. D., Chowdhury, K., and Gruss, P., 1985, Structural analysis of murine genes containing homeobox sequences and their expression in embryonal carcinoma cells, *Nature (Lond.)* **314:**713–718.

Cohnheim, J., 1875, Congenitales, querqestreiftes Muskelsarkom der Nieren, *Arch. Pathol. Anat. Physiol. Klin. Med.* **65:**64–69.

Cohnheim, J., 1889, *Lectures on General Pathology* (A. B. McKee, transl.), The New Sydenham Society, London [in English].

Cooper, D., Schermer, A., and Sun, T-T, 1985, Classification of human epithelia and their neoplasms using monoclonal antibodies to keratins: Strategies, applications, and limitations, *Lab. Invest.* **52:**243–256.

Corbeil, L. B., 1967, Differentiation of rhabdomyosarcoma and neonatal muscle cells in vitro, *Cancer* **20:**572–578.

Crocker, J. F. S., and Vernier, R. L., 1972, Congenital nephroma of infancy: Induction of renal structures by organ culture, *J. Pediatr.* **80:**69–73.

Daenen, S., Vellenga, E., van Dobbenburgh, O. A., and Halie, M. R., 1986, Retinoic acid as antileukemic therapy in a patient with acute promyelocytic leukemia and *Aspergillus* pneumonia, *Blood* **67:**559–561.

Damjanov, I., Damjanov, A., and Solter, D., 1987, Production of teratocarcinomas from embryos transplanted to extra uterine sites, in: *Teratocarcinomas and Embryonic Stem Cells: A Practical Approach* (E. J. Robertson, ed.), pp. 1–18, IRL Press Ltd., Oxford.

Dawe, C. J., and Law, L. W., 1959, Morphologic changes in salivary-gland tissue of the newborn mouse exposed to parotid-tumor agent in vitro, *J. Natl. Cancer Inst.* **23:**1157–1177.

Deschamps, J., DeLaaf, R., Joosen, L., Meijlink, F., and Destree, O., 1987, Abundant expression of homeobox genes in mouse embryonal carcinoma cells correlates with chemically induced differentiation, *Proc. Natl. Acad. Sci. USA* **84:**1304–1308.

Dewey, M. J., Martin, D. W., Martin, G. R., and Mintz, B., 1977, Mosaic mice with teratocarcinoma-derived mutant cells deficient in hypoxanthine phosphoribosyltransferase, *Proc. Natl. Acad. Sci. USA* **74:**5564–5568.

Dexter, D. L., 1977, *N,N*-Dimethylformamide-induced morphological differentiation and reduction in tumorigenicity in cultured mouse rhabdomyosarcoma cells, *Cancer Res.* **37:**3136–3140.

Dexter, D. L., Hager, J. C., Gold, D., Miller, F., Fligiel, Z., and Calabresi, P., 1978, Induction of maturation and loss of tumorigenicity in human colon carcinoma cells by polar solvents, *Clin. Res.* **26:**434A.

Dexter, D. L., Barbosa, J. A., and Calabresi, P., 1979, *N,N*-dimethylformamide-induced alteration of cell culture characteristics and loss of tumorigenicity in cultured human colon carcinoma cells, *Cancer Res.* **39:**1020–1025.

Dexter, D. L., Konieczny, S. F., Lawrence, J. B., Shaffer, M., Mitchell, P., and Coleman, J. R., 1981, Induction by butyrate of differentiated properties in cloned murine rhabdomyosarcoma cells, *Differentiation* **18:**115–122.

Dexter, D. L., Spremulli, E. N., Matook, G. M., Diamond, I., and Calabresi, P., 1982, Inhibition of the growth of human colon cancer xenografts by polar solvents, *Cancer Res.* **42:**5018–5022.

DiBerardino, M. A., 1980, Genetic stability and modulation of metazoan nuclei transplanted into eggs and oocytes, *Differentiation* **17:**17–30.

DiBerardino, M. A., 1987, Genomic potential of differentiated cells analyzed by nuclear transplantation, *Am. Zool.* **27:**623–644.

DiBerardino, M. A., and Hoffner, N. J., 1980, The current status of cloning and nuclear reprogramming in amphibian eggs, in: *Differentiation and Neoplasia* (R. G. McKinnell, M. A. DiBerardino, M. Blumenfeld, and R. D. Bergad, eds.), pp. 53–64, Springer-Verlag, Berlin.

DiBerardino, M. A., and Hoffner, N. J., 1983, Gene reactivation in erythrocytes: Nuclear transplantation in oocytes and eggs of Rana, *Science* **219**:862–864.

DiBerardino, M. A., King, T. J., and McKinnell, R. G., 1961, Embryonic development and the frog renal adenocarcinoma, in: *Frog, Kidney Adenocarcinoma Conference* (W. R. Duryee and L. Warner, eds.), pp. 81–90, National Institutes of Health, Bethesda.

DiBerardino, M. A., King, T. J., and McKinnell, R. G., 1963, Chromosome studies of a frog renal adenocarcinoma line carried by serial intraocular transplantation, *J. Natl. Cancer Inst.* **31**:769–789.

DiBerardino, M. A., Mizell, M., Hoffner, N. J., and Friesendorf, D. G., 1983, Frog larvae cloned from nuclei of pronephric adenocarcinoma, *Differentiation* **23**:213–217.

DiBerardino, M. A., Orr, N. H., and McKinnell, R. G., 1986, Feeding tadpoles cloned from *Rana* erythrocyte nuclei, *Proc. Natl. Acad. Sci. USA* **83**:8231–8234.

Dixon, F. J., and Moore, R. A., 1953, Testicular tumors, *Cancer* **6**:427–454.

Doolittle, R. F., Hunkapiller, M. W., Hood, L. E., Devare, S. G., Robbins, K. C., Aaronson, S. A., and Antoniades, H. N., 1983, Simian sarcoma virus *onc* gene, v-*sis*, is derived from the gene (or genes) encoding a platelet-derived growth factor, *Science* **221**:275–277.

Downward, J., Yarden, Y., Mayes, E., Scarce, G., Totty, N., Stockwell, P., Ullrich, A., Schlessinger, J., and Waterfield, M. D., 1984, Close similarity of epidermal growth factor receptor and v-*erb*-B oncogene protein sequences, *Nature (Lond.)* **307**:521–527.

Dudouet, B., Robine, S., Huet, C., Sahuquillo-Merino, L., Blair, E., Coudrier, E., and Louvard, D., 1987, Changes in villin synthesis and subcellular distribution during intestinal differentiation of HT29-18 clones, *J. Cell Biol.* **105**:359–369.

Earle, E. D., and Torrey, J. G., 1965, Morphogenesis in cell colonies grown from *Convolvulus* cell suspensions plated on synthetic media, *Am. J. Bot.* **52**:891–899.

Edwards, M. K., and McBurney, M. W., 1983, The concentration of retinoic acid determines the differentiated cell types formed by a teratocarcinoma cell line, *Dev. Biol.* **98**:187–191.

Einat, M., Resnitzky, D., and Kimchi, A., 1985, Close link between reduction of c-*myc* expression by interferon and G_0/G_1 arrest, *Nature (Lond.)* **313**:597–600.

Ellis, Z., Schaefer, A., and Koch, G., 1987, Changes in intracellular pH and cell volume during the early phase of DMSO-induced differentiation of Friend erythroleukemia cells. *Experientia* **43**:914–916.

Eto, Y., Tsuji, T., Takezawa, M., Takano, S., Yokogawa, T., and Shibai, H., 1987, Purification and characterization of erythroid differentiation factor (EDF) isolated from human leukemia cell line THP-1, *Biochem. Biophys. Res. Commun.* **142**:1095–1103.

Fekete, E., and Ferrigno, M. A., 1952, Studies on a transplantable teratoma of the mouse, *Cancer Res.* **12**:438–440.

Fenderson, B. A., Andrews, P. W., Nudelman, E., Clausen, H., and Hakomori, S-I, 1987, Gycolipid core structure switching from globo- to lacto- and ganglio- series during retinoic acid-induced differentiation of TERA-2-derived human embryonal carcinoma cells, *Dev. Biol.* **122**:21–34.

Filmus, J., and Buick, R. N., 1985, Relationship of c-*myc* expression to differentiation and proliferation of HL-60 cells, *Cancer Res.* **45**:822–825.

Francis, G. E., and Pinsky, C. M., 1988, Clinical trials of differentiation therapy: Current trials and future prospects, in: *The Status of Differentiation Therapy of Cancer* (S. Waxman, G. B. Rossi, and F. Takaku, eds.), pp. 331–347, Raven, New York.

Freshney, R. I., 1985, Induction of differentiation in neoplastic cells, *Anticancer Res.* **5**:111–130.

Friend, C., 1978, The phenomenon of differentiation in murine erythroleukemic cells, *Harvey Lect.* **72**:253–281.

Friend, C., 1980, The regulation of differentiation in murine virus-induced erythroleukemic cells, in: *Differentiation and Neoplasia* (R. G. McKinnell, M. DiBerardino, M. Blumenfeld, and R. D. Bergad, eds.), pp. 202–212, Springer-Verlag, Berlin.

Friend, C., Patuleia, M. C., and de Harven, E., 1966, Erythrocytic maturation *in vitro* of murine (Friend) virus-induced leukemia cells, *Natl. Cancer Inst. Monog.* **22**:505–522.

Friend, C., Scher, W., Holland, J. G., and Sato, T., 1971, Hemoglobin synthesis in murine virus induced leukemia cells *in vitro* stimulation of erythroid differentiation by dimethyl sulfoxide, *Proc. Natl. Acad. Sci. USA* **68**:378–382.

Fry, R. J. M., 1985, Principles of cancer biology: Physical carcinogenesis, in: *Cancer, Principles and Practice of Oncology*, 2nd ed. (V. T. DeVita, Jr., S. Hellman, and S. A. Rosenberg, eds.), pp. 101–112, Lippincott, Philadelphia.

Fujii, Y., Yuki, N., Takeichi, N., Kobayashi, H., and Miyazaki, T., 1987, Differentiation therapy of a myelomonocytic leukemia (c-WRT-7) in rats by injection of lipopolysaccharide and daunomycin, *Cancer Res.* **47**:1668–1673.

Gabrilove, J. L., 1986, Differentiation factors, *Semin. Oncol.* **13**:228–233.

Gallagher, R. E., Lurie, K. J., Leavitt, R. D., and Wiernik, P. H., 1987, Effects of interferon and retinoic acid on the growth and differentiation of clonogenic leukemic cells from acute myelogenous leukemia patients treated with recombinant leukocyte-αA interferon, *Leuk. Res.* **11**: 609–619.

Glazer, R. I., Cohen, M. B., Hartman, K. D., Knode, M. C., Lim, M. I., and Marquez, V. E., 1986, Induction of differentiation in the human promyelocytic leukemia cell line HL-60 by the cyclopentenyl analogue of cytidine, *Biochem. Pharmacol.* **35**:1841–1848.

Goldstein, M. N., and Plurad, S., 1980, Drug-induced differentiation of human neuroblastoma: Transformation into ganglion cells with mitomycin-C, in: *Differentiation and Neoplasia* (R. G. McKinnell, M. A. DiBerardino, M. Blumenfeld, and R. D. Bergad, eds.), pp. 259–264, Springer-Verlag, Berlin.

Goldstein, M. N., Burdman, J. A., and Journey, L. J., 1964, Long-term tissue culture of neuroblastomas. II. Morphologic evidence for differentiation and maturation, *J. Natl. Cancer Inst.* **32**:165–199.

Gotti, C., Sher, E., Corbrini, D., Bondiolotti, G., Wanke, E., Mancinelli, E., and Clementi, F., 1987, Cholinergic receptors, ion channels, neurotransmitter synthesis, and neurite outgrowth are independently regulated during the in vitro differentiation of a human neuroblastoma cell line, *Differentiation* **34**:144–155.

Goustin, A. S., Leof, E. B., Shipley, G. D., and Moses, H. L., 1986, Growth factors and cancer, *Cancer Res.* **46**:1015–1029.

Grant, S., Bhalla, K., Weinstein, B., Pestka, S., Mileno, M. D., and Fisher, P. B., 1985, Recombinant human interferon sensitizes resistant myeloid leukemic cells to induction of terminal differentiation, *Biochem. Biophys. Res. Commun.* **130**:379–388.

Greenberg, M. E., and Ziff, E. B., 1984, Stimulation of 3T3 cells induces transcription of the c-*fos* proto-oncogene, *Nature (Lond.)* **311**:433–438.

Grobstein, C., 1955, Inductive interaction in the development of the mouse metanephros, *J. Exp. Zool.* **130**:319–339.

Gross, L., 1951, "Spontaneous" leukemia developing C3H mice following inoculation, in infancy, with AK-leukemic extracts or AK-embryos, *Proc. Soc. Exp. Biol. Med.* **76**:27–32.

Gross, L., 1983, *Oncogenic Viruses*, 3rd ed., Pergamon, Elmsford, New York.

Grosso, L. E., and Pitot, H. C., 1984, Modulation of c-*myc* expression in the HL-60 cell line, *Biochem. Biophys. Res. Commun.* **119**:473–480.

Gurdon, J. B., 1987, Embryonic induction—Molecular prospects, *Development* **99**:285–306.

Harrison, F. W., Zambernard, J., and Cowden, R. R., 1975, Fluorescent cytochemistry of calid and algid normal and Lucké tumor-bearing kidneys, *Acta Histochem. (Jena)* **54**:295–306.

Harshbarger, J. C., Charles, A. M., and Spero, P. M., 1981, Collection and analysis of neoplasms in sub-homeothermic animals from a phyletic point of view, in: *Phyletic Approaches to Cancer* (C. J. Dawe, J. C. Harshbarger, S. Kondo, T. Sugimura, and S. Takayama, eds.), pp. 357–384, Japan Scientific Societies Press, Tokyo.

Hauser, C. A., Joyner, A. L., Klein, R. D., Learned, T. K., Martin, G. R., and Tjian, R., 1985, Expression of homologous homeo-box-containing genes in differentiated human teratocarcinoma cells and mouse embryos, *Cell* **43**:19–28.

Hay, E., 1974, Cellular basis of regeneration, in: *Concepts of Development* (J. Lash and J. R. Whittaker, eds.), pp. 404–428, Sinauer Associates, Stamford, Connecticut.

Hennen, S., 1970, Influence of spermine and reduced temperature on the ability of transplanted nuclei to promote normal development in eggs of *Rana pipiens*, *Proc. Natl. Acad. Sci. USA* **66**: 630–637.

Hickman, J. A., and Friedman, J. A., 1988, Pharmacology and mechanisms of action of differentiation agents, in: *The Status of Differentiation Therapy of Cancer* (S. Waxman, G. B. Rossi, and F. Takaku, eds.), pp. 231–252, Raven, New York.

Holtfreter, J., 1947, Neural induction in explants which have passed through sublethal cytolysis, *J. Exp. Zool.* **106**:197–222.

Holtfreter, J., 1948, Concepts on the mechanisms of embryonic induction and their relation to parthenogenesis and malignancy, *Symp. Soc. Exp. Biol.* **2**:17–48.

How, S. W., and Snell, K. C., 1967, Skin tumors induced in rats by the dietary administration of N,N^1-2, 7-Fluroenylenebisacetamide, *J. Natl. Cancer Inst.* **38**:407–434.

Huet, C., Sahuquillo-Merino, C., Coudrier, E., and Louvard, D., 1987, Adsorptive and mucus-secreting subclones isolated from a multipotent intestinal cell line (HT-29) provides new models for cell polarity and terminal differentiation, *J. Cell Biol.* **105**:345–357.

Ichikawa, Y., 1969, Differentiation of a cell line of myeloid leukemia, *J. Cell Physiol.* **74**:223–234.

Illmensee, K., and Mintz, B., 1976, Totipotency and normal differentiation of single teratocarcinoma cells cloned by injection into blastocysts, *Proc. Natl. Acad. Sci. USA* **73**:549–553.

Ishikawa, T., Prince Masahito, Matsumoto, J., and Takayama, S., 1978, Morphologic and biological characterization of erythrophoromas in goldfish (*Carassius auratus*), *J. Natl. Cancer. Inst.* **61**: 1461–1470.

Jimenez, J. J., and Yunis, A. A., 1987, Tumor cell rejection through terminal cell differentiation, *Science* **238**:1278–1280.

Kalil, M., and Hildebrandt, A. C., 1981, Pathology and distribution of plant tumors, in: *Neoplasms—Comparative Pathology of Growth in Animals, Plants, and Man* (H. E. Kaiser, ed.), pp. 813–821, Williams & Wilkins, Baltimore.

Kaplinsky, C., Estrov, Z., Freedman, M. H., and Cohen, A., 1986, Induction of differentiation in HL-60 promyelocytic cells: A comparative study of two sublines, *Blood Cells* **11**:459–468.

Kasukabe, T., Honma, Y., Hozumi, M., Suda, T., and Nishii, Y., 1987, Control of Proliferating potential of myeloid leukemia cells during long-term treatment with vitamin D$_3$ analogues and other differentiation inducers in combination with antileukemic drugs: *In vitro* and *in vivo* studies, *Cancer Res.* **47**:567–572.

Kelly, K., Cochran, B. H., Stiles, C. D., and Leder, P., 1983, Cell-specific regulation of the c-myc gene by lymphocyte mitogens and platelet-derived growth factor, *Cell* **35**:603–610.

Kim, Y. S., 1988, Differentiation of normal and cancerous colon cells, in: *The Status of Differentiation Therapy of Cancer* (S. Waxman, G. B. Rossi, and F. Takaku, eds.), pp. 29–44, Raven, New York.

King, T. J., and DiBerardino, M. A., 1965, Transplantation of nuclei from the frog renal adenocarcinoma. I. Development of tumor nuclear-transplant embryos, *Ann. NY Acad. Sci.* **126**:115–126.

King, T. J., and McKinnell, R. G., 1960, An attempt to determine the developmental potentialities of the cancer cell nucleus by means of transplantation, in: *Cell Physiology of Neoplasia*, pp. 591–617, University of Texas Press, Austin.

Kline, G., and Kline, E., 1986, Conditioned tumorigenicity of activated oncogenes, *Cancer Res.* **46**: 3211–3224.

Kleinsmith, L. J., and Pierce, G. B., 1964, Multipotentiality of single embryonal carcinoma cells, *Cancer Res.* **24**:1544–1551.

Kohlhepp, E. A., Condon, M. E., and Hamburger, A. W., 1987, Recombinant human interferon α enhancement of retinoic-acid-induced differentiation of HL-60 cells, *Exp. Hematol.* **15**:414–418.

Kordower, J. H., Nolter, M. F. D., Yeh, H. M., and Gash, D. M., 1987, An *in vivo* and *in vitro* assessment of differentiated neuroblastoma cells as a source of donor tissue for transplantation, *Ann. NY Acad. Sci.* **495**:606–622.

Kos, Z., Pavelic, L., Pekic, B., and Pavelic, K., 1987, Reversal of human myeloid leukemia cells into normal granulocytes and macrophages: Activity and intracellular distribution of catalase, *Oncology* **44**:245–247.

Kroll, A. J., Kuwabara, T., and Howard, G. M., 1963, Electron microscopy of rhabdomyosarcoma of the orbit, *Invest. Ophthalmol.* **2**:523–537.

Kruijer, W., Cooper, J. A., Hunter, T., and Verma, I. M., 1984, Platelet-derived growth factor induces rapid but transient expression of the c-*fos* gene and protein, *Nature (Lond.)* **312**:711–716.

Lachman, H. M., and Skoultchi, A. I., 1984, Expression of c-myc changes during differentiation of mouse erythroleukemia cells, Nature (Lond.) **310**:592–594.

Langdon, S. P., and Hickman, J. A., 1987a, Alkylformamides as inducers of tumour cell differentiation—A mini-review, Toxicology **43**:239–249.

Langdon, S. P., and Hickman, J. A., 1987b, Correlation between the molecular weight and potency of polar compounds which induce the differentiation of HL-60 human promyelocytic leukemia cells, Cancer Res. **47**:140–144.

Larsson, L-G., Gray, H. E., Tötterman, T., Pettersson, U., and Nilsson, K., 1987, Drastic increased expression of MYC and FOS protooncogenes during in vitro differentiation of chronic lymphocytic leukemia cells, Proc. Natl. Acad. Sci. USA **94**:223–227.

Levin, I., and Levine, M., 1920, Malignancy of the crown-gall and its analogy to animal cancer, J. Cancer Res. **5**:243–260.

Levine, J. M., and Flynn, P., 1986, Cell surface changes accompanying neural differentiation of an embryonal carcinoma cell line, J. Neurosci. **6**:3374–3384.

Lim, R., Hicklin, D. J., Ryken, T. C., Han, X-M., Liu, K-N., Miller, J., and Baggenstoss, B. A., 1986, Suppression of glioma growth in vitro and in vivo by glia maturation factor, Cancer Res. **46**:5241–5247.

Lippincott, J. A., and Lippincott, B. B., 1981, Crown gall, and "malignant plant tumor," in: Neoplasms—Comparative Pathology of Growth in Animals, Plants, and Man, (H. E. Kaiser, ed.), pp. 833–839, Williams & Wilkins, Baltimore.

Lotem, J., and Sachs, L., 1978, In vivo induction of normal differentiation in myeloid leukemia cells, Proc. Natl. Acad. Sci. USA **75**:3781–3785.

Lotem, J., and Sachs, L., 1981, In vivo inhibition of the development of myeloid leukemia by injection of macrophage- and granulocyte-inducing protein, Int. J. Cancer **28**:375–386.

Lotem, J., and Sachs, L., 1984, Control of in vivo differentiation of myeloid leukemic cells. IV. Inhibition of leukemia development by myeloid differentiation-inducing protein, Int. J. Cancer **33**:147–154.

Lucké, B., 1934, A neoplastic disease of the kidney of the frog, Rana pipiens, Am. J. Cancer **20**:352–379.

Lucké B., and Schlumberger, H., 1949, Induction of metastasis of frog carcinoma by increase of environmental temperature, J. Exp. Med. **89**:269–278.

Maddox, A-M., and Haddox, M. K., 1988, Characteristics of cyclic AMP enhancement of retinoic acid induction of increased transglutaminase activity in HL-60 cells, Exp. Cell Biol. **56**:49–59.

Mareel, M., Bruyneel, R., Tweedell, K., McKinnell, R., and Tarin, D., 1985, Temperature-dependence of PNKT-4B frog renal carcinoma cell invasion in vitro, in: Treatment of Metastasis: Problems and Prospects (K. Hellmann and S. A. Eccles, eds.), pp. 335–338, Taylor and Francis, London.

Markert, C. L., 1968, Neoplasia: A disease of cell differentiation, Cancer Res. **28**:1908–1914.

Marks, P. A., 1985, Introduction, in: Genetics, Cell Differentiation, and Cancer (P. A. Marks, ed.), pp. 1–11, Academic, New York.

Marks, P. A., and Rifkind, R. A., 1978, Erythroleukemic differentiation, Annu. Rev. Biochem. **47**:419–448.

Marks, P. A., Sheffery, M., and Rifkind, R. A., 1987, Induction of transformed cells to terminal differentiation and the modulation of gene expression, Cancer Res. **47**:659–666.

Marx, J. L., 1987, Human trials of new cancer therapy begin, Science **236**:778–779.

Matsumoto, J., Ishikawa, T., Prince Masahito, Takayama, S., 1980, Permanent cell lines from erythrophoromas in goldfish (Carassius auratus), J. Natl. Cancer Inst. **64**:879–890.

Matsumoto, J., Ishikawa, T., Prince Masahito, Oikawa, A., and Takayama, S., 1981, Multiplicity in phenotypic expression of fish erythrophoroma and irido-melanophoroma cells in vitro, in: Phyletic Approaches to Cancer (C. J. Dawe, J. C. Harshbarger, S. Kondo, T. Sugimura, and S. Takayama, eds.), pp. 253–266, Japan Scientific Societies Press, Tokyo.

Matsumoto, J., Lynch, T. J., Grabowski, S., Richards, C. M., Lo, S. L., Clarke, C., Kern, D., Taylor, J. D., Tchen, T. T., Ishikawa, T., Prince Masahito, and Takayama, S., 1983, Fish tumor pigment cells: Differentiation and comparison to their normal counterparts, Am. Zool. **23**:569–580.

Matsumoto, J., Akiyama, T., Taylor, J. D., and Tchen, T. T., 1985, Modification of differentiation programs of goldfish erythrophoroma cells by dual application of different inducing agents: A problem of blast (stem) cells, in: *Pigment Cell 1985: Biological, Molecular, and Clinical Aspects of Pigmentation* (J. T. Bagnara, S. N. Klaus, E. Paul, and M. Schartl, eds.), pp. 333–340, Tokyo University Press, Tokyo.

Mattsson, M. E. K., Ruusala, A. I., and Pahlman, S., 1984, Changes in inducibility of ornithine decarboxylase activity in differentiating human neuroblastoma cells, *Exp. Cell Res.* **155**:105–112.

Matzner, Y., Gavison, R., Rachmilewitz, E. A., and Fibach, E., 1987, Expression of granulocytic functions by leukemic promyelocytic HL-60 cells: Differential induction by dimethylsulfoxide and retinoic acid, *Cell Diff.* **21**:261–269.

McKinnell, R. G., 1972, Nuclear transfer in *Xenopus* and *Rana* compared, in: *Cell Differentiation* (R. Harris, P. Allin, and D. Viza, eds.), pp. 61–64, Munksgaard, Copenhagen.

McKinnell, R. G., 1973a, The Lucké frog kidney tumor and its herpesvirus, *Am. Zool.* **13**:97–114.

McKinnell, R. G., 1973b, Nuclear transplantation, in: *Seventh National Cancer Conference Proceedings*, pp. 65–72, Lippincott, Philadelphia.

McKinnell, R. G., 1978, *Cloning, Nuclear Transplantation in Amphibia*, University of Minnesota Press, Minneapolis.

McKinnell, R. G., 1979, The pluripotential genome of the frog renal tumor cell as revealed by nuclear transplantation, *Int. Rev. Cytol.* **9**(Suppl.):179–188.

McKinnell, R. G., 1984, Lucké tumor of frogs, in: *Diseases of Amphibians and Reptiles* (G. L. Hoff, F. L. Frye, and E. R. Jacobson, eds.), pp. 581–605, Plenum, New York.

McKinnell, R. G., 1985, *Cloning of Frogs, Mice and Other Animals*, University of Minnesota Press, Minneapolis.

McKinnell, R. G., 1989, Expression of differentiated function in neoplasms, in: *The Pathobiology of Neoplasia* (A. E. Sirica, ed.), pp. 435–460, Plenum, New York.

McKinnell, R. G., and Cunningham, W. P., 1982, Herpesviruses in metastatic Lucké renal adenocarcinoma, *Differentiation* **22**:41–46.

McKinnell, R. G., and Duplantier, D. P., 1970, Are there renal adenocarcinoma-free populations of leopard frogs?, *Cancer Res.* **30**:2730–2735.

McKinnell, R. G., and Tarin, D., 1984, Temperature-dependent metastasis of the Lucké renal carcinoma and its significance for studies on mechanisms of metastasis, *Cancer Metast. Rev.* **3**:373–386.

McKinnell, R. G., and Tweedell, K. S., 1970, Induction of renal tumors in triploid leopard frogs, *J. Natl. Cancer. Inst.* **44**:1161–1166.

McKinnell, R. G., Deggins, B. A., and Labat, D. D., 1969, Transplantation of pluripotential nuclei from triploid frog tumors, *Science* **165**:394–396.

McKinnell, R. G., Steven, L. M., Jr., and Labat, D. D., 1976, Frog renal tumors are composed of stroma, vascular elements, and epithelial cells. What type nucleus programs for tadpoles with the cloning procedure?, in: *Progress in Differentiation Research* (N. Muller-Berat, ed.), pp. 319–330, North-Holland, Amsterdam.

McKinnell, R. G., DeBruyne, G. K., Mareel, M. M., Tarin, D., and Tweedell, K. S., 1984, Cytoplasmic microtubules of normal and tumor cells of the leopard frog: Temperature effects, *Differentiation* **26**:231–234.

McKinnell, R. G., Bruyneel, E. A., Mareel, M. M., Seppanen, E. D., and Mekala, P. R., 1986, Invasion *in vitro* by explants of Lucke renal carcinoma cocultured with normal tissue is temperature dependent, *Clin. Exp. Metast.* **4**:237–243.

McKinnell, R. G., Bruyneel, E. A., Mareel, M. M., Tweedell, K. S., and Mekala, P., 1988a, Temperature-dependent malignant invasion *in vitro* by frog renal carcinoma-derived PNKT-4B cells, *Clin. Exp. Metast.* **6**:49–59.

McKinnell, R. G., De Bruyne, G., Mareel, M., Rollins-Smith, L., Lust, J., and Kane, D., 1988b, Differentiation of Lucké renal carcinoma induced by nuclear transplantation: Temperature effects on cytoplasmic microtubules, *Proc. Am. Assoc. Cancer Res.* **29**:34.

Meins, F., Jr., 1974, Mechanisms underlying tumor transformation and tumor reversal in crown-gall, a neoplastic disease at higher plants, in: *Developmental Aspects of Carcinogenesis and Immunity* (T. J. King, ed.), pp. 23–39, Academic, New York.

Mendelsohn, N., Michl, J., Gilbert, H. S., Acs, G., and Christman, J. K., 1980, L-Ethionine as an inducer of differentiation in human promyelocytic leukemia cells (HL-60), *Cancer Res.* **40:** 3206–3210.

Metcalf, D., 1985, The granulocyte-macrophage colony-stimulating factors, *Science* **229:**16–22.

Metcalf, D., 1986, The molecular biology and functions of the granulocyte–macrophage colony-stimulating factors, *Blood* **67:**257–267.

Mintz, B., 1978, Gene expression in neoplasia and differentiation, *Harvey Lect.* **71:**193–246.

Mintz, B., and Illmensee, K., 1975, Normal genetically mosaic mice produced from malignant teratocarcinoma cells, *Proc. Natl. Acad. Sci. USA* **72:**3585–3589.

Moore, E. E., Mitra, N. S., and Moritz, E. A., 1986, Differentiation of F9 embryonal carcinoma cells. Differences in the effects of retinoic acid, 5-bromodeoxyuridine, and N-N^1dimethylacetamide, *Differentiation* **31:**183–190.

Muir, W. H., Hildebrandt, A. C., and Riker, A. J., 1958, The preparation, isolation, and growth in culture of single cells from higher plants, *Am. J. Bot.* **45:**589–597.

Mukherjee, A. B., Czirbik, R. J., Parsa, N. Z., and Testa, J. R., 1985, Induction of terminal differentiation and nuclear appendage(s) formation in a human myeloid leukaemia cell line (HL-60), *Cytobios* **44:**109–118.

Mulcahy, L. S., Smith, M. R., and Stacey, D. W., 1985, Requirement for *ras* proto-oncogene function during serum-stimulated growth of NIH 3T3 cells, *Nature (Lond.)* **313:**241–243.

Müller, R., and Verma, I. M., 1984, Expression of cellular oncogenes, *Curr. Topics Microbiol. Immunol.* **112:**73–116.

Muller, R., Bravo, R., Burckhardt, J., and Curran, T., 1984, Induction of c-*fos* gene and protein by growth factors precedes activation of c-*myc*, *Nature (Lond.)* **312:**716–720.

Murray, M. R., and Stout, A. P., 1947, Distinctive characteristics of the sympathicoblastoma cultivated in vitro, *Am. J. Pathol.* **23:**429–441.

Naegele, R. F., Granoff, A., and Darlington, R. W., 1974, The presence of the Lucke herpesvirus genome in induced tadpole tumors and its oncogenicity: Koch–Henle postulates fulfilled, *Proc. Natl. Acad. Sci. USA* **71:**830–834.

Nakayasu, M., Terada, M., Tamura, G., and Sugimura, T., 1980, Induction of differentiation of human and murine myeloid leukemia cells in culture by tunicamycin, *Proc. Natl. Acad. Sci. USA* **77:**409–413.

Nameroff, M. A., Reznik, M., Anderson, P., and Hansen, J. L., 1970, Differentiation and control of mitosis in a skeletal muscle tumor, *Cancer Res.* **30:**596–600.

Needham, J., 1936, New advances in the chemistry and biology of organized growth, *Proc. R. Soc. Med.* **29:**1577–1626.

Needham, J., 1942, *Biochemistry and Morphogenesis*, Cambridge University Press, Cambridge.

Nicola, N. A., 1987, Granulocyte colony-stimulating factor and differentiation-induction in myeloid leukemic cells, *Int. J. Cell Cloning* **5:**1–15.

Oberling, C., 1944, *The Riddle of Cancer*, Yale University Press, New Haven, Connecticut.

Ogilvie, D. J., McKinnell, R. G., and Tarin, D., 1984, Temperature-dependent elaboration of collagenase by the renal adenocarcinoma of the leopard frog, *Rana pipiens*, *Cancer Res.* **44:**3438–3441.

Ohashi, Y., Ueda, K., Hayaishi, O., Ikai, K., and Niwa, O., 1984, Induction of murine teratocarcinoma cell differentiation by suppression of poly(ADP-ribose) synthesis, *Proc. Natl. Acad. Sci. USA* **81:**7132–7136.

Okada, Y. K., 1938, Neural induction by means of inorganic implantation, *Growth* **2:**49–53.

Orr, N. H., DiBerardino, M. A., and McKinnell, R. G., 1986, The genome of frog erythrocytes displays centuplicate replication, *Proc. Natl. Acad. Sci. USA* **83:**1369–1373.

Papaioannou, V. E., McBurney, M. W., Gardner, R. L., and Evans, M. J., 1975, Fate of teratocarcinoma cells injected into early mouse embryos, *Nature (Lond.)* **258:**70–73.

Papaioannou, V. E., Gardner, R. L., McBurney, M. W., Babinet, C., and Evans, M. J., 1978, Participation of cultured teratocarcinoma cells in mouse embryogenesis, *J. Embryol. Exp. Morphol.* **44:** 81–92.

Paterno, G. D., and McBurney, M. W., 1985, X chromosome inactivation during induced differentiation of a female mouse embryonal carcinoma cell line, *J. Cell Sci.* **75:**149–163.

Paukovits, J. B., Paukovits, W. R., and Laerum, O. D., 1986, Identification of a regulatory peptide

distinct from normal granulocyte-derived hemoregulatory peptide produced by human promyelocytic HL-60 leukemia cells after differentiation induction with retinoic acid, *Cancer Res.*
46:4444–4448.

Paul, J., 1978, Cell differentiation and cancer—A summary, in: *Cell Differentiation and Neoplasia*
(G. F. Saunders, ed.), pp. 525–532, Raven, New York.

Pierce, G. B., 1961, Teratocarcinomas, *Can. Cancer Conf.* **4**:119–137.

Pierce, G. B., 1974, The benign cells of malignant tumors, in: *Developmental Aspects of Carcinogenesis and Immunity* (T. J. King, ed.), pp. 3–22, Academic, New York.

Pierce, G. B., and Dixon, F. J., 1959, Testicular teratomas. I. The demonstration of teratogenesis by
metamorphosis of multipotential cells, *Cancer* **12**:573–583.

Pierce, G. B., and Speers, W. C., 1988, Tumors as caricatures of the process of tissue renewal:
Prospects for therapy by directing differentiation, *Cancer Res.* **48**:1996–2004.

Pierce, G. B., and Wallace, C., 1971, Differentiation of malignant to benign cells, *Cancer Res.* **31**:
127–134.

Pierce, G. B., Dixon, F. J., Jr., and Verney, E., 1959, Testicular teratomas, I. Demonstration of
teratogenesis by metamorphosis of multipotential cells, *Cancer* **12**:573–583.

Pierce, G. B., Dixon, F. J., Jr., and Verney, E. L., 1960, Teratocarcinogenic and tissue-forming
potentials of the cell types comprising neoplastic embryoid bodies, *Lab. Invest.* **9**:583–602.

Pierce, G. B., Shikes, R., and Fink, L. M., 1978, *Cancer, A Problem of Developmental Biology*,
Prentice-Hall, Englewood Cliffs, New Jersey.

Pierce, G. B., Arechaga, J., Muro, C., and Wells, R. S., 1988, Differentiation of ICM cells into
trophectoderm, *Am. J. Pathol.* **132**:356–364.

Pietra, G., Spencer, K., and Shubik, P., 1959, Response of newly born mice to a chemical car-
cinogen, *Nature (Lond.)* **183**:1689.

Pilz, R. B., Van den Berghe, G., and Boss, G. R., 1987, Induction of HL-60 differentiation by
starvation for a single essential amino acid but not by protein synthesis inhibitors, *J. Clin.
Invest.* **79**:1006–1009.

Pimentel, E., 1986, *Oncogenes*, CRC Press, Boca Raton, Florida.

Pitot, H. C., 1985, Principles of cancer biology: Chemical carcinogenesis, in: *Cancer, Principles and
Practice of Oncology*, 2nd ed. (V. T. DeVita, Jr., S. Hellman, and S. A. Rosenberg, eds.), pp. 79–
99, Lippincott, Philadelphia.

Podesta, A. H., Mullins, J., Pierce, G. B., and Wells, R. S., 1984, The neurula-stage mouse embryo in
control of neuroblastoma, *Proc. Natl. Acad. Sci. USA* **81**:7608–7611.

Potter, V. R., 1969, Recent trends in cancer biochemistry: The importance of studies on fetal tissue,
Can. Cancer Conf. **8**:9–30.

Potter, V. R., 1987, Blocked ontogeny, *Science* **237**:964.

Prasad, K. N., 1982, Maturation of neuroblastoma, in: *Prolonged Arrest of Cancer* (B. A. Stoll, ed.),
pp. 281–308, Wiley, Chichester.

Prescott, D. M., and Flexer, A. S., 1986, *Cancer, The Misguided Cell*, 2nd ed., Sinauer Associates,
Sunderland, Massachusetts.

Ranzi, S., and Tamini, E., 1939, Die Wirkung von NaSCN auf die Entwicklung von Froschembry-
onen, *Naturwissenschaften* **27**:566–567.

Rather, L. J., 1978, *The Genesis of Cancer*, Johns Hopkins Press, Baltimore.

Razin, A., Webb, C., Szyf, M., Yisraeli, J., Rosenthal, A., Neven-Many, T., Scicky-Gallili, N., and
Cedar, H., 1984, Variations in DNA methylation during mouse differentiation in vivo and in
vitro, *Proc. Natl. Acad. Sci. USA* **81**:2275–2279.

Ream, L. W., and Gordon, M. P., 1982, Crown gall disease and prospects for genetic manipulation
in plants, *Science* **218**:854–859.

Reitsma, P. H., Rothberg, P. G., Astrin, S. M., Trial, J., Bar-Shavit, Z., Hall, A., Teitelbaum, S. L., and
Kahn, A. J., 1983, Regulation of *myc* gene expression in HL-60 leukemia cells by a vitamin D
metabolite, *Nature (Lond.)* **306**:492–494.

Rose, S. M., 1949, Transformed cells, *Sci. Am.* **181**(6):22–24.

Rose, S. M., 1970, *Regeneration: Key to Understanding Normal and Abnormal Growth and Development*, Appleton & Lange, E. Norwalk, Connecticut.

Rose, S. M., and Wallingford, H. M., 1948, Transformation of renal tumors of frogs to normal tissue in regenerating limbs of salamanders, *Science* **107**:457.

Rowley, P. T., and Skuse, G. R., 1987, Oncogene expression in myelopoiesis, *Int. J. Cell Cloning* **5**: 255–266.

Ruben, L. N., 1955, The effects of implanting anuran cancer into non-regenerating and regenerating larval urodele limbs, *J. Exp. Zool.* **128**:29–51.

Ruben, L. N., 1956, The effects of implanting anuran cancer into regenerating adult urodele limbs. I. Simple regenerating systems, *J. Morphol.* **98**:389–403.

Ruben, L. N., and Balls, M., 1964, The implantation of lymphosarcoma of *Xenopus laevis* into regenerating and non-regenerating forelimbs of that species, *J. Morphol.* **115**:225–238.

Rudnicki, M. A., and McBurney, M. W., 1987, Cell culture methods and induction of differentiation of embryonal carcinoma cell lines, in: *Teratocarcinomas and Embryonic Stem Cells: A Practical Approach* (E. J. Robertson, ed.), pp. 19–49, IRL Press Ltd., Oxford.

Sachs, L., 1978, The differentiation of myeloid leukemic cells: New possibilities for therapy, *Br. J. Haematol.* **40**:509–517.

Sachs, L., 1980, Activation of normal differentiation genes and the origin and development of myeloid leukemia, in: *Differentiation and Neoplasia* (R. G. McKinnell, M. A. DiBerardino, M. Blumenfeld, and R. D. Bergad, eds.), pp. 213–216, Springer-Verlag, Berlin.

Sachs, L., 1986, Growth, differentiation and the reversal of malignancy, *Sci. Am.* **254**:40–47.

Sachs, L., 1987a, Cell differentiation and bypassing of genetic defects in the suppression of malignancy, *Cancer Res.* **47**:1981–1986.

Sachs, L., 1987b, The molecular regulators of normal and leukaemic blood cells, *Proc. R. Soc. Lond. [Biol.]*, **231**:289–312.

Samara, A., Yagen, B., Agranat, I., Rachmilewitz, E. A., and Fibach, E., 1987, Induction of differentiation in human myeloid leukemic cells by T-2 toxin and other trichothecenes, *Toxicol. Appl. Pharmacol.* **89**:418–428.

Sanchez, S. S., and Barbieri, F. D., 1988, Extracellular materials and determination of neuroectoblast in amphibian gastrula, *Exp. Cell Biol.* **56**:60–66.

Saxén, L., and Toivonen, S., 1962, *Primary Embryonic Induction*, Prentice-Hall, Englewood Cliffs, New Jersey.

Saxén, L., Toivonen, S., and Nakamura, O., 1978, Primary embryonic induction: An unsolved problem, in: *Organizer—A Milestone of a Half Century from Spemann* (O. Nakamura and S. Toivonen, eds.), pp. 315–320, Elsevier/North-Holland, Amsterdam.

Schilperoort, R. A., Klapwijk, P. M., Hooykaas, P. J. J., Koekman, B. P., Ooms, G., Otten, L. A. B. M., Wurzer-Figurelli, E. M., Wullems, G. J., and Rorsch, A., 1978, A. tumefaciens plasmids as vectors for genetic transformation of plant cells, in: *Frontiers of Plant Tissue Culture 1978* (T. A. Thorpe, ed.), pp. 85–94, International Association for Plant Tissue Culture, Calgary.

Schlumberger, H. G., and Lucké, B., 1948, Tumors of fishes, amphibians and reptiles, *Cancer Res.* **8**: 657–754.

Schubert, D., Humphreys, S., Baroni, C., and Cohn, M., 1969, *In vitro* differentiation of a mouse neuroblastoma, *Proc. Natl. Acad. Sci. USA* **64**:316–323.

Schubert, D., Humphreys, S., DeVitry, F., and Jacob, F., 1971, Induced differentiation of a neuroblastoma, *Dev. Biol.* **25**:514–546.

Schwartz, E. L., Brown, B. J., Nierenburg, M., Marsh, J. C., and Sartorelli, A. C., 1983, Evaluation of some anthracycline antibiotics in an *in vivo* model for studying drug-induced human leukemia cell differentiation, *Cancer Res.* **43**:2725–2730.

Scott, R. E., Edens, M., Estervig, D. N., Filipac, M., Hoerl, B. J., Hsu, B. M., Maercklein, P. B., Minoo, P., Tzen, C-Y., Wilke, M. R., and Zschunke, M. A., 1988, Cellular differentiation and the prevention and treatment of cancer, in: *The Status of Differentiation Therapy of Cancer* (S. Waxman, G. B. Rossi, and F. Takaku, eds.), pp. 3–16, Raven, New York.

Seeds, N. W., Gilman, A. G., Amano, T., and Nirenberg, M. W., 1970, Regulation of axon formation by clonal lines of a neural tumor, *Proc. Natl. Acad. Sci. USA* **66**:160–167.

Seilern-Aspang, F., and Kratochwil, K., 1962, Induction and differentiation of an epithelial tumour in the newt (*Triturus cristatus*), *J. Embryol. Exp. Morphol.* **10**:337–356.

Seppanen, E. D., McKinnell, R. G., Tarin, D., Rollins-Smith, L. A., and Hanson, W., 1984, Temperature-dependent dissociation of Lucké renal adenocarcinoma cells, *Differentiation* **26:**227–230.

Shen, S. C., 1939, A quantitative study of amphibian neural tube induction with a water-soluble hydrocarbon, *J. Exp. Biol.* **16:**143–149.

Sheremetieva, E. A., 1965, Spontaneous melanoma in regenerating tails of axolotls, *J. Exp. Zool.* **158:**101–122.

Sherr, C. J., Rettenmier, C. W., Sacca, R., Roussei, M. F., Look A. T., and Stanley, E. R., 1985, The c-*fms* proto-oncogene product is related to the receptor for the mononuclear phagocyte growth factor, CSF-1, *Cell* **41:**665–676.

Shumway, W., 1940, Stages in the normal development of *Rana pipiens*. I. External form, *Anat. Rec.* **78:**139–147.

Sidell, N., 1981, Retinoic acid induced growth inhibition and morphologic differentiation of human neuroblastoma cells *in vitro, J. Natl. Cancer. Inst.* **68:**589–596.

Siminovitch, L., and Axelrad, A. A., 1963, Cell-cell interactions *in vitro:* Their relation to differentiation and carcinogenesis, *Can. Cancer Conf.* **5:**149–165.

Simon-Assmann, P., Bouziqes, F., Daviaud, D., Haffen, K., and Kedinger, M., 1987, Synthesis of glycosaminoglycans by undifferentiated and differentiated HT29 human colonic cancer cells, *Cancer Res.* **47:**4478–4484.

Slungaard, A., Confer, D. L., and Schubach, W. H., 1987, Rapid transcriptional down-regulation of c-*myc* expression during cyclic adenosine monophosphate-promoted differentiation of leukemic cells, *J. Clin. Invest.* **79:**1542–1547.

Smith, H. H., 1988, The inheritance of genetic tumors in *Nicotiana* hybrids, *J. Hered.* **79:**277–283.

Smith, E. F., and Townsend, C. O., 1907, A plant-tumor of bacterial origin, *Science* **25:**671–673.

Smithers, D. W., 1969, Maturation in human tumours, *Lancet* **2:**949–952.

Solter, D., Skreb, N., and Damjanov, I., 1970, Extrauterine growth of mouse egg-cylinder results in malignant teratoma, *Nature (Lond.)* **227:**503–504.

Souza, L. M., Boone, T. C., Gabrilove, J., Lai, P. H., Zsebo, K. M., Murdock, D. C., Chazin, V. R., Bruszewski, J., Lu, H., Chen, K. K., Barendt, J., Platzer, E., Moore, M. A. S., Mertelsmann, R., and Welte, K., 1986, Recombinant human granulocyte colony-stimulating factor: Effects on normal and leukemic myeloid cells, *Science* **232:**61–65.

Spemann, H., 1938, *Embryonic Development and Induction*, Yale University Press, New Haven, Connecticut.

Spemann, H., and Mangold, H., 1924, Uber Induktion von Embryonalanlagen durch Implantation artfremder Organisatoren, *Arch. Mikrosc. Anat. Entwicklungsmech.* **100:**599–638.

Spiegelman, B. M., Lopata, M. A., and Kirschner, M. W., 1979, Aggregation of microtubule initiation sites preceeding neurite outgrowth of mouse neuroblastoma cells, *Cell* **16:**253–263.

Sporn, M. B., Roberts, A. B., and Driscoll, J. S., 1985, Principles of cancer biology: Growth factors and differentiation, in: *Cancer, Principles and Practice of Oncology* (V. T. DeVita, Jr., S. Hellman, and S. A. Rosenberg, eds.), pp. 49–65, Lippincott, Philadelphia.

Sporn, M. B., and Roberts, A. B., 1984, Role of retinoids in differentiation and carcinogenesis, *J. Natl. Cancer Inst.* **73:**1381–1387.

Stevens, L. C., 1960, Embryonic potency of embryoid bodies derived from a transplantable testicular teratoma of the mouse, *Dev. Biol.* **2:**285–297.

Stevens, L. C., 1970*a*, The development of transplantable teratocarcinomas from intertesticular grafts of pre- and post-implantation mouse embryos, *Dev. Biol.* **21:**364–382.

Stevens, L. C., 1970*b*, Experimental production of testicular teratomas in mice of strains 129, A/He and their F_1 hybrids, *J. Natl. Cancer Inst.* **44:**923–929.

Stevens, L. C., 1973, A new inbred subline of mice (129/terSv) with a high incidence of spontaneous testicular teratoma, *J. Natl. Cancer Inst.* **50:**235–242.

Stevens, L. C., 1981, Genetic influences on the development of gonadal tumors in mice with emphasis on teratomas, in: *Neoplasms—Comparative Pathology of Growth in Animals, Plants, and Man* (H. E. Kaiser, ed.), pp. 467–474, Williams & Wilkins, Baltimore.

Steward, F. C., Mapes, M. O., Kent, A. E., and Holsten, R. D., 1964, Growth and development of cultured plant cells, *Science* **143:**20–27.

Steward, F. C., Ammirato, P. V., and Mapes, M. O., 1970, Growth and development of totipotent cells, *Ann. Bot.* **34**:761–787.

Stewart, T. A., and Mintz, B., 1981, Successive generations of mice produced from an established culture line of euploid teratocarcinoma cells, *Proc. Natl. Acad. Sci. USA* **78**:6314–6318.

Strickland, S., and Mahdavi, V., 1978, The induction of differentiation in teratocarcinoma stem cells by retinoic acid, *Cell* **15**:393–403.

Sugimoto, T., Kato, T., Sawada, T., Horii, Y., Kemshead, J. T., Hino, T., Morioka, H., and Hosoi, H., 1988, Schwannian cell differentiation of human neuroblastoma cell lines *in vitro* induced by bromodeoxyuridine, *Cancer Res.* **48**:2531–2537.

Takebe, I., Labib, G., and Melchers, G., 1971, Regeneration of whole plants from isolated mesophyll protoplasts of tobacco, *Naturwissenschaften* **58**:318–320.

Takeda, K., Minowada, J., Bloch, A., 1982, Kinetics of appearance of differentiation associated characteristics of ML-1, a line of human myeloblastic leukemia cells, after treatment with TPA, DMSO, or Ara-C, *Cancer Res.* **42**:5152–5158.

Tarin, D., 1973, Histochemical and enzyme digestion studies on neural induction in *Xenopus laevis*, *Differentiation* **1**:109–126.

Thomashow, M. F., Hugly, S., Buchholz, W. G., and Thomashow, L. S., 1986, Molecular basis for auxin-independent phenotype of crown gall tumor tissues, *Science* **231**:616–618.

Timofejewsky, A. D., and Benewolenskaya, S. W., 1929, Neue Beobachtungen an lymphoiden Zellen der myeloiden und lymphatischen Leukamie in Explantations-Versuchens, *Arch. Exp. Zellforsch.* **8**:1–34.

Torrey, J. G., 1985, The development of plant biotechnology, *Am. Sci.* **73**:354–363.

Trinchieri, G., Rosen, M., and Perussia, B., 1987, Retinoic acid cooperates with tumor necrosis factor and human interferon in inducing differentiation and growth inhibition of the human promyelocytic leukemic cell line HL-60, *Blood* **69**:1218–1224.

Triolo, V. A., 1965, Nineteenth century foundations of cancer research advances in tumor pathology, nomenclature, and theories of oncogenesis, *Cancer Res.* **25**:75–106.

Tsokos, M., Scarpa, S., Ross, R. A., Triche, T. J., 1987, Differentiation of human neuroblastoma recapitulates neural crest development, *Am. J. Pathol.* **128**:484–496.

Tsonis, P. A., 1984, Limb regeneration in newts with spontaneous skin cancer, *Can. J. Zool.* **62**: 2681–2685.

Turgeon, R., Wood, H. N., and Braun, A. C., 1976, Studies on the recovery of crown gall tumor cells, *Proc. Natl. Acad. Sci. USA* **73**:3562–3564.

Tweedell, K. S., 1965, Cytopathology of a frog renal adenocarcinoma *in vitro* with fluorescence microscopy, *Ann. NY Acad. Sci.* **126**:170–187.

Tweedell, K. S., 1967, Induced oncogenesis in developing frog kidney cells, *Cancer Res.* **27**:2042–2052.

Tweedell, K. S., 1978, Pronephric tumor cell lines from herpesvirus transformed cells, *Int. Agency Res. Cancer Sci. Pub.* **24**:609–616.

Unsworth, B., and Grobstein, C., 1970, Induction of kidney tubules in mouse metanephrogenic mesenchyme by various embryonic mesenchymal tissues, *Dev. Biol.* **21**:547–556.

Vasil, V., and Hildebrandt, A. C., 1965, Differentiation of tobacco plants from single, isolated cells in microculture, *Science* **150**:889–892.

Waddington, C. H., 1935, Cancer and the theory of organizers, *Nature (Lond.)* **135**:606–608.

Wartiovaara, J., and Rechardt, L., 1985, Neural differentiation in embryonal carcinoma cells, *Prog. Clin. Biol. Res.* **171**:3–13.

Waterfield, M. D., Scrace, G. T., Whittle, N., Stroobant, P., Johnsson, A., Wasteson, A., Westermark, B., Heldin, C. H., Huang, J. S., and Devel, T. F., 1983, Platelet-derived growth factor is structurally related to the putative transforming protein p28*sis* of simian sarcoma virus, *Nature (Lond.)* **304**:35–39.

Watson, B., Currier, T. C., Gordon, M. P., Chilton, M. D., and Nester, E. W., 1975, Plasmid required for virulence of *Agrobacterium tumefaciens*, *J. Bacteriol.* **123**:255–264.

Waxman, S., and Takaku, F., 1988, The incorporation of differentiation induction into the design of cancer therapy, in: *The Status of Differentiation Therapy of Cancer* (S. Waxman, G. B. Rossi, and F. Takaku, eds.), pp. 407–419, Raven, New York.

Waxman, S., Scher, W., and Scher, B. M., 1986, Basic principles for utilizing combination differentiation agents, *Cancer Detect. Prev.* **9**:395–407.

Webb, M., Graham, C., and Walsh, F., 1986, Neuronal differentiation of cloned human teratoma cells in response to retinoic acid *in vitro, J. Neuroimmunol.* **11**:67–86.

Wells, R. S., and Miotto, K. A., 1986, Widespread inhibition of neuroblastoma cells in the 12- to 17-day old mouse embryo, *Cancer Res.* **46**:1659–1662.

Westin, E. H., Wong-Staal, F., Gelman, E. P., Dalla Favera, R., Papas, T. S., Lautenberger, J. A., Eva, A., Premkumar Reddy, E., Tronick, S. R., Aaronson, S. A., and Gallo, R. C., 1982, Expression of cellular homologues of retroviral *onc* genes in human hematopoietic cells, *Proc. Natl. Acad. Sci. USA* **79**:2490–2494.

Whitehead, R. H., Jones, J. K., Gabriel, A., and Lukies, R. E., 1987, A new colon carcinoma cell line (LIM 1863) that grows as organoids with spontaneous differentiation into crypt-like structures *in vitro, Cancer Res.* **47**:2683–2689.

Wiemann, M., Alexander, P., and Calabresi, P., 1988, Combination differentiation therapy, in: *The Status of Differentiation Therapy of Cancer* (S. Waxman, G. B. Rossi, and F. Takaku, eds.), pp. 299–314. Raven, New York.

Willis, R. A., 1962, *The Borderland of Embryology and Pathology,* 2nd ed., Butterworths, London.

Witschi, E., 1948, Migration of the germ cells of human embryos from the yolk sac to the primitive gonadal folds, *Contrib. Embryol. Carnegie Inst. Wash.* **32**:67–80.

Witschi, E., 1956, *Development of Vertebrates,* W. B. Saunders, Philadelphia.

Yoshida, M., Normura, S., and Beppu, T., 1987, Effects of trichostatins on differentiation of murine erythroleukemia cells, *Cancer Res.* **47**:3688–3691.

Yunis, A. A., Arimura, G. K., Wu, F-M., and Wu, M-C., 1987, Differentiation of cultured promyelocytic leukemia cells (HL-60) induced by endotoxin-treated human lung conditioned medium, *Leuk. Res.* **11**:673–679.

Yuspa, S. H., and Roop, D. R., 1988, Alterations in control of terminal differentiation in premalignant and malignant mouse keratinocytes, in: *The Status of Differentiation Therapy of Cancer* (S. Waxman, G. B. Rossi, and F. Takaku, eds.), pp. 17–28, Raven, New York.

Zimmerman, B. T., and Speers, W. C., 1988, Induction of differentiation of murine embryonal carcinoma cells by ouabain, *Differentiation* **36**:164–173.

Zweibaum, A., Pinto, M., Chevalier G., Dussaulx, E., Triadou, N., Lacroix, B., Haffen, K., Brun, J-L., and Rousset, M., 1985, Enterocytic differentiation of a subpopulation of the human colon tumor cell line HT-29 selected for growth in sugar-free medium and its inhibition by glucose, *J. Cell Physiol.* **122**:21–29.

Index